CHIPLESS RFID SENSORS

CHIPLESS RFID SENSORS

NEMAI CHANDRA KARMAKAR
EMRAN MD AMIN
JHANTU KUMAR SAHA

Published by John Wiley & Sons, Inc., Hoboken, New Jersey
Published simultaneously in Canada

For general information on our other products and services or for technical support, please contact our Customer Care Department within the United States at (800) 762-2974, outside the United States at (317) 572-3993 or fax (317) 572-4002.

Wiley also publishes its books in a variety of electronic formats. Some content that appears in print may not be available in electronic formats. For more information about Wiley products, visit our web site at www.wiley.com.

Library of Congress Cataloging-in-Publication Data applied for:

ISBN: 9781118936009

Typeset in 10/12pt TimesLTStd by SPi Global, Chennai, India

Printed in the United States of America

10 9 8 7 6 5 4 3 2 1

1 2016

The book is dedicated

To my beloved wife, Shipra, and daughters, Antara and Ananya
— N. C. K.

To my parents, beloved wife, Mysha and son, Ayman
— E. M. A.

To my mother, beloved wife, Sima, and son, Dhritisundar, and daughter, Joyeeta
— J. K. S.

VISIONARY STATEMENT

Deliver a technology that would replace optical barcodes with low-cost, compact, printable and highly sensitive chipless RFID sensors. This will promote green technology and pollution-free disposable sensor nodes for pervasive sensing. Such low-cost ubiquitous sensing technology can uniquely identify and monitor each and every physical object through with Internet of Things (IoT).

CONTENTS

PREFACE

RFID AND RF SENSORS

Radio frequency identification (RFID) is an emerging wireless technology for automatic identifications, access controls, tracking, security and surveillance, database management, inventory control, and logistics. However, the application-specific integrated circuits (ASICs) in the chipped RFID tags make the tag costly and hinder their applications in mass tagging. Chipless RFID tags are voids of these microchips. Some chipless RFID are fully printable passive microwave and mm-wave circuits. They can be produced very cheaply. Integration of physical parameter sensors with chipless RFID will open up a new domain for energy-efficient housing, control and monitoring of perishable items, equipment, and people. In the new millennium, low-cost ubiquitous tagging and sensing of objects, homes, and people will make the system efficient, reduce wastage, and lower the healthcare budget. This book presents various sensing techniques incorporated in the chipless RFID systems.

The RFID has two main components—a tag and a reader. The reader sends an interrogating radio signal to the tag. In return, the tag responds with a unique identification code to the reader. The reader processes the returned signal from the tag into a meaningful identification code. Some tags coupled with RF sensors can also provide data of surrounding environment such as temperature, relative humidity, pressure or impact, moisture content, and location.

The tags are classified into active, semi-active, and passive tags based on their onboard power supplies. An active tag contains onboard battery to energize the processing chip and amplify signals. A semiactive tag contains a battery as well, but the battery is used only to energize the chip, hence yielding better longevity compared to the active tag. A passive tag does not have a battery. It scavenges power for its processing chip from the interrogating signal from the tag, hence last forever. However, the processing power and reading distance is limited by the transmitted power of the reader.

SIGNIFICANCE OF CHIPLESS RFID

As stated earlier, the main constraint of mass deployment of RFID tags for low-cost item tagging is the cost of the tag. The main cost comes from the microchip of the tag. If the chip can be removed without losing functionality of the tag, then the tag can be produced in subcents and has the potential to replace the optical barcode.

The optical barcode has limitations in operation such as each barcode is individually read, needs human intervention, and has less data handling capability. Soiled barcodes cannot be read and barcodes need line-of-sight operation. Despite these limitations, the low-cost benefit of the optical barcode makes it very attractive as it is printed almost without any extra cost. Therefore, there is a pressing need to remove the ASIC from the RFID tag to make it competitive in deployment to coexist or replace trillions of optical barcodes printed each year. The solution is to remove the ASIC from the RFID tag. The tag should be fully printable on low-cost substrates such as paper and plastics similar to the optical barcodes. A reliable prediction by the respected RFID research organization IDTechEx advocates [1] that 60% of the total tag market will be occupied by the chipless tag if the tag can be made in less than one cent.

However, removal of signal processing ASIC from the tag is not a trivial task. It needs tremendous investigation and investment in designing low-cost but robust passive microwave circuits and antennas using conductive ink on low-cost substrates. However, obtaining high fidelity response from low-cost lossy materials is very difficult. To overcome these challenges, new materials characterization and fabrication processes are to be innovated for chipless RFID tags and sensors. In the interrogation and decoding sides of the RFID system is the development of the RFID reader, which is capable to read the chipless RFID tag. The authors' group has tremendous progress in this frontier developing multiple chipless RFID tag readers in 2.45, 24, and 60 GHz frequency bands. Currently, only a few fully printable chipless RFID tags, which are in the inception stage, are reported in the literature. They are a capacitive gap coupled dipole array [2], a reactively loaded transmission line [3], a ladder network [4], and finally, a piano and a Hilbert curve fractal resonators [5]. These tags are in prototype stage and no further development in commercial grade is reported so far. Only commercially successful chipless RFID is RF-SAW, but they are not printable and expensive [6]. There is much stride to develop thin-film transistor circuit (TFTC) chipless tags to attract huge market of high-frequency (HF) tags [7]. However, they are complex circuits and need complex fabrication processes. To fill up the gap in the literature of the potential chipless RFID field, the author's chipless RFID research team has been working on the paradigm chipless RFID tag since 2004. The designed tag has mainly targeted to tag Australian polymer banknotes, library access cards, and apparels [8–11].

Significant successes have been achieved to tag not only the polymer banknotes but also many low-cost items such as books, postage stamps, secured documents, bus tickets, and hung-tags for apparels. The technology relies on encoding spectral signatures and decoding the amplitude and phase of the spectral signature. The other methods are phase encoding of backscattered spectral signals and time-domain delay lines. So far as many as more than 10 chipless RFID tags [8–11] and three generations of readers [12] are designed. The proof-of-concept technology is being transferred to the banknote polymer and paper for low-cost item tagging. These tags have potential to coexist or replace trillions of optical barcodes printed each year [9]. To this end, it is imperative to invest in low-loss conducting ink, high-resolution printing process, and characterization of laminates on which the tag will be printed. The design needs to push in higher frequency bands to accommodate and increase the number of bits in the chipless tag to compete with optical barcodes. The book has addressed all these issues in 11 chapters.

WHY INCORPORATION OF SENSING ELEMENTS IN CHIPLESS RFID TAGS—THE HYPOTHESIS OF THE BOOK?

While successes are achieved in very low-cost multibit chipless tag design, there are pressing needs to extend the functionality for real-time wireless sensing and monitoring of physical parameters such as temperature, relative humidity, pressure or impact, moisture content, sensing of noxious gases, light intensity, and location of objects [13–16]. In these pursuits, various sensing materials that are compatible with the printable RF/microwave electronics are also investigated. Various smart materials that are identified for low-cost chipless RFID sensor fabrication are (i) ionic plastic crystals, whose ionic conductivity changes due to organic molecule defects and the movement of crystals; (ii) conductive polymers (PEDOTs), whose conductivity increases with frequency; (iii) composite/conjugate polymer, mixed with conductive and nonconductive polymers [17]; and (iv) nanostructured metal oxides that exhibit multifunctional properties and are very susceptible to external environmental changes, such as pressure, temperature, and electric fields [18]. Implementation of these smart materials in fully printable multibit chipless RFID tags brings many new innovations in areas such as new chipless RFID tag design, metamaterial-based high-Q resonator design for sensing purposes, microwave and mm-wave frequency characterization of smart materials, fabrication of integrated chipless RFID sensors, and finally evaluations of such sensing devices in various ambient environments. The book aims to address all these issues mentioned above to make the chipless RFID sensors a viable commercial product for mass deployment. The book covers all these materials in five sections: (i) Introduction to chipless RFID sensors; (ii) RFID sensors design; (iii) smart materials; (iv) fabrication, integration, and testing; and finally (v) applications. The book presents many new designs, concepts, and results in the new field. The authors believe the book will create a significant impact in the research community.

OVERALL OBJECTIVE

In recent decades, RFID has been revolutionizing supply chain management, security, and access controls by tagging items and personnel. The mandate of tagging manufactured items by vendors of retail giant Walmart has accelerated the impact of using RFID [19]. However, RFID has not become a low-cost item tagging device like optical barcodes due to its high cost per tag. Mass deployment of RFID technology will only be possible if the tag is made chipless and fully printable like the barcode. There are a few books on conventional chipped tags in the market. A couple of books on chipless RID tags and readers have been published by the author's group in recent years.

Adding sensing capabilities with the chipless RFID tags will open up many new application areas such as agriculture, construction, health care, energy sectors, retails, public transportations, logistics, and supply chain management.

No book on chipless RFID sensors has been published yet. This will be the first effort to publish a book in the niche area of the chipless RFID sensors based on the outcomes of fundamental research conducted by the author's research group from 2009. Once the chipless RFID tag sensors are made fully printable similar to the optical barcode, it will revolutionize the mass market of low-cost and perishable item tagging and sensing.

REFERENCES

1. P. Harrop and R. Das. *Printed and Chipless RFID Forecasts, Technologies & Players 2011–2021 [Online]*. Available: http://www.idtechex.com/research/reports/printed-and-chipless-rfid-forecasts-technologies-and-players-2011-2021-000254.asp.
2. I. Jalaly and I. D. Robertson, "RF barcodes using multiple frequency bands," in *Microwave Symposium Digest, 2005 IEEE MTT-S International*, 2005, p. 4.
3. S. Shrestha, J. Vemagiri, M. Agarwal, and K. Varahramyan, "Transmission line reflection and delay-based ID generation scheme for RFID and other applications," *International Journal of Radio Frequency Identification Technology and Applications,* vol. 1, pp. 401–416, 2007.
4. S. Mukherjee, "System for identifying radio-frequency identification devices," US20070046433.
5. J. McVay, A. Hoorfar, and N. Engheta, "Theory and experiments on Peano and Hilbert curve RFID tags", *Proceedings of SPIE*, 6248, Wireless Sensing and Processing, 624808, doi: 10.1117/12.666911, 2006.
6. S. Preradovic, N. C. Karmakar, and I. Balbin, "RFID Transponders," *IEEE Microwave Magazine*, vol. 9, pp. 90–103, 2008.
7. R. Das and P. Harrop. *Chip-less RFID Forecasts, Technologies & Players 2006–2016 [Online].* Available: http://www.idtechex.com/products/en/view.asp?productcategoryid=96.
8. S. Preradovic, "Chipless RFID System for Barcode Replacement," Doctor of Philosophy, Department of Electrical and Computer Systems Engineering, Monash University, 2009.
9. S. Preradovic and N. C. Karmakar, "Chipless RFID: bar code of the future," *IEEE Microwave Magazine*, vol. 11, pp. 87–97, 2010.
10. S. Preradovic, I. Balbin, N. C. Karmakar, and G. F. Swiegers, "Multiresonator-based chipless RFID system for low-cost item tracking," *IEEE Transactions on Microwave Theory and Techniques*, vol. 57, pp. 1411–1419, 2009.

11. M. A. Islam and N. Karmakar, "Compact printable chipless RFID system using polarization diversity," Monash University, 2011.
12. N. C. Karmakar, S. M. Roy, and M. S. Ikram, "Development of smart antenna for RFID reader," in *RFID, 2008 IEEE International Conference on*, 2008, pp. 65–73.
13. E. M. Amin and N. Karmakar, "Development of a chipless RFID temperature sensor using cascaded spiral resonators," presented at the IEEE SENSORs 2011, 2011.
14. E. M. Amin, N. Karmakar, and S. Preradovic, "Towards an intelligent EM barcode," in *Electrical & Computer Engineering (ICECE), 2012 7th International Conference on*, 2012, pp. 826–829.
15. E. M. Amin and N. Karmakar, "Partial discharge monitoring of high voltage equipment using chipless RFID sensor," presented at the Asia-Pacific Microwave Conference, Melbourne, Australia, 2011.
16. E. M. Amin, S. Bhuiyan, N. Karmakar, and B. Winther-Jensen, "A novel EM barcode for humidity sensing," in *RFID (RFID), 2013 IEEE International Conference on*, 2013, pp. 82–87.
17. J. R. Terje and A. Skotheim, *Conjugated Polymers: Processing and Applications*, 3rd ed: CRC Press, 2007, p. 207.
18. J. Xia, C. Sui, H. Wang, T. Xu, B. Yan, and Y. Liu, "Optical temperature sensor based on ZnO thin film's temperature-dependent optical properties," *Review of Scientific Instruments*, vol. 82, pp. 084901–3, 2011.
19. M. Roberti. (2005). *Wal-Mart Begins RFID Process Changes*. Available: http://www.rfidjournal.com/article/view/1385.

ACKNOWLEDGMENTS

I would like to thank Ms Kari Capone, Content Capture Manager, Wiley, for her invitation to write a book on *Chipless RFID Sensors*. Thanks also go to the reviewers who reviewed the book proposal and chapters outline. I must acknowledge Brett Kurzman, Editor, Ms Divya Narayanan, Project Editor, and Alex Castro, Editorial Assistant of Wiley for their continuous support and patience throughout the preparation, submission, and reviewing processes of the manuscript.

Emran Md Amin would like to acknowledge *AutoID Lab, Massachusetts Institute of Technology (MIT)*, USA, and *EISLAB, Lulea University of Technology*, Sweden, for providing him an excellent opportunity and scholarship.

Emran Md Amin and Jhantu K. Saha also would like to highly acknowledge for technical support and guidance of Assoc. Prof. Bjorn Jensen, Materials Engineering Department, Monash University.

Finally, the research funding supports from Australian Research Council's Discovery Project Grants and Linkage Project Grants are highly acknowledged.

The book project is my vision and I guided my PDF and PhD student coauthors to fulfill the vision throughout the whole project. Therefore, the book is a culmination of hard work of the two dedicated authors Emran Md Amin and Dr. Jhantu K. Saha. Without their continuous motivation, dedication, and perseverance, the book would not have shaped in this high-quality piece of scientific artwork.

Assoc. Prof. Nemai Chandra Karmakar

Department of Electrical and Computer System Engineering
Monash University
April 2015

ABBREVIATIONS

ASICs Application-specific integrated circuits
BJT Bipolar junction transistors
BST Barium strontium titanate
CdS Cadmium sulfide
CMOS Complementary metal-oxide semiconductor
CPW Coplanar waveguide
CST Computer simulation technology
DAC Digital to analog converter
EBG Electronic band gap
EIS Electrochemical impedance spectroscopy
EIRP Equivalent isotropically radiated power
ELC Electric inductor capacitor
EM Electromagnetic
FDTD Finite difference time domain
FMCW Frequency modulated continuous wave
FSS Frequency selective surface
GPD Gain phase detector
HF High frequency
HV High voltage
IC Integrated circuit
ICT Information and communication technology

IDT Inter-digital transducer
IF Intermediate frequency
IoT Internet of things
ITO Indium tin oxide
LED Light emitting diode
LH Left hand
LO Local oscillator
MEMS Micro-electromechanical systems
MWS Microwave studio
OLA Overlap add algorithm
PAni Polyanniline
pC Picocoulomb
PD Partial discharge
PEDOT Poly(3,4-ethylenedioxythiophene)
PMMA Poly(methyl methacrylate)
PPM Pulse position modulation
PVA Polyvinyl alcohol
PVC Polyvinyl chloride
RCS Radar cross section
RF Radio frequency
RFID Radio-frequency identification
RH Relative humidity
SAW Surface acoustic wave
SIR Stepped impedance resonator
STFT Short time frequency transform
SW-BNT Single wall carbon nanotube
TDR Time domain reflectometry
THF Tetrahydrofuran
UHF Ultra high frequency
UIR Uniform impedance resonators
USN Ubiquitous sensing network
UWB Ultra-wide band
VCO Voltage control oscillator
VNA Vector network analyzer
WISP Wireless identification and sensing platform
WLAN Wireless local area network

SYMBOLS

α Absorption coefficient

c Speed of light

Eg Optical band gap

E Electric field

f_r Resonant frequency

Δf_{SN} Maximum frequency shift band for slot resonator

φ Phase of incident signal

λ Wavelength of EM signal

L_0 Physical length of slot resonator

$\varepsilon_r{}'$ Real part of electric permittivity

$\varepsilon_r{}''$ Imaginary part of electric permittivity

K Cell constant

hv Photon energy

$K(k_0)$ Modulus of complete elliptic integrals

$P_{\min}$ Minimum power at resonance

μ Carrier mobility

v Carrier drift velocity

q_1, q_2 Partial filling factor

R Hop size for STFT analysis

S_{21} Insertion loss

S_{11} Reflection loss

σ Conductivity

S_{fr} Sensitivity for resonant frequency

Sens_cal_data Sensor calibrated data

T_c Transition temperature of sublimate material

$\tan\delta$ Dielectric loss tangent

ϑ_T Electrical length of stepped resonator

T_{cd} Temperature coefficient of delay

γ_n Carrier phase for reflected signal

Δt Time resolution for STFT analysis

$w(n)$ Window function

$X_m(\omega)$ DTFT of windowed signal

Z_s Equivalent impedance of SIR resonator

1

INTRODUCTION

1.1 TRACKING ID TECHNOLOGY

A feature of modern society is the increasing use of machine-reading techniques in everyday life. From recording commercial transactions to monitoring logistics, machine reading is increasingly pervasive and important in the modern economy. Although a variety of different technologies have been used, today two technologies are competing for dominance: (i) barcodes and (ii) radio-frequency identification (RFID). Figure 1.1 shows how tracking and tracing ID technology in industry has evolved significantly from the first self-adhesive label manufactured in the 1930s. In the following sections, each of the technologies is discussed.

1.1.1 Barcoding

Barcode identification is a line-of-sight technology that involves scanning a printed pattern comprising light and dark (mark and space) elements with a laser-reader apparatus. The laser beam is either reflected or absorbed by the elements, with the resulting pattern being detected by the reader and converted into digital data according to a conversion protocol. Barcoding was first commercially used in 1974. Today, billions of barcodes are printed yearly for numerous applications related to product tracking, management, and logistics [1].

The key advantage barcodes provide is their information density and low cost; they can be printed for fractions of a cent each. Barcode technology, nevertheless, has important disadvantages. The line-of-sight (LOS) requirement means that a human operator must usually be present to direct the reading process, or at least verify it.

Chipless RFID Sensors, First Edition. Nemai Chandra Karmakar, Emran Md Amin and Jhantu Kumar Saha.

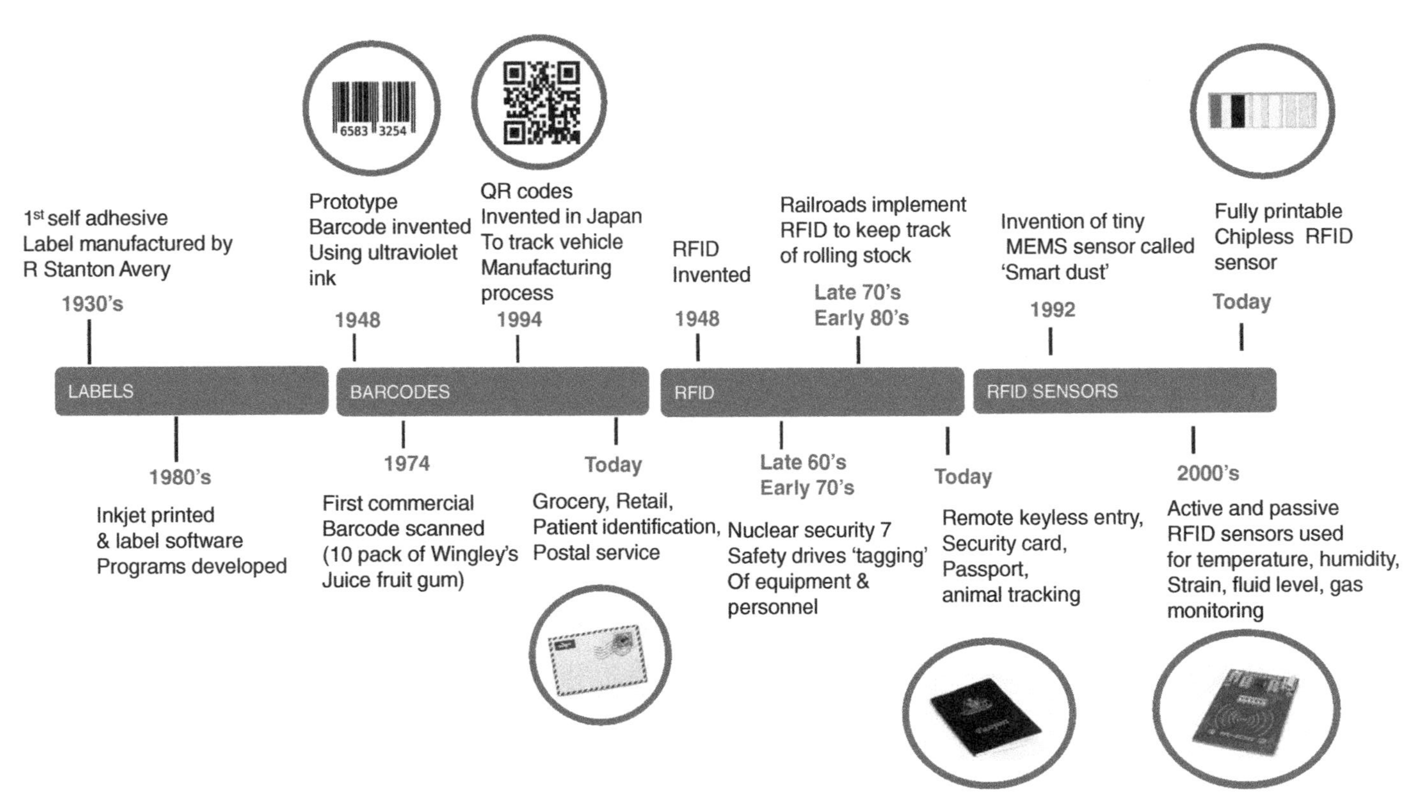

Figure 1.1 Evolution of tracking ID technology

Scanning is usually limited to a single item at a time, making it a slow process compared to purely electronic systems.

1.1.2 Radio-Frequency Identification

RFID technology usually consists of three components: (i) a small and mobile tag unit (or transponder) that is attached to items of interest and (ii) a reader (or transceiver) whose location is generally fixed and which contains (iii) an attached antenna (Figure 1.2). Signals are broadcast by the reader via its attached antenna. The tag receives these signals and responds by either reading or writing the data, or by replying with another signal containing some data, such as an identity code or a measurement value. The tag may also rebroadcast the original signal received from the reader, sometimes with a time delay.

RFID is a rapidly developing revolutionary wireless data collection technology for automatic identification, asset tracking, access control, security surveillance, electronic toll collection, car immobilizers, and smart logistics. The concept of RFID technology germinated in the early 1950s and evolved through the 1980s with the rapid advent of very low-power and application-specific integrated circuits (ASICs). To date, the RFID market has surpassed \$10 billion [2] through its omnipresence in communications and transport, banking systems, retail, distribution logistics, hospital management, and automotive systems. Major retailers such as Wal-Mart in the United States [3] and Coles Myer in Australia [4] are increasingly using RFID and have reduced their reliance on optical barcodes in their businesses. However, RFID has not fully replaced barcodes due to the high tag price. While optical barcodes can be directly printed on packed items with almost no extra cost, RFID tags need special procedures for application to items. However, for expensive items, this cost barrier is mitigated by the extra benefits and flexibility in operation the RFID provides. For example, Gillette has tagged razors costing over \$20 with high-frequency (HF) tags. Optical barcodes have many limitations: (i) they need LOS reading, (ii) barcodes cannot be read in sunlight and dark, (iii) soiled tags cannot be read, and finally, (iv) barcodes have low data capacity. In contrast, RFIDs can be read non-LOS without any human intervention and item-level tagging is possible due to their immense data capacity. For most logistical purposes such as supply chain management in the

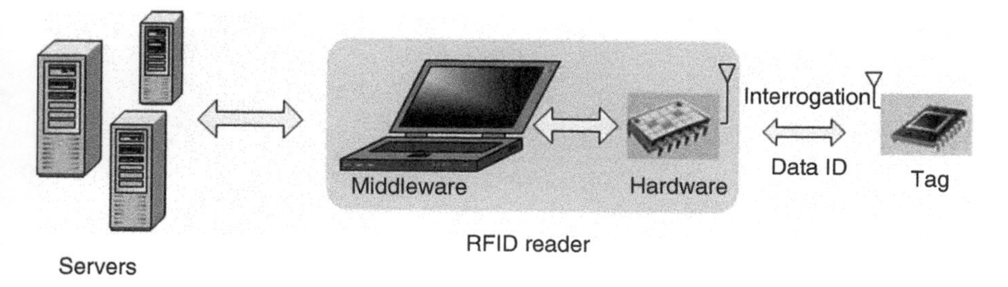

Figure 1.2 Block diagram of generic RFID system

retail sector, RFID tags must offer at least 1 billion unique permutations (known as addresses). This corresponds to a 30-bit binary code with 2^{30} permutations. The only way to produce such high data bits on an RFID transponder at present is by the incorporation of a silicon chip. The cost of fabricating and affixing such a chip is substantial (>10¢), even when the chip is produced in quantities of billions. A printable, chipless RFID system with multibit capacity is needed to overcome the cost limitation and make RFID competitive with optical barcodes.

1.1.3 Chipless RFID

The chipless RFID tag is a breakthrough in overcoming the limitations of conventional RFID technology as it removes the cost associated with the silicon IC chip in the tag circuit. Moreover, the tag is fully printable and passive and is thus resistant to extremely harsh environments and weather conditions. Chipless RFID tags can even be printed on metals and bottles containing liquids [5]. The potential advantages of these unique features permit chipless RFID in unique applications that could not be achieved previously with both barcodes and chipped RFIDs. Some examples include low-cost item tagging such as for banknotes, ID cards, books, aluminum cans, drink bottles, and consumer goods.

1.1.4 Chipless RFID Sensors

Emerging challenges in tracking ID technology demand ubiquitous sensing together with tagging of an object. Figure 1.3 shows the advances in application areas for tracking ID technology over time. Clearly, supply chain management, logistics, transport, and storage of goods have become sophisticated with the advent of the Internet of things (IoT) platform. Here, the primary goal is to connect every object to

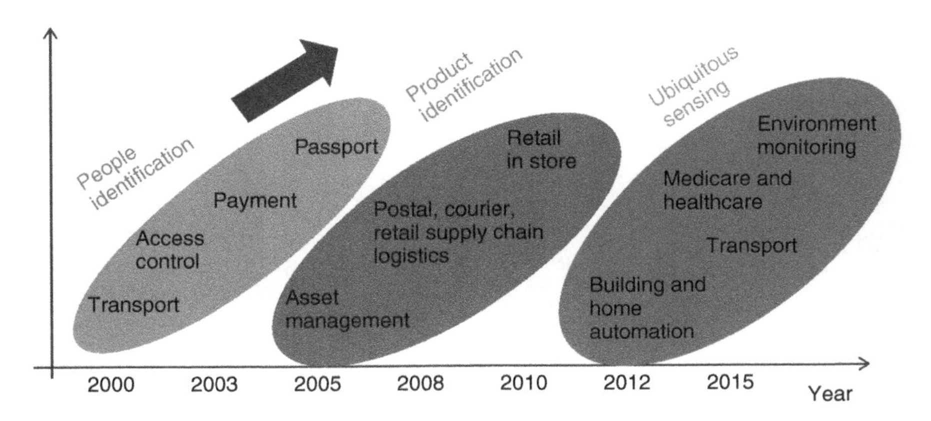

Figure 1.3 Advancement of application areas of tracking ID technology with time

"cloud data" and monitor critical information. However, machine-to-machine communication creates a significant burden on the cost budget of a system due to the high cost of each sensor node.

Alternatively, chipless RFID tag sensors provide identification data and monitor a number of physical parameters of tagged objects without having an active sensor in the tag circuitry [6]. The chipless RFID sensor has benefits over traditional sensors because of its lower cost, longer storage life, robustness, and lower radiated power.

The aim of chipless RFID tags is to replace the existing barcodes used for item-level tagging. The functional evolution of our proposed RFID sensor is shown in Figure 1.4. First, an analogy between the optical barcode (Figure 1.4(a)) and the radar cross section (RCS)-based chipless RFID tag (Figure 1.4(b)) is presented. A chipless RFID tag can be realized by placing conductive and dielectric strips of different shapes adjacent to each other, similar to barcodes [7]. The different combinations of the strips have distinct frequency responses in the backscattered RCS spectrum when interrogated with a polarized E-field. This unique response can attribute to encoded data bits in frequency and phase spectrum. However, migrating from optical barcodes to this crude chipless RFID tag has major advantages. The chipless tag has a longer reading range, it is not limited to LOS reading, and, most importantly, it can be used as an anticounterfeiting solution.

Furthermore, a chipless RFID sensor can be realized by modifying one of the metallic/dielectric strips with a smart sensing material (refer to Figure 1.4(c)). The

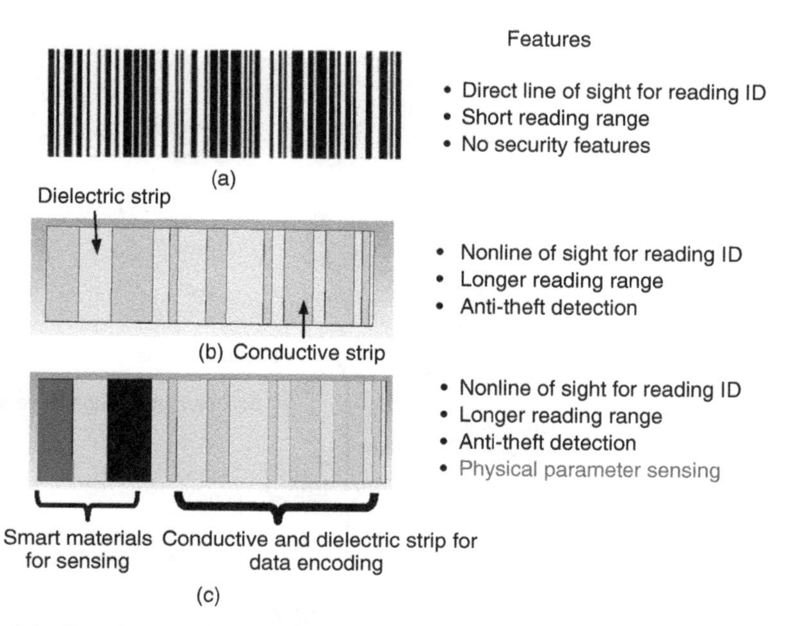

Figure 1.4 Functional evolution and features of tracking ID technology (a) optical barcode, (b) illustration of chipless RFID tag, and (c) illustration of chipless RFID sensor

RCS response of this particular strip corresponds to certain environmental parameters (i.e., humidity, temperature, light, gas, and pH). Hence, a chipless sensor has the features of both data encoding and sensing in a single platform. Moreover, the sensing strips can be modified to include a number of smart materials. Each material responds independently to the change of a particular physical parameter. Here, the data encoding scheme is independent of the sensing mechanism, which entails that multiple parameter sensing is possible in a chipless RFID tag.

1.2 CHIPLESS RFID SENSOR SYSTEM

Our proposed tag sensor consists of a number of RCS backscatterers or resonators that emit a distinct frequency signature when illuminated by an ultra-wide band (UWB) signal. Here an UWB signal is interrogated by a chipless RFID reader (see Figure 1.5).

There are two types of backscatterers within the tag. The first set of scatterers carries the data ID of the tag, and the second set of scatterers carries the sensing information. Here, each scatterer gives a unique spectral response for ID generation and sensing. The resonant frequencies depend on the equivalent circuit parameters of the individual scatterer. In addition, the variation of a particular scatterer's structural parameters does not affect the other resonant frequencies. By tuning the resonant properties of sensing scatterers individually using smart polymer materials, we can develop a single chipless RFID tag with multiple physical parameter-sensing capabilities. The uniqueness of our chipless sensor compared with existing reported studies

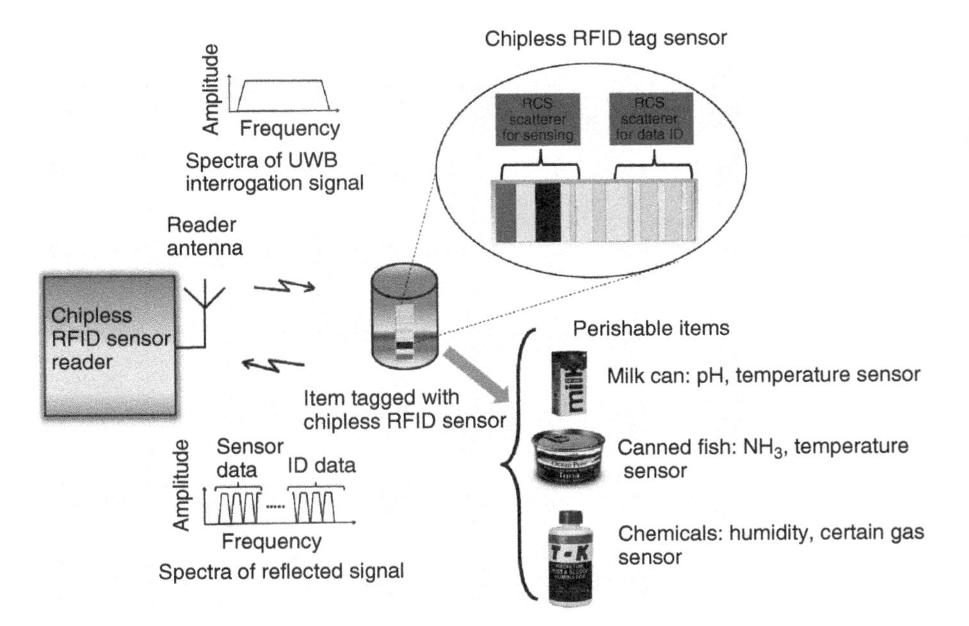

Figure 1.5 Generic block diagram of proposed chipless RFID sensor system

are as follows: (i) data ID and sensing information are coded in both magnitude and phase spectrum, (ii) a single tag has multiple parameter sensing capabilities, and (iii) RF sensing is incorporated using smart materials rather than external sensors or lumped components in the tag circuitry. In addition, the primary application for this chipless RFID sensor is short range (up to 50 cm). Therefore, its application is not affected by UWB power limitation regulation, which entails maximum transmitted equivalent isotropically radiated power (EIRP) of −45 dB m (outdoors) and −55 dB m (indoors) over the UWB microwave frequency band from 3 to 10.6 GHz.

So far, the definition and significance of chipless RFID sensors have been presented. In the next section, the aims of this book and chapter outline are presented.

1.3 PROPOSED CHIPLESS RFID SENSOR

There is a tremendous market push to develop a very low-cost, printable, passive single-node multiparameter chipless RFID sensor for ubiquitous sensing. To address this market demand, an interdisciplinary research project is conducted in two key areas: (i) microwave passive circuit design using metamaterials and (ii) identification, characterization, and fabrication of smart materials for microwave sensing of physical parameters. Integration of smart sensing materials with state-of-the-art passive microwave circuit design provides new fully printable chipless RFID sensors for ubiquitous tagging and sensing of low-cost items. The outcomes are novel high data density and highly sensitive passive RFID sensors, which have numerous real-world applications. Three potential RF sensing applications are addressed in the book: (i) noninvasive radiometric partial discharge (PD) detection and localization, (ii) microwave sensors for environment monitoring, and (iii) nonvolatile microwave memory sensors for event detection. Each application involves the physical layer design of the sensor tag, prototype fabrication, and experimentation. Finally, future research direction is presented in the areas of nanofabrication techniques to make the tag sensor fully printable. Also, a chipless RFID reader architecture is proposed, which is capable of reading data ID and sensing information from the chipless sensor. The book concludes with numerous potential case studies for many emerging applications such as food safety and security, health care, logistics, transportations, smart cities, agriculture, infrastructures, and homeland security.

Figure 1.6 shows an overview of topics covered in the book.

1.4 CHAPTER OVERVIEW

1.4.1 Chapter 1: Introduction

This chapter presents an outline to the overall book on chipless RFID sensors.

1.4.2 Chapter 2: Literature Review

In the first section of this chapter, a comprehensive review of conventional RFID sensors is presented. The objective is to present the classification of RFID sensors

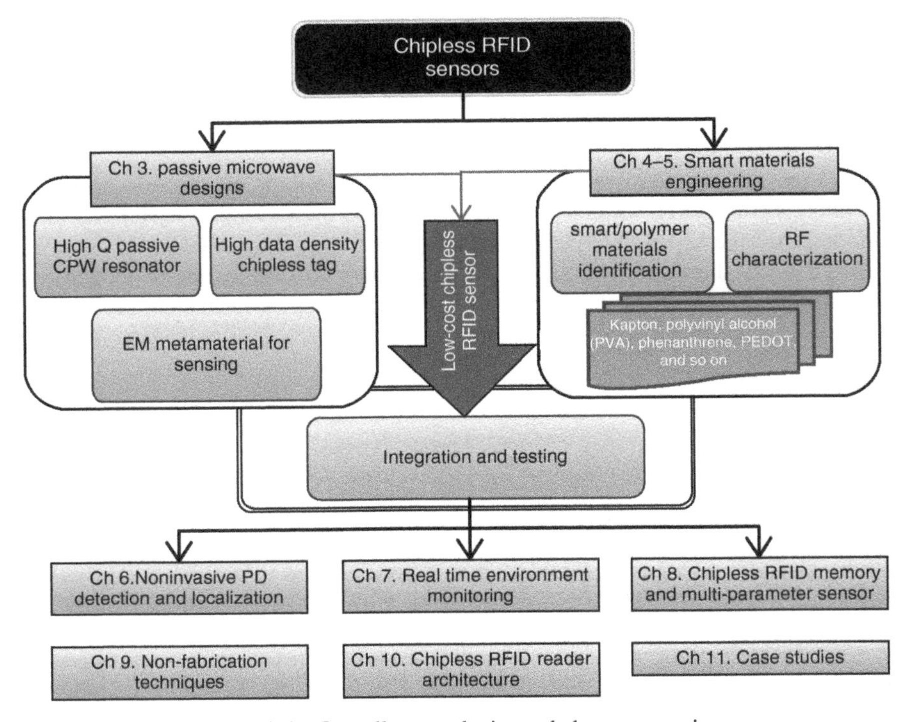

Figure 1.6 Overall research aim and chapter overview

by sensing principle, power requirement, functionality, and applications. This provides a broad overview of the key challenges of current RFID sensor technologies in item-level tagging and sensing.

In the second section, a comprehensive review of state-of-the-art chipless RFID sensors is presented. A comparative summary of various chipless RFID sensor technologies is presented to highlight ongoing research trends. The objective is to establish the aims and challenges in realizing a maintenance-free, batteryless ubiquitous RFID sensor in three application areas considered in this book.

1.4.3 Chapter 3: Passive Microwave Designs

This chapter presents passive RF designs to achieve a compact, high data density, highly sensitive tag sensor for a number of real-world ubiquitous sensing applications. Here, we present three types of passive microwave components: (i) cascaded multiresonator, (ii) RCS backscatterer, and (iii) UWB antenna. A high Q stepped impedance resonator (SIR) is presented for noninvasive radiometric PD detection and localization. A high data density and highly sensitive RCS scatterer is also presented for real-time environment monitoring and event detection. This design includes a slot-loaded patch and an ELC resonator. Here we present the theory, design guidelines, layout, simulation, and measured results for each passive resonator.

1.4.4 Chapter 4: Smart Materials for Chipless RFID Sensors

Smart materials exhibit large and sharp physical and/or chemical changes in response to small physical or chemical stimuli. These materials have great potential for integration with RF devices for environment sensing. This chapter provides a classification of smart materials based on sensing physical parameters (i.e., humidity, temperature, pH, gas, strain, and light). For each class of smart materials, dielectric or conductive property variations in millimeter and microwave frequency are presented. In this chapter, smart materials for microwave sensing application are reviewed and their microwave characteristics in the influence of physical parameters are also explored.

1.4.5 Chapter 5: Characterization of Smart Materials

This chapter presents various novel analysis and characterization techniques including microstructural and surface morphology (X-ray diffraction, XRD; Raman; secondary ion mass spectrometer, SIMS; Fourier-transform infrared reflection, FTIR; atomic force microscopy, AFM; scanning electron microscope, SEM; transmission electron microscopy, TEM; spectroscopic ellipsometry, SE, etc.), optical (UV–vis, SE), electrical and thermal (DC conductivity, stability, etc.), and microwave scattering parameters such as complex permittivity, dielectric loss, and reflection loss in the gigahertz range for sensing materials. This chapter also describes various characterization procedures for dielectric, loss tangent, and conductivity (attenuation) measurement for unknown dielectric and conductive materials. Finally, RF characterization of materials is conducted to investigate their sensitivity and dynamic range.

1.4.6 Chapter 6: Chipless RFID Sensor for Noninvasive PD Detection and Localization

In this chapter, a passive chipless RFID sensor is presented for noninvasive radiometric PD detection in HV equipment. The sensor is a multiresonator-based passive circuit with an antenna for capturing PD signals. The low-cost RF sensor is installed in an individual HV sensing unit for monitoring multiple small units. The sensor block captures the PD signal, processes it with distinct spectral signatures as identification data bits, and transmits the signal to a single sampling channel. From the captured RF signal, both the PD level and source identification can be retrieved. The proposed passive sensor addresses both aspects of PD monitoring, PD detection and faulty source identification. Moreover, in identifying the data bits, time–frequency analysis is utilized for superior detectability. This analysis enables the detection of multiple PD events separated by a defined time delay although they occur at the same instance. Therefore, the sensor can distinguish simultaneous PD occurrences from multiple sources. The proposed sensor system provides low-cost, automated, and battery-free condition monitoring of HV units in a distribution substation.

1.4.7 Chapter 7: Chipless RFID Sensor for Real-Time Environment Monitoring

This chapter presents a chipless RFID sensor for environment monitoring. Two smart materials, Kapton and PVA, are compared for humidity sensitivity in the ultrahigh-frequency (UHF) range. This analysis also determines a number of RF sensing parameters for calibrating humidity in real environments. Next, a chipless RFID humidity sensor is developed and tested for real-time humidity monitoring. Detailed measurement results and sensitivity curves are presented to verify the repeatability and time response of the sensor.

1.4.8 Chapter 8: Chipless RFID Temperature Memory and Multiparameter Sensor

This chapter first presents a chipless RFID memory sensor for temperature threshold detection. This sensor acts as a nonvolatile memory device that is triggered only once. The aim of this memory sensor is to detect and store a particular event when a threshold is exceeded. We use Phenanthrene sublimate material to realize temperature threshold sensing. Detailed experiments are reported to verify the memory effect of our sensor. Finally, a chipless RFID sensor with both humidity and temperature sensing capability is presented.

1.4.9 Chapter 9: Nanofabrication Techniques for Chipless RFID Sensor

This chapter first presents an overview various fabrication techniques that can be used for the development of various chipless RFID sensors. It includes a review on innovative micro- and nanofabrication technologies suitable for roll to roll chipless RFID sensor printing. In addition to the various printing facilities, state-of-the-art micro-/nanofabrication processes, such as hand casting, spin coating, electrodeposition, wet chemical, physical and chemical vapor deposition, laser ablation, direct pattern writing by photolithography/electron beam lithography/ion beam lithography, nanoimprint lithography and etching, and surface and bulk micromachining, suitable for chipless RFID sensor fabrication are described. A general survey and comparison of the different fabrication techniques are also given. The aim is to highlight the limitations of conventional fabrication process and their solutions for on-demand, high-speed printing for flexible, robust, mass productivity of chipless RFID sensors. Moreover, printing, imaging, and characterization procedures of micro- and nanostructures and their integration into RF sensing devices are presented. Printing of microwave passive design on polymers and organic materials has great research potential.

1.4.10 Chapter 10: Chipless RFID Reader Architecture

This chapter details development of a novel chipless RFID reader for decoding chipless RFID sensor's ID information and sensory data. It presents overall architecture of the reader and its operation.

1.4.11 Chapter 11: Case Studies

The chipless RFID sensor has tremendous potential in regard to technological breakthroughs and its social and environmental impacts. It has a number of innovative features such as fully printable, passive, sub-cent, and environment friendly. The potential advantages of these unique features permit chipless RFID sensors in unique applications that could not be achieved previously with both traditional RFID sensors. This section presents a detailed case study on various application areas suitable for chipless RFID sensors. These include retail, pharmaceutical, logistics, power, and construction industries just to name a few.

REFERENCES

1. M. Bellis. (2013). *Bar Codes*. Available: http://inventors.about.com/od/bstartinventions/a/Bar-Codes.htm.
2. D. P. Harrop and R. Das. (17 May). *Printed and Chipless RFID Forecasts,Technologies & Players 2009–2029*. Available: http://media2.idtechex.com/pdfs/en/R9034K8915.pdf.
3. M. Roberti. (2005). *Wal-Mart Begins RFID Process Changes*. Available: http://www.rfidjournal.com/article/view/1385.
4. (2004). Intel Assists Coles Myer With Trial Of RFID Technology. Available: http://www.idg.com.au/mediareleases/4433/intel-assists-coles-myer-with-trial-of-rfid-techno/.
5. S. Preradovic, "Chipless RFID system for barcode replacement," Doctor of Philosophy, Department of Electrical and Computer Systems Engineering, Monash University, 2009.
6. R. Bhattacharyya, C. Floerkemeier, and S. Sarma, "Low-Cost, Ubiquitous RFID-Tag-Antenna-Based Sensing," *Proceedings of the IEEE*, vol. 98, pp. 1593–1600, 2010.
7. S. Kofman, Y. Meerfeld, M. Sandler, S. Dukler, and V. Alchanatis, *Radio Frequency Identification System and Data Reading Method*, Google Patents, 2012.

2

LITERATURE REVIEW

2.1 INTRODUCTION

Radio-frequency identification (RFID) is emerging as a successful alternative to traditional barcodes due to its embedded security, larger reading distance irrespective of line-of-sight (LOS), greater storage capacity, and improved reliability [1]. The RFID tags are made robust so that they can be used in harsh environments and temperatures. Each tag has a unique ID that is used to track or localize goods. The tag uses modulated backscattering of RF signals to communicate with the reader.

Apart from tracking an object, RFID tags can monitor the surrounding environmental conditions and act as sensors of physical parameters, such as temperature, pressure, relative humidity, and gas content [2–4] in addition to the identification function. An object providing its own condition and identification simultaneously simplifies the infrastructure and enhances the quality of the information, thereby creating a link between the *information and communication technology (ICT)* and the *physical worlds* [5]. The addition of physical parameter-sensing capability to RFIDs extends their applications in the retail, pharmaceutical, logistics, biotechnology, and construction industries. Real-time monitoring of perishable items will offer flexible expiry dates of those items and save billions of dollars annually.

In the new era of ubiquitous sensing networks (USNs), where every object connects through Internet of things (IoT), there is a pressing demand for low-cost sensor nodes. This chapter's primary focus is to present a comprehensive literature review of traditional RFID sensor technology available on the market and reported in peer-reviewed articles. This chapter summarizes the challenges of achieving

Chipless RFID Sensors, First Edition. Nemai Chandra Karmakar, Emran Md Amin and Jhantu Kumar Saha.

a low-cost, printable RFID sensor on flexible materials. Finally, a novel chipless RFID sensor for multidimension application areas is presented.

2.2 TRADITIONAL RFID SENSORS

RFID sensors can be classified broadly into two groups: active and passive, according to their working mechanism and power requirements. These sensors have different features and applications, which are discussed in the following sections.

2.2.1 Active RFID Sensors

Active sensors are a subclass of sensors that use batteries to power their communication circuitry, microcontroller, and sensor. These sensor tags operate over a wide range and can achieve high data rates. However, the power supply required in the tag circuitry increases device complexity, cost, and weight.

In Figure 2.1, the basic block diagram of an active RFID sensor tag is shown [6]. The main part of the tag is the microcontroller, which connects the sensors to the RF protocol. The sensors are connected to the microcontroller through the sensor bus and these sensors (S_1, S_2, … , S_n) can be a temperature sensor, pressure sensor, humidity sensor, or any sensor that detects change in physical parameters. The microcontroller and sensors are driven by a power source. The microcontroller has a memory unit as well as a logic unit. It is connected to the RF protocol block, which includes all the logic required to support the commands and drive the RF front end. The RF front end works as a transceiver connecting the analog electrical domain coming from the tag antenna and the digital domain needed for the RF protocol block. The antenna essentially maintains communication with the reader.

A brief description of some widely used active RFID sensors follows.

VarioSens is a smart card temperature sensor incorporating active RFID developed by KSW Microtec AG [7]. It is a very flexible credit card size product, with

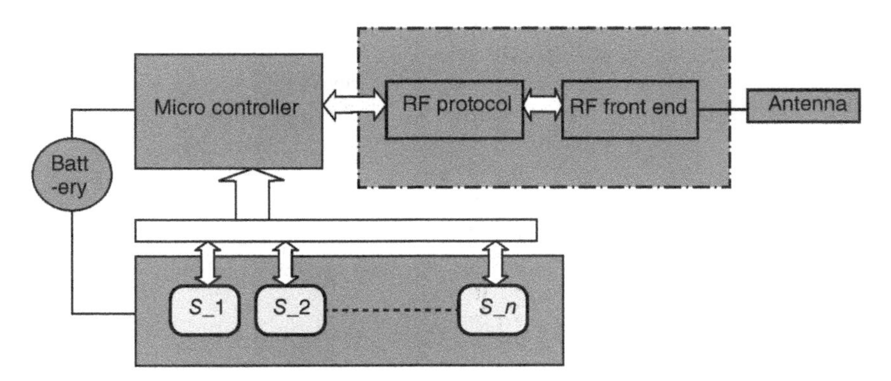

Figure 2.1 Basic block diagram of active RFID sensor tag [6]

superior performance to the conventional temperature sensor tags. It is widely used in shipping products for its high-speed reading and data capture capabilities and light weight. It works in a temperature range from −20 to 50 °C with an accuracy of about 1 °C. It also has a reading range up to 25 cm.

GAO Tek Inc. has developed a temperature sensor tag that collects real-time temperature and transmits to the reader periodically. It is useful in cold storage and medicine transport because of the time information of the data. It has a wide reading range up to 100 m and a temperature range of −50 to +150 °C with 0.7 °C accuracy [8].

RFIT-Pro T 40028 is a temperature and humidity sensor tag supporting long-range operation. It has additional features including tilt and rotation detection, light emitting diode (LED) visual indication, and alarm setup. It works on a temperature range of −40 to +123.8 °C with 0.5 °C accuracy and a humidity range of 0–100% [9].

G2 Microsystems has developed a system on a chip that uses IEEE 802.11 (Wi-Fi) for wireless communication. In addition to asset tracking and identification, G2-based tags can monitor and record various parameters such as temperature, pressure, shock, motion, and security [10]. The external sensor interface is used to monitor the ambient temperature of the environment over a range of −40 to +55 °C with 1 °C accuracy. Moreover, it can send a warning message over Wi-Fi if the temperature exceeds a threshold value.

Alien-BAP is a long-range battery-powered tag developed by Alien Technology [11]. It uses the backscatter power to communicate with the reader operating in 2.45 GHz. The tag has the benefit of interfacing with any external sensors, such as temperature sensors, tamper detection, and shock sensors. It has been used in cold chain management and military applications.

The active and semiactive sensors have onboard power sources and microcontrollers to drive the sensing circuits. They are very costly, bulky, and rigid in operation, have finite lifetimes, and need maintenance. To overcome these problems, researchers have developed passive RFID sensors.

2.2.2 Passive RFID Sensors

Passive RFID sensors obtain their operating power from the RFID reader. The tag circuitry is energized by the EM power radiated from the reader antenna, which then either retransmits or backscatters the modulated signal. Therefore, the reader captures and demodulates the signal to retrieve the tag information. Passive sensors have benefits due to their lower cost, weight, and volume. As the tags do not require an onboard power source, they have longer lifetimes than active tags. Moreover, it is possible to permanently embed the tags in objects for sensing purposes. According to the working principle and circuitry, passive tags can be subdivided broadly into chipless and chip-based tags.

2.2.2.1 Chip-Based Passive RFID Sensors These have an onboard silicon chip or integrated circuit (IC), making them costly and nonplanar. The presence of an

IC in the tag circuit also makes it nonprintable, hindering its application in robust environments. A number of chipped passive sensor technologies have been reported in the literature. These are outlined in the following section.

Complementary metal-oxide semiconductor (CMOS) analog circuit-based passive sensors have been widely used for developing remote sensing networks for real-time security applications due to their low cost and high sensitivity. The tag circuit extracts power from the RF link transmitted from the reader section, and thus requires no onboard power supply. These sensors utilize the fundamental behavior of bipolar junction transistors (BJTs) and integrated resistors.

The basic architecture of this kind of sensor comprises of an analog front end, a digital controller, and a sensing current generator [12–14]. The analog front end maintains communication with the RF link and the digital controller encodes and decodes data, controls memory, and other main operations within the chip.

Intel Research Seattle has developed the **Wireless Identification and Sensing Platform (WISP)** to explore sensor-enabled RFID applications. Besides operating as an RFID tag, it has the capability to sense parameters such as light, temperature, acceleration, strain, and liquid level, and to investigate embedded security [15]. It is reported to be the first fully programmable computing platform that uses power transmitted from a long-range RFID reader and transfers random data in a single response. Some key features of WISP include a read range of 10 ft; include light, temperature, and motion sensor; and it is EPC class 1 Gen2 compatible [16].

Passive LC resonator-based sensors rely on the frequency variation output, which is due to the permittivity change of sensitive material. Electrical capacitance/inductance change is the principle behind this type of sensor. As the physical properties of temperature-sensitive material change with temperature, the equivalent impedance changes, causing a frequency deviation denoting temperature shift. These sensors have the drawbacks of being less sensitive, have complex circuitry, and are nonplanar [17].

Bhattacharyya *et al.* [18, 19] have developed a prototype passive RFID sensor for monitoring the change in temperature or other environmental effects. The sensors use the variation of **reader threshold transmitted power** and tag backscatter power to quantify sensing parameters. There has been significant research on the development of an RFID sensor platform in flexible substrates [20–22]. Although these fully passive sensors can be used without a battery, the IC used in the tag circuit introduces a cost overhead. Some of the key players who are developing RFID sensors are IBM Research, IDENTEC SOLUTIONS, MicroStrain, Permasense, and Senceive. Rida *et al.* [23] propose **paper/polymer-based chipped RFID** sensors that use a 600 dpi inkjet printer. They integrate a single-wall carbon nanotube (SW-CNT) for ammonia gas sensing. However, due to the lumped components and microcontroller, the solution is costly.

2.2.3 Low-Cost Chipless RFID Sensors

The chipless RFID tag sensor provides identification data and monitors a number of physical parameters of tagged objects without having an active sensor in the tag circuitry [18]. The fundamental difference of these sensor tags compared with chip-based tags is the absence of a silicon chip in the tag circuit. The entire cost of the RFID system largely depends on the cost of individual tags. However, an onboard IC is the principal element in the overhead cost of the tag [24]. Moreover, the chipless RFID sensor has benefits over traditional sensors because of its lower cost, longer storage life, robustness, and lower radiated power. Several approaches have been proposed to realize a chipless RFID tag with an integrated sensor. The challenge for designing chipless RFID sensors is to incorporate data encoding and sensing in passive operation. In response to this challenge, three general types of RFID sensors can be categorized: time-domain reflectometry (TDR)-based; frequency modulation-based, and phase-encoded chipless RFID sensors. Figure 2.2 shows a classification of current chipless RFID sensor technologies, depending on their operating principle.

TDR-based chipless RFID tags are energized by sending a UWB signal from the reader in the form of a pulse and capturing the reflected pulse sent by the tag. An array of pulses is hence created, which can be used to encode data. A number of RFID tags have been reported using TDR-based technology for data encoding. These can be classified as nonprintable and printable tags.

Surface acoustic wave (SAW)-based RFID tags have TDR-based operation and **nonprintable** circuitry. SAW tags have been efficiently used to localize tag position, direction of travel, and tag temperature [3, 25–27]. In addition, SAW tags can endure security and safety-related processes that involve elevated operating temperatures in a robust environment.

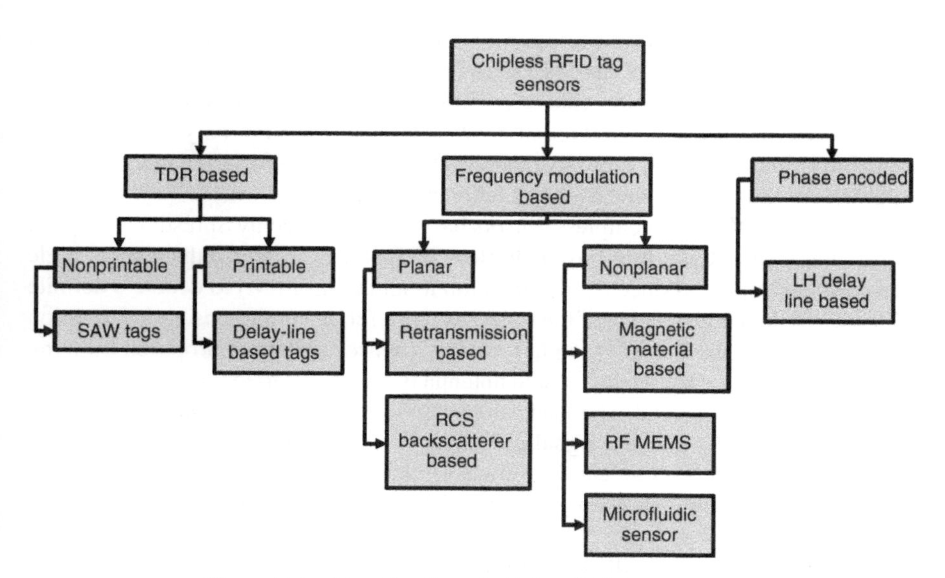

Figure 2.2 Classification of chipless RFID tag sensors

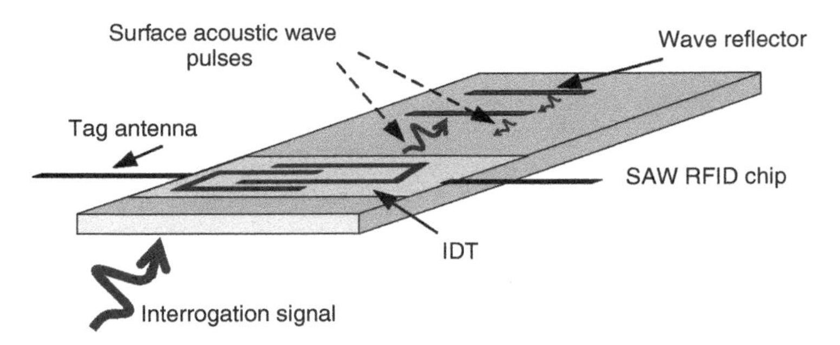

Figure 2.3 Generic SAW-based RFID tag [28]

In a SAW-based RFID system, the reader antenna illuminates a high-frequency EM probe signal, which is picked up by the tag antenna (Figure 2.3) [28]. An inter-digital transducer (IDT) then converts the received signal into a SAW through the converse piezoelectric effect. The SAW propagates from the IDT toward the reflectors arranged in a designated pattern and is partially reflected at each reflector. Moreover, the temperature of the SAW tag can be measured by calculating the time dilation or contraction of the tag response. The temperature coefficient of delay, T_{cd}, at a current temperature with respect to a reference temperature is a measure of difference in time dilation of the impulse responses divided by the delay time at the reference temperature. Hence, T_{cd} is a mathematical function that relates the relative time delay of the response and the tag temperature [29, 30].

After returning to the IDT, the acoustic wave packets are converted back into electrical signals and retransmitted to the receiver antenna. This signal contains information about the number and location of reflectors, together with the propagation characteristics of the SAW. In SAW-based sensors, a physical or chemical influence changes the propagation characteristics of the SAW and accordingly changes the response of the device. Although SAW sensor tags are the only commercially available chipless sensors, they have the disadvantages of being nonplanar, and have complex circuitry and an inconvenient fabrication process.

Printable TDR-based chipless sensors have been reported by Shrestha *et al.* [31]. Here, a microstrip **delay-line-based chipless RFID** tag is presented that uses a single transmission line to produce a pulse position modulation (PPM) to denote a definite ID code and the phase change of the reflected pulse for detecting sensory information [32] (Figure 2.4). In Ref. [31], a platform for a sensor system is proposed utilizing delay-line configuration. Here, a patch antenna is connected to the transmission line, which has a load at the end. The load can be any sensor impedance that generates the phase change of the reflected signal according to the sensing parameters. However, the length of the delay line can be varied to create different ID codes. Therefore, from the reader's end, calculating the time delay and phase change provides the ID and sensory data.

Frequency modulation-based chipless tag sensors encode data into the frequency spectrum using resonant structures or RCS scatterers. These have advantages

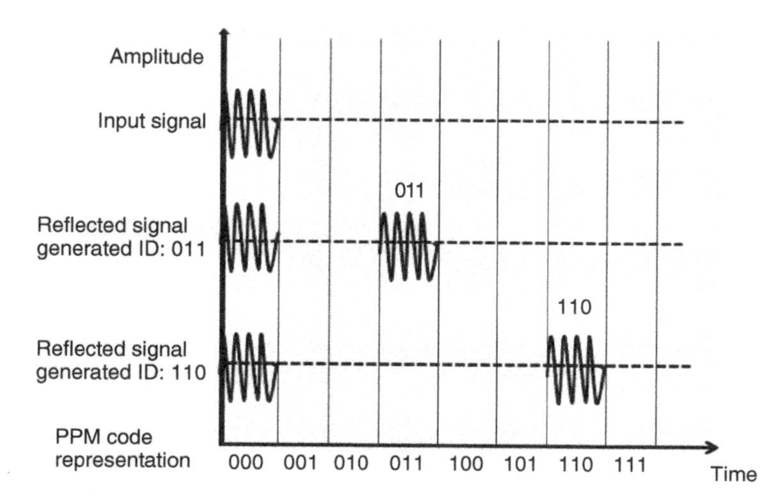

Figure 2.4 Interrogation and coding of delay-line-based chipless tag [32]

over TDR-based tags in regard to their compact size, high data density, and lower cost, and tag reading is less prone to environmental noise. To date, five types of frequency modulation-based sensors have been reported. These are classified into planar and nonplanar circuits.

Planar chipless RFID sensors are advantageous as they are printable on flexible substrates. These tags can be retransmission-based and RCS scatterer-based. **Retransmission-based** tag sensors use a number of cascaded high Q resonators for data encoding and UWB antennas for data transmission and reception (Figure 2.5). In contrast, **RCS scatterer-based** tags have EM scatterers to perform both signal transmission and data encoding. Spectral signature-based chipless RFID tag-integrated

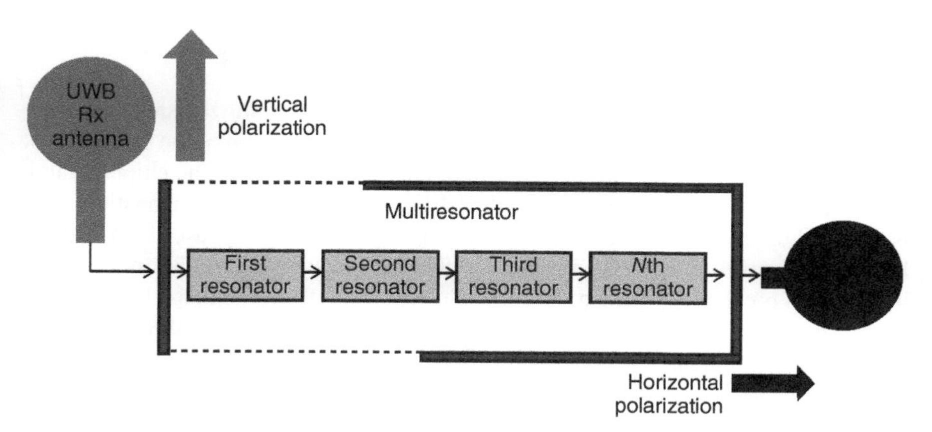

Figure 2.5 Block diagram of retransmission-based chipless RFID tag

sensors are presented in Refs. [4, 33, 34]. Here, a number of microstrip resonator/RCS scatterers modify the frequency spectra of interrogated UWB signals according to their resonant behavior. The unique signature in the backscattered signal is used for data encoding. In addition, the resonators can be designed to resemble specific sensing parameters [35]. The wireless tag sensor has the potential to integrate data encoding and multisensing mechanisms.

Nonplanar chipless sensors use multiple layers of dielectric materials to create a capacitive sensor. **A magnetic material-based** chipless RFID temperature sensor has been proposed by Fletcher and Gershenfeld [36]. These sensors have three different layers of magnetic materials, which change the magnetic spectrum with temperature. Three types of temperature sensors have been developed: (i) nonreversible threshold sensors, (ii) reversible temperature threshold sensors, and (iii) continuous readout temperature sensors using a 30%/70% Ni–Fe layer as modulation element and amorphous alloy strips as signal and bias elements. The continuous readout temperature sensor can be used for real-time monitoring of temperature by tracking the resonant frequency shift. The operating temperature range is 20–70 °C, and this range can be adjusted by varying the Curie temperature of the magnetic material.

Mahmood *et al.* [37] reported an **RF microelectromechanical system (MEMS) sensor** in using capacitive loading of an evanescent-mode cavity resonator. The sensing mechanism is achieved through a rectangular array of actuated microcantilever beams on a silicon substrate. A temperature change causes the microcantilevers to bend, altering the parasitic capacitance of the cavity, therefore causing a shift of resonant frequency [38]. A **microfluidic** sensing mechanism is used in Ref. [39]. Here, a temperature-dependent dielectric fluid is channeled underneath a microstrip structure. Hence, modification of planar capacitance is achieved using dielectric fluid expansion. Neither the RF MEMs nor the microfluid sensor has been explored for data encoding. Furthermore, these structures have complex fabrication methods that increase sensor cost.

Phase-encoded chipless RFID sensors require less bandwidth than the previous TDR and frequency modulation-based sensors. Here data encoding is performed by varying the phase of the backscattered signal. The sensing element is connected as RF loading with the antenna structure. Hence, it modifies the "antenna mode" response of the reflected signal.

Left-hand (LH) delay-line loading of the chipless tags utilizes analog components for phase modulation and enhances the RCS response time of the tag using the slow-wave effect of delay lines [40], which reduces the tag size. The chipless tag is energized by a band-limited pulse emitted from the reader. The interrogation pulse is received by the tag antenna and transmits through a number of cascaded LH delay lines. The received interrogation signal is reflected once it reaches a discontinuity and information is encoded by the differential phase of the reflected signal with respect to a reference phase. As shown in Figure 2.6, an incident wave travels through a delay line with reflection sections. At each reflector, the phase changes φ_1, φ_2, and φ_3. Finally, the carrier phase (γ_n) is the sum of each reflected signal phase.

In Ref. [41], a chipless RFID temperature sensor using a Barium strontium titanate (BST) thick-film temperature-sensitive capacitor is presented. The chip

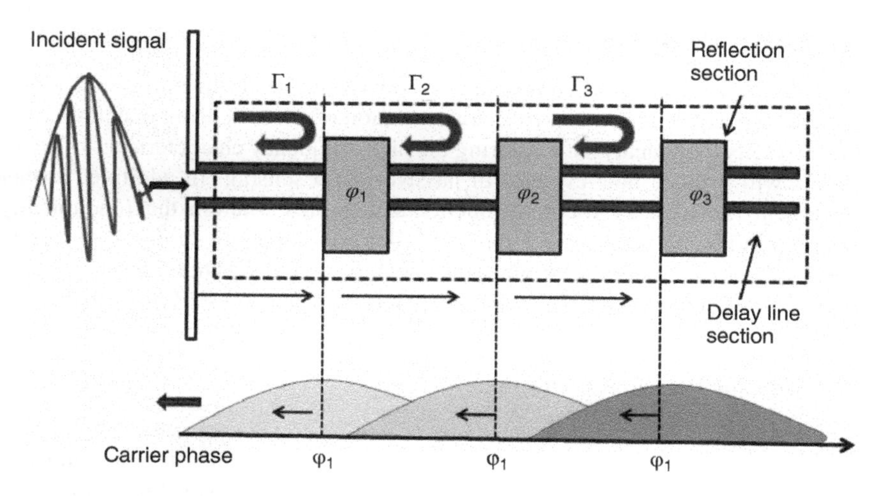

Figure 2.6 Operating principle of LH delay-line-based chipless RFID tag [40]

has a temperature-sensitive circuit integrated with a phase-encoded chipless tag for temperature sensing and data identification. The drawbacks of this tag sensor are its bulkiness and the low data capacity compared to frequency modulation-based tag sensors.

Various chipless RFID sensors that have been reported in open literature have been presented in this section. The limitations in achieving low-cost ubiquitous sensors have also been analyzed. These challenges and limitations are summarized as follows.

2.3 CHALLENGES AND LIMITATIONS OF CURRENT CHIPLESS RFID SENSORS

Chipless RFID sensors are yet to reach the commercial stage as there are a number of challenges researchers need to address. Details of these challenges are discussed in this section.

2.3.1 Fully Printable

The primary goal of chipless RFID tags is for them to be printable on flexible substrates such as polymer, paper, cartons, and boxes [24]. This unique feature of chipless tags has helped them to penetrate the market for low-cost item tagging. However, incorporating sensing feature in the chipless domain hinders printability. To date, existing chipless sensor technologies are not compatible with roll-to-roll printing. Researchers need to investigate novel fabrication techniques such as laser printing or combined chemical etching and nanofabrication to overcome this barrier.

2.3.2 Smart Sensing Materials

Most smart materials have been investigated for DC or low-frequency applications. There is a major gap in microwave characterization of smart sensing materials, synthesis, and sensitivity analysis. Exploring the high-frequency characteristics of smart materials will provide new avenues in passive RFID sensing. In addition, further development is necessary for multifunctional materials wherein the functionality arises from individual nanoscale components that can impart electrical conductivity or responsiveness to light and other physical parameters. These materials will incorporate multiple parameter-sensing mechanisms with enhanced sensitivity.

2.3.3 Multiple Parameter Sensing

A decisive goal of low-cost chipless RFID sensors is to tag each and every perishable item for checking expiry dates. Perishable products typically have a number of physical parameters (temperature, humidity, certain gas, pH, etc.) to indicate spoilage. Therefore, to accurately detect expiry dates, it is desirable to have multiple parameters sensing in a single chipless RFID tag. For example, as milk spoils, it emits carbon monoxide and its pH changes. Monitoring both these parameters will give precise monitoring of the condition of milk. However, this is a fundamental challenge with traditional chipless technologies as the sensing mechanism is linked with the data encoding. Therefore, a chipless RFID with multiple parameter sensing features has great potential in the tagging industry.

2.3.4 Chipless RFID Sensor Systems

Chipless RFID sensors are still in the research phase. There has not been a complete system level implementation of this technology for commercial use. This entails (i) developing readers in regard to wireless communication standards, (ii) calibration of sensors for reference data, (iii) power budget analysis, (iv) sensing data reception and signal processing, and (v) study of the repeatability and stability of sensory data. These steps will ensure a complete system solution for chipless RFID sensor implementation.

2.3.5 Applications

From previous research, we find very streamlined applications of chipless RFID sensors in the real world. Most reported studies have focused on generic environmental parameter monitoring, such as temperature, humidity, strain, cracking, and NH_3 gas. However, their applications can be broadened. For example, it is possible to envisage low-cost disposable sensors for perishable products, chipless sensors for event/threshold detection, sensors for structural health monitoring, and smart homes. Each application requires specific standards and designs of the tag sensors that developers need to take into account. Exploring multidimensional applications of chipless RFID sensor technology is another research domain yet to be explored.

2.4 MOTIVATION FOR A NOVEL CHIPLESS RFID SENSOR

RFID sensors are the fundamental component of the IoT that connects each and every physical object to the cloud database for exchange of information and status. Therefore, the development of low-cost, efficient, high data density, printable sensor nodes is a fundamental research domain in today's high-technology era.

This review of the literature on RFID sensors indicates that there is a pressing demand for chipless sensors for multidimensional applications. The main motivation for this research is therefore to develop a chipless RFID sensor with the features of

- High data density
- Superior sensitivity
- Multiple parameter sensing
- Printable on flexible substrates
- Suitable for versatile real-world applications.

2.5 PROPOSED CHIPLESS RFID SENSOR

The authors aim to develop a low-cost, printable, high data density chipless RFID tag platform for multidimensional sensing applications. In this research, chipless RFID sensing potential has been investigated in three distinct domains for specific applications.

- Noninvasive RF level detection and localization
- Real-time environment monitoring
- Nonvolatile memory sensor for event detection.

2.5.1 Noninvasive PD Detection and Localization

A chipless RFID sensor platform is developed for noninvasive partial discharge (PD) detection and localization of faulty HV equipment. Here, a proof-of-concept sensor prototype for PD detection is developed. The sensor is a multiresonator-based passive circuit with an antenna for capturing PD signals. The sensor node captures the PD signal, processes it with distinct spectral signatures as identification data bits, and transmits the signal to a single sampling channel. From the captured RF signal, both the PD level and source identification can be retrieved. Moreover, in identifying the data bits, time–frequency analysis is utilized for superior detectability. This analysis enables detection of multiple PD events separated by a defined time delay although they occur at the same instant. Therefore, the sensor can distinguish simultaneous PD occurrences from multiple sources. The proposed sensor system provides low-cost, automated, and battery-free condition monitoring of HV units in a distribution substation.

2.5.2 Real-Time Environment Monitoring

Here, a chipless RFID sensor is proposed for environment monitoring. The microwave sensor tag has frequency selective surface (FSS) resonators to encode data. Each resonator operates as an electromagnetic band gap (EBG) structure to attenuate a certain frequency bandwidth. By modifying a particular EBG structure using a dielectric-sensitive material to resemble an environment physical parameter, the frequency characteristics of that resonator can be calibrated to measurable sensing quantities. As a result, a single tag can have multiple FSS structures to carry identification data as well as other designated structures to convey sensory data. This sensor tag will provide real-time information on the environment, such as temperature, relative humidity, pressure, light exposure, and presence of noxious gases, together with data ID when scanned using an RF reader.

2.5.3 Nonvolatile Memory Sensor for Event Detection

The third application extends the chipless RFID sensor platform to realize a non-volatile memory sensor for event detection. The principle of this memory sensor is to detect and store a particular event when a threshold is exceeded. The sensor acts as a nonvolatile memory device that is triggered only once. This sensing mechanism within the chipless RFID platform can revolutionize RFID implementation in supply chain management. In this research, a chipless RFID threshold sensor is developed using the irreversible dielectric property change of certain chemicals. This sensor permanently changes its spectral response when a threshold temperature is exceeded and does not exhibit its actual response, if the temperature goes lower than the threshold once triggered. Therefore, it acts as a memory sensor to store the event of threshold violation. This sensor tag has a potential market in cold chain management, the transport of perishable products, and the storage of explosive chemicals.

2.5.4 Single-Node Multiparameter Chipless RFID Sensor

In addition to the three main application areas of chipless RFID sensors mentioned in the previous section, we present a chipless RFID for multiple parameter sensing. This chipless RFID sensor combines both real-time humidity monitoring and temperature threshold sensing. This feature is a novel finding for wireless sensor platform and has not been reported earlier. Multiparameter sensing in a single chipless RFID tag has great potential in applications where more than one parameter is essential for decision making.

2.6 CONCLUSION

This chapter presents a detailed review of existing RFID sensor technology. Firstly, it discusses the functionality and limitations of chip-based active and passive sensors. Next, a classification of chipless RFID sensor technologies is presented. The aim is to critically analyze existing limitations in realizing a low-cost, fully printable, high data density tag sensor. We found that current chipless RFID sensor technology has limited data capacity, low sensitivity and is not suitable for various real-world applications. This analysis establishes the research gap considered in this book. We aim to fulfill the research gaps with passive microwave design, smart material characterization, fabrication, measurement, and testing. The chipless sensor proposed in this book has a wide range of applications for ubiquitous sensing.

The next chapter presents the first step in chipless RFID sensor development. It details the design, fabrication, and measurement of a number of passive resonators and RCS scatterers for realizing a frequency modulation-based tag sensor.

REFERENCES

1. K. Finkenzeller. *Introduction*: John Wiley & Sons, Ltd, 2010.
2. Y. Li, Z. Rongwei, D. Staiculescu, C. P. Wong, and M. M. Tentzeris, "A Novel Conformal RFID-Enabled Module Utilizing Inkjet-Printed Antennas and Carbon Nanotubes for Gas-Detection Applications," *IEEE Antennas and Wireless Propagation Letters*, vol. 8, pp. 653–656, 2009.
3. W. Buff, S. Klett, M. Rusko, J. Ehrenpfordt, and M. Goroli, "Passive remote sensing for temperature and pressure using SAW resonator devices," *IEEE Transactions on Ultrasonics, Ferroelectrics and Frequency Control*, vol. 45, pp. 1388–1392, 1998.
4. E. Perret, A. Vena, S. Tedjini, D. Kaddour, A. Potie, and T. Baron, "A compact chipless RFID tag with environment sensing capability," in *IEEE International Microwave Symposium*, Montreal, Quebec, Canada, 2012.
5. R. Want, "Enabling Ubiquitous Sensing with RFID," *Computer*, vol. 37, pp. 84–86, 2004.
6. M. Hanhikorpi, A. Ruhanen, F. Bertuccelli, A. Colonna, W. Malik, D. Ranasinghe, T. Sánchez López, N. Yan, and M. Tavilampi. *Sensor-Enabled RFID Tag Handbook* [Online] (accessed 1 October 2013).
7. KSW – VarioSens Basic. Available: http://www.ksw-microtec.de/index.php?ILNK=Active_RFID_VarioSens&iL=2&PHPSESSID=78fa3c304ef36f145d8bc2486bc872f5.
8. GAO Tek Inc. Available: http://www.gaotek.com/index.php?main_page=product_info&products_id=1260.
9. RFIT Temperature & Humidity Sensor Tag Pro T 40028. Available: http://www.rfidinfotek.com/detail/rfid-temperature-and-humidity-sensor-tag/422.html (accessed 1 October 2013).
10. G2 Microsystems – G2C501. Available: http://www.g2microsystems.com/html/products.htm#temp (accessed 1 October 2013).
11. ALIEN – Battery Assisted Passive Tag. Available: http://www.alientechnology.com/docs/AT_DS_BAP.pdf.
12. K. Opasjumruskit, T. Thanthipwan, O. Sathusen, P. Sirinamarattana, P. Gadmanee, E. Pootarapan, *et al.*, "Self-Powered Wireless Temperature Sensors Exploit RFID Technology," *IEEE Pervasive Computing*, vol. 5, pp. 54–61, 2006.
13. S. Hongwei, L. Lilan, and Z. Yumei, "Fully integrated passive UHF RFID tag with temperature sensor for environment monitoring," in *ASIC 2007 (ASICON '07), 7th International Conference on*, 2007, pp. 360–363.

14. D. Pardo, A. Vaz, S. Gil, J. Gomez, A. Ubarretxena, D. Puente, *et al.*, "Design criteria for full passive long range UHF RFID sensor for human body temperature monitoring," in *RFID, 2007. IEEE International Conference on*, 2007, pp. 141–148.
15. A. P. Sample, D. J. Yeager, P. S. Powledge, A. V. Mamishev, and J. R. Smith, "Design of an RFID-based battery-free programmable sensing platform," *IEEE Transactions on Instrumentation and Measurement*, vol. 57, pp. 2608–2615, 2008.
16. Wireless Identification and Sensing Platform (WISP). Available: http://www.seattle.intel-research.net/wisp/.
17. Y. Wang, Y. Jia, Q. Chen, and Y. Wang, "A Passive Wireless Temperature Sensor for Harsh Environment Applications," *Sensors*, vol. 8, pp. 7982–7995, 2008 (accessed 1 October 2013).
18. R. Bhattacharyya, C. Floerkemeier, and S. Sarma, "Low-Cost, Ubiquitous RFID-Tag-Antenna-Based Sensing," *Proceedings of the IEEE*, vol. 98, pp. 1593–1600, 2010.
19. R. Bhattacharyya, C. Floerkemeier, S. Sarma, and D. Deavours, "RFID tag antenna based temperature sensing in the frequency domain," in *RFID (RFID), 2011 IEEE International Conference on*, 2011, pp. 70–77.
20. A. Rida, L. Yang, and M. Tentzeris, *Design and Development of Radio Frequency Identification (RFID) and RFID-Enabled Sensors on Flexible Low Cost Substrates*: Morgan & Claypool Publishers, 2009.
21. R. Vyas, V. Lakafosis, A. Rida, N. Chaisilwattana, S. Travis, J. Pan, *et al.*, "Paper-Based RFID-Enabled Wireless Platforms for Sensing Applications," *IEEE Transactions on Microwave Theory and Techniques*, vol. 57, pp. 1370–1382, 2009.
22. A. Oprea, N. Barsan, U. Weimar, M. L. Bauersfeld, D. Ebling, and J. Wollenstein, "Capacitive Humidity Sensors on Flexible RFID Labels," in *Solid-State Sensors, Actuators and Microsystems Conference, 2007. TRANSDUCERS 2007. International*, 2007, pp. 2039–2042.
23. A. Rida, Y. Li, R. Vyas, S. Bhattacharya, and M. M. Tentzeris, "Design and integration of inkjet-printed paper-based UHF components for RFID and ubiquitous sensing applications," in *Microwave Conference, 2007. European*, 2007, pp. 724–727.
24. S. Preradovic and N. C. Karmakar, "Chipless RFID: Bar Code of the Future," *IEEE Microwave Magazine*, vol. 11, pp. 87–97, 2010.
25. V. K. Varadan, *et al.*, "Design and development of a smart wireless system for passive temperature sensors," *Smart Materials and Structures*, vol. 9, p. 379, 2000.
26. J. Dowling, M. M. Tentzeris, and N. Beckett, "RFID-enabled temperature sensing devices: a major step forward for energy efficiency in home and industrial applications?," in *Wireless Sensing, Local Positioning, and RFID, 2009. IMWS 2009. IEEE MTT-S International Microwave Workshop on*, 2009, pp. 1–4.
27. A. Stelzer, S. Scheiblhofer, S. Schuster, and R. Teichmann, "Wireless sensor marking and temperature measurement with SAW-identification tags," *Measurement*, vol. 41, pp. 579–588, 2008.
28. W. E. Bulst, G. Fischerauer, and L. Reindl, "State of the art in wireless sensing with surface acoustic waves," in *Industrial Electronics Society, 1998. IECON '98. Proceedings of the 24th Annual Conference of the IEEE*, 1998, pp. 2391–2396, Vol. 4.
29. R. J. Barton, T. F. Kennedy, R. M. Williams, P. W. Fink, P. H. Ngo, and R. R. Ingle, "Detection, identification, location, and remote sensing using SAW RFID sensor tags," in *Aerospace Conference, 2010 IEEE*, 2010, pp. 1–19.

30. R. Fachberger, G. Bruckner, R. Hauser, and L. Reindl, "Wireless SAW based high-temperature measurement systems," in *International Frequency Control Symposium and Exposition, 2006 IEEE*, 2006, pp. 358–367.
31. S. Shrestha, M. Balachandran, M. Agarwal, V. V. Phoha, and K. Varahramyan, "A Chipless RFID Sensor System for Cyber Centric Monitoring Applications," *IEEE Transactions on Microwave Theory and Techniques*, vol. 57, pp. 1303–1309, 2009.
32. J. Vemagiri, A. Chamarti, M. Agarwal, and K. Varahramyan, "Transmission Line Delay-Based Radio Frequency Identification (RFID) Tag," *Microwave and Optical Technology Letters*, vol. 49, pp. 1900–1904, 2007.
33. E. Md. Amin and N. Karmakar, "Development of a chipless RFID temperature sensor using cascaded spiral resonators," presented at *the IEEE SENSORs 2011*, 2011.
34. S. Preradovic and N. Karmakar, "Chipless RFID tag with integrated sensor," in *Sensors, 2010 IEEE*, 2010, pp. 1277–1281.
35. Y. Duroc and S. Tedjini, "From radiator to signal processing antenna," presented at *the APMC 2011*, Melbourne, Australia, 2011.
36. R. R. Fletcher and N. A. Gershenfeld, "Remotely Interrogated Temperature Sensors Based on Magnetic Materials," *IEEE Transactions on Magnetics*, vol. 36, pp. 2794–2795, 2000.
37. A. Mahmood, H. H. Sigmarsson, H. Joshi, W. J. Chappell, and D. Peroulis, "An evanescent-mode cavity resonator based thermal sensor," in *Sensors, 2007 IEEE*, 2007, pp. 950–953.
38. T. T. Thai, J. M. Mehdi, H. Aubert, P. Pons, G. R. DeJean, M. M. Tentzeris, *et al.*, "A novel passive wireless ultrasensitive RF temperature transducer for remote sensing," in *Microwave Symposium Digest (MTT), 2010 IEEE MTT-S International*, 2010, pp. 473–476.
39. S. Bouaziz, F. Chebila, A. Traille, P. Pons, H. Aubert, and M. M. Tentsiris, "A new millimeter-wave micro-fluidic temperature sensor for wireless passive radar interrogation," in *Sensors, 2012 IEEE*, 2012, pp. 1–4.
40. M. Schueler, C. Mandel, M. Puentes, and R. Jakoby, "Metamaterial Inspired Microwave Sensors," *IEEE Microwave Magazine*, vol. 13, pp. 57–68, 2012.
41. C. Mandel, H. Maune, M. Maasch, M. Sazegar, X. Schu, *et al.*, "Passive wireless temperature sensing with BST-based chipless transponder," in *Microwave Conference (GeMIC), 2011 German*, 2011, pp. 1–4.

3

PASSIVE MICROWAVE DESIGN

3.1 INTRODUCTION

A passive microwave component is a circuit that operates without any active electronics component such as transistors or integrated circuits. Examples of commonly used passive microwave components are resistors, capacitors, antennas, and filters. These are vital in a chipless radio-frequency identification (RFID) sensor system. Particularly, in a chipless tag circuitry, there are no lumped or active components. Hence, passive components are inevitable for tag operation. Passive components in a chipless sensor receive interrogation signal and transmit backscatter signal encoded with tag ID and sensing information.

The primary aim of this book is to design passive microwave circuits for implementing a chipless RFID sensor. Microwave circuits are the fundamental component of realizing a fully printable, passive, planar chipless sensor. This chapter introduces the different types of passive microwave circuits used in this book.

3.2 CHAPTER OVERVIEW

Figure 3.1 gives an overview of passive microwave components and different types of sensors presented in this book. The components can be categorized into three types according to operation: (i) cascaded multiresonator, (ii) backscatterer, and (iii) ultra-wide band (UWB) antenna. Based on the application, one or more of this different types of passive components are used to realize chipless RFID sensor. These components are integrated to construct two types of chipless sensor in this book.

Chipless RFID Sensors, First Edition. Nemai Chandra Karmakar, Emran Md Amin and Jhantu Kumar Saha.

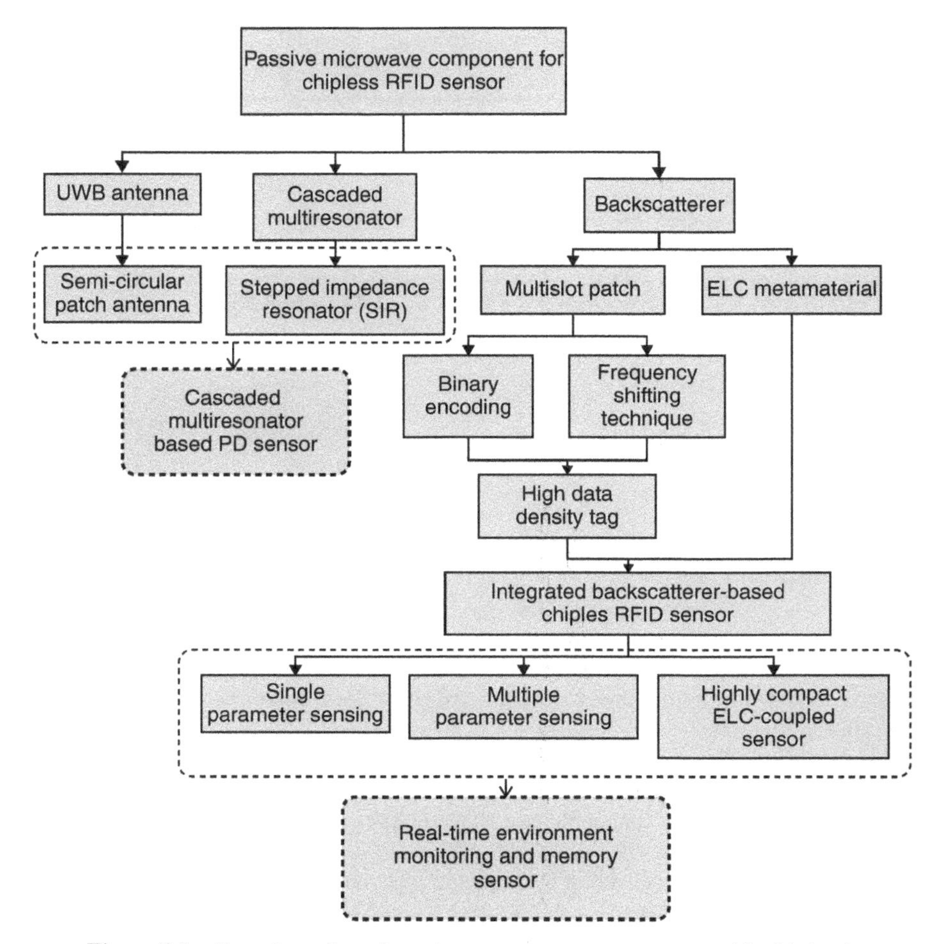

Figure 3.1 Overview of passive microwave components presented in this book

The first type of sensor is cascaded multiresonator-based partial discharge (PD) sensor. It incorporates a semicircular patch antenna and a number of stepped impedance resonator (SIR) connected in series for PD level detection and faulty source identification.

The second type of sensor is based on planar backscatterer. It incorporates a highly compact multislot patch resonator for data encoding and electric inductive–capacitive (ELC) metamaterial for physical parameter sensing. Here, data encoding is performed in two topologies: (i) binary encoding and (ii) frequency shifting technique. Moreover, ELC resonator encompasses certain smart material to perform dielectric sensing. Finally, we present three distinct backscatterer chipless RFID sensors integrating multislot resonator and ELC resonator as shown in Figure 3.1. These sensors are used for real-time humidity and temperature threshold sensing.

The organization of this chapter is as follows:

- Section 3.3 presents detailed theory of various types of passive components considered in this book. Later, this section presents the theory of operation for two types of chipless RFID sensor developed in this book.
- Section 3.3.1 presents circuit layout for various passive microwave components and integrated sensor. It shows generic design guideline for realizing a fully passive, planar, backscatterer-based chipless RFID sensor. It also describes the fabrication method of microwave components.
- Section 3.4 presents detailed simulation and measured results for the designed passive microwave components and integrated sensor.
- Finally, the conclusion section outlines the significant outcomes of this chapter.

3.3 THEORY

The organization of theory section is shown in Figure 3.2. It is divided broadly in two sections. In Section 3.3.1, four passive microwave components considered in this are presented. Next, integration of these components to realize chipless RFID sensors is presented in Section 3.3.2.

3.3.1 Passive Microwave Components

The following sections present the four passive microwave components considered in this book. These are (i) tri-step SIR, (ii) semicircular patch antenna, (iii) multislot patch, and (iv) ELC resonator. The tri-step resonator and UWB patch antenna are designed for developing a PD sensor. Moreover, multislot patch and ELC resonator are two radar cross-section (RCS)-based passive resonators and designed for an integrated chipless RFID sensor for environment monitoring and memory sensor.

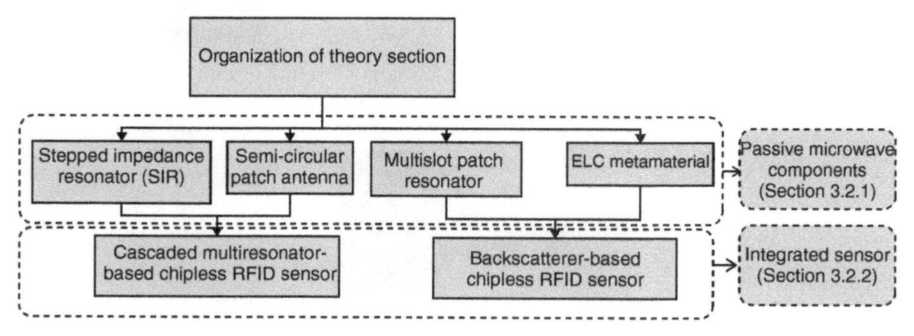

Figure 3.2 Organization of theory section

3.3.1.1 Stepped Impedance Resonator SIRs are transmission line resonators utilizing quasi-TEM modes. The SIR has advantages over uniform impedance resonators (UIR) in terms of the wide degree of freedom in design, their compact size, and ease of fabrication. These structures can be used in various frequency bands from RF and millimeter wave as well as in applications such as filters, mixers, and oscillators [1, 2].

A tri-step SIR bandstop filter is considered in this book for high Q stopband application. A tri-step SIR gives additional size reduction compared with the two-step SIR or UIR [3]. The basic structure of a three-element half-wave SIR is shown in Figure 3.3(a). Similar to the two-step SIR, this structure is symmetric at the mid center and comprises two cascaded quarter-wave tri-step SIRs, as shown in Figure 3.3(b). The characteristic impedance of the three steps is Z_1, Z_2, and Z_3 with electrical lengths θ_1, θ_2, and θ_3. The equivalent impedance of tri-step SIR is given in Ref. [4] by Equation 3.1.

$$Z_s = \frac{j\left(Z_1Z_2\tan\theta_2 + Z_1Z_3\tan\theta_3 + Z_1{}^2\tan\theta_1 - \frac{Z_1{}^2Z_2}{Z_3}\tan\theta_1\tan\theta_2\tan\theta_3\right)}{\left(Z_1 - Z_2\tan\theta_1\tan\theta_2 - Z_3\tan\theta_1\tan\theta_3 - \frac{Z_1Z_2}{Z_3}\tan\theta_2\tan\theta_3\right)} \tag{3.1}$$

Taking $a = Z_1/Z_2$ and $b = Z_3/Z_2$, from the resonance condition for the SIR filter, the electrical length of Z_3 can be found as

$$\theta_3 = \tan^{-1}\left(\frac{1 - a\tan^2\theta_1}{P\tan\theta_1}\right)$$

where $P = (b + a/b)$.

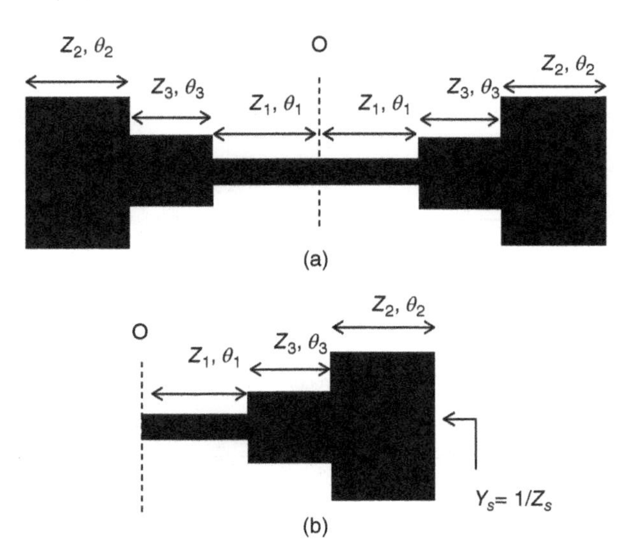

Figure 3.3 (a) Proposed $\lambda/2$ type tri-section SIR structure with two quarter wavelength tri-step SIRs as in (b) (here O is the short terminal)

Therefore, the overall length of the SIR is

$$\theta_T = 2\theta_1 + \theta_3 \tag{3.2}$$

The overall length is a function of a, b, and θ_1. From Equation 3.2, for a set of values of (a, b), we get a distinct value for θ_1 that makes θ_T minimum. Performing a parametric sweep over $\theta_{1,}$ it is found that for the range $a > 1 > b$, the overall tri-step SIR length was minimum. This range is selected for designing the tri-step SIR filter [3, 4]. Design of coplanar waveguide (CPW) tri-step SIR is detailed in Section 3.4.1.

The main reasons for choosing a tri-step quarter-wave SIR for cascaded multiresonator-based PD sensor are as follows:

- The function of SIRs in a PD sensor is to attenuate UWB signals at certain frequency to generate frequency-modulated signal for faulty source identification. In this regard, high Q passive SIR is a suitable microwave bandstop filter.
- SIRs have advantage over other passive filters in terms of higher degree of freedom to achieve sharp selectivity and narrow stopband [4].
- Tri-step SIRs have a larger footprint area to dissipate heat while operating adjacent to HV power apparatus.

3.3.1.2 Semicircular Slot-Loaded Patch Antenna on Air Gap Layer A semicircular patch antenna developed by the Monash Microwave Antennas RFID and Sensors Research group is used for chipless RFID PD sensors. This antenna is developed under the Australian Research Council's (ARC) Linkage project grant LP0989355, "RFID-based sensor to detect partial discharge from faulty power apparatus." The antenna is L-shaped slot-loaded semicircular patch on air gap layer (see Figure 3.4). This type of antenna has been studied earlier by Ray and Krishna and has been shown to have a wide bandwidth [5–7]. Air was used as the substrate of the antenna in order to reduce the weight and cost of fabrication. Design specification and measured results of the antenna is presented and discussed in Section 3.4.2. It needs to be mentioned that this antenna element is not planar and used to verify proof of concept

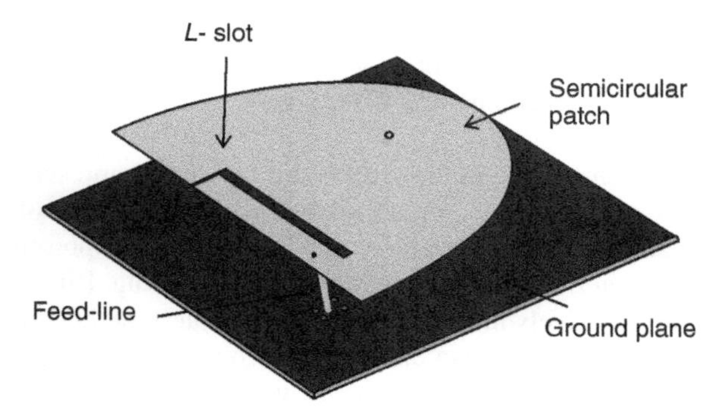

Figure 3.4 Layout of a semicircular patch antenna on air gap layer

for the PD sensor. In future, the antenna will be replaced by a planar, CPW monopole antenna for wider bandwidth.

3.3.1.3 Backscatterer A backscatterer or RCS scatterer is a "filter" for plane waves. These are narrow-band two-dimensional frequency selective surface (FSS) structures for which transmission and reflection coefficients are dependent on the frequency of operation, polarization, and incident angle of the transmitted wave. These were initially developed to control the transmission and reflection characteristics of an incident EM wave. An RCS scatterer can be designed to have a much smaller size compared with wavelength along with the increased bandwidth.

Here, we propose two backscatterers: (i) a multiple-slot patch resonator for high data density tags and (ii) an ELC resonator for dielectric sensing. Detailed theory and operation are presented in the following sections.

3.3.1.4 Multislot Patch Resonator Data bit encoding in chipless RFID tags is generally accomplished by generating spectral signatures utilizing different resonating scatterers. Space-filling curves [8], split microstrip dipoles [9], and half-wave slot resonators [10] are some commonly reported resonating scatterers for chipless RFID applications. As a number of slot resonators can be accommodated in a small footprint, they are the most space-efficient [10]. For a slot resonator with a physical length L_0, the resonant frequency or the frequency whereby the signature will be produced can be predicted using Equation 3.3 [11]:

$$f_r = \frac{c}{2L_0}\sqrt{\frac{2}{1+\varepsilon_r}} \tag{3.3}$$

where c denotes the speed of light and ε_r denotes the relative permittivity of the substrate.

A rectangular patch with N number of concentric U-shaped slots has been investigated in this research for data encoding. Figure 3.5 shows a generic illustration of a rectangular patch with N slots. The slots $S_1, S_2, S_3, \ldots, S_N$ have a total physical length of $L_1, L_2, L_3, \ldots, L_N$, and from Equation 3.3, each slot will resonate at distinct frequency. As the patch is illuminated by a plane wave, it shows a corresponding resonance in the backscattered RCS spectrum. Each slot will produce a magnitude "dip" resulting in N number of resonances in the frequency spectrum (Figure 3.5). Depending on the tag design, this multislot rectangular patch can encode data in two configurations: (i) binary encoding and (ii) frequency shifting.

Binary Encoding In this configuration, each slot resonator represents a single data bit. The number of slots has 1:1 correspondence with the number of encoded bits. The presence or absence of a magnitude "dip" or "null" is represented by "0" or "1". This means that an N-slot patch can encode N bits of data. Binary encoding has been studied comprehensively in Ref. [12]. It is the simplest method of data encoding in frequency modulation-based data encoding. Encoded data bits can be accurately retrieved by using an RFID reader compared with frequency shifting-based tags.

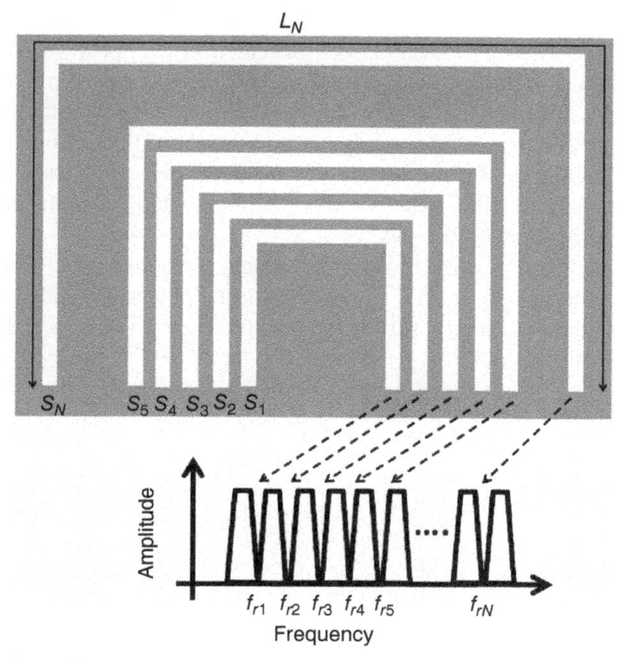

Figure 3.5 Illustration of *N*-slot rectangular patch

However, binary encoded tags have limited data capacity and require a large footprint area. Also, mutual coupling between adjacent slots put limit in the number of slots and frequency signatures in a defined band. Therefore, a second method called frequency shifting technique is proposed in the next section.

Frequency Shifting to Encode Data Bits In this configuration, a different approach for data bit encoding has been adopted and the method enables encoding of multiple bits by using a single resonator. The proposed method relies on the absence or presence of resonance as well as on frequency shifts of resonance to encode data bits. From Equation 3.3, it is evident that the length L_0 of the resonator has an inverse relationship with the resonant frequency. Therefore, a slight variation in L_0, which can be accomplished by increasing the length of the slot or using a metallic filling element to shorten the length of the slot, will correspondingly cause the resonant frequency (f_r) to shift to a lower or higher frequency, respectively. As an increase or decrease in the resonator length correspondingly gives rise to a left shift (to a lower frequency) or to a right shift (to a higher frequency) in the resonance frequency, the variation in resonator length plays a critical role in the proposed data encoding method and has been referred as "*frequency shifting parameter.*" Resonance frequencies slightly shifted to left or right, but originating from a single resonator, can be linked to represent different states of a data bit. Here, unlike the encoding methods used in Refs. [13–15], where only two bit states were generated from a single resonator, the total number of states will be more than two. A similar approach of encoding data bits using the length variation of dipoles was presented in Ref. [16]. In addition, a preliminary concept of data encoding with multislot resonators was presented in Ref. [17]. With a sensitive reader, it will be possible to exactly determine the position of the resonance and, hence, accurately retrieve the ID of the designed chipless tag. The resonant frequency shifting technique enhances the overall data density of the slot-loaded rectangular patch-based chipless RFID tag and offers improvement over the chipless tag reported in Ref. [18].

Figure 3.6 shows a generic illustration of data encoding using frequency shifting technique. The slots $S_1, S_2, S_3, \ldots, S_N$ have initial physical lengths of $L_{i1}, L_{i2}, L_{i3}, \ldots, L_{iN}$ and variable frequency shifting parameters $x_1, x_2, x_3, \ldots, x_N$. This gives the realized length of each slot as

$$L_{f1} = L_{i1} - 2x_1$$
$$L_{f2} = L_{i2} - 2x_2$$
$$L_{f3} = L_{i3} - 2x_3$$
$$L_{fN} = L_{iN} - 2x_N$$

Now, for the initial length L_{iN} of the Nth slot, we obtain a resonant dip at f_{rN} in the frequency spectrum. Therefore, a variable frequency shifting parameter x_N can map the resonant frequency within a band Δf_{SN}. Here, Δf_{SN} is the *maximum frequency shift band* allocated for the Nth slot. If there is M number of frequency channels having bandwidths $\Delta f_{SN1}, \Delta f_{SN2}, \Delta f_{SN3}, \ldots, \Delta f_{SNM}$ within Δf_{SN}, then we can have a total of 2^{M+1} states represented by each slot. Allocated bandwidth for each frequency channel or state is referred to as *frequency channel band* Δf_{SNM}.

Here, the $(M+1)$th state is the condition where the slot is detuned by shorting. This gives the total number of bits encoded by an N-slot monopole $N \times \log_2 (M+1)$. Compared with the binary data encoding configuration, frequency shifting gives $\log_2 (M+1)$ times more data bits in the same footprint area. This gives a high data density, compact tag for chipless RFID realization. In Section 3.4.4, we discuss the detailed design of an $N=3$ slot resonator patch encoding data by using the frequency shifting technique.

3.3.1.5 ELC Metamaterial

Metamaterial for RF Sensing Metamaterials are artificial scaled structures that resonate at a frequency smaller than the wavelength of the interrogation signal. They exhibit a strong electric field enhancement in a confined space and can be used for optimized, highly sensitive tools for dielectric sensing. Based on this property, a number of novel applications of metamaterials have been proposed recently. For instance, surface Plasmon sensors based on metamaterials are proposed in Ref. [18]. In Ref. [19], metamaterials are utilized for high-frequency sensors. Huang and Yang [20] analyzed the performance of metamaterial sensors and concluded that metamaterials can significantly enhance sensor resolution and sensitivity. Zheludev [21] presented

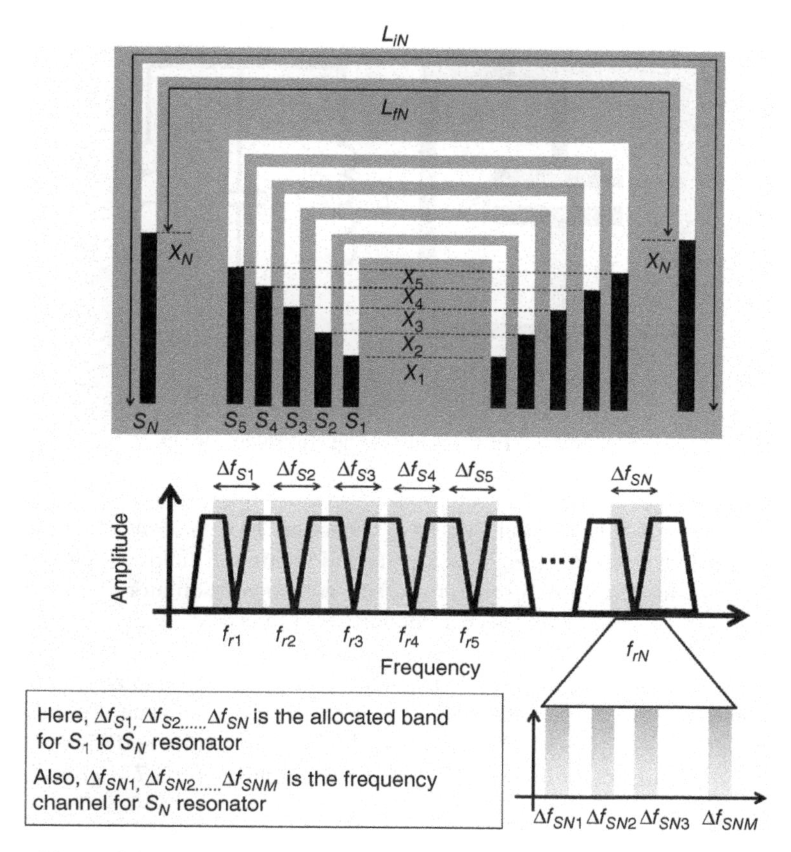

Figure 3.6 Illustration of frequency shifting technique for data encoding

future trends of metamaterials applications in sensing. These create new degrees of freedom in RF sensor design, which have the potential to enhance sensitivity and expedite readout.

A point to note that sensing devices can detect minute changes, subject to the following four conditions: first, the sensors need to operate at a frequency low enough so that it is not affected by background and substrate absorption. This presents a major challenge as traditional sensing devices have a limited footprint area, and such compact space tends to increase the operating frequency of sensors. Hence, it is desirable to retain a small layout of the sensor while decreasing its operating frequency. Second, it is essential for the sensors to produce high measurable readout data with high Q resonant characteristics to accurately detect the shift in transmission spectra. The third condition relates to the linearity of sensing, which depends on the quality of sensors. The fourth condition is the sensitivity of sensors. For high-sensitivity sensors, we can resolve minute shifts in transmission spectra using the limited number of data points in one frequency scan of the RFID reader.

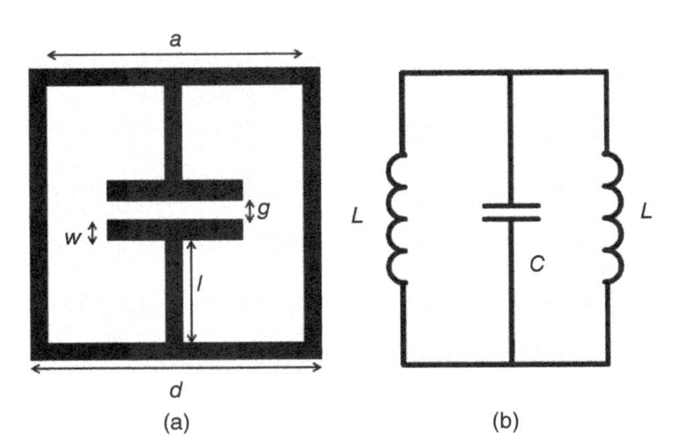

Figure 3.7 ELC resonator and equivalent circuit

ELC Resonator In this research, we have chosen an ELC resonator with a fundamental mode that couples strongly to a polarized incident E-field and marginally to a uniform H-field. The ELC resonator is a self-contained oscillator that is robust in preserving its resonant properties near boundaries or interfaces. Figure 3.7 shows the physical layout and the equivalent circuit model of an ELC resonator. As a plane wave illuminates the resonator, the middle capacitor-like structure couples to the E-field and is connected to two parallel loops that provide the inductance. Therefore, the structure resonates at a frequency determined by its equivalent L and C components. The resonant frequency of the ELC resonator shown in Figure 3.7 is given by (3.4) [22],

$$f_0 = \frac{1}{2\pi}\sqrt{\frac{2}{LC}} \tag{3.4}$$

In Equation 3.4, L and C are the equivalent inductance and capacitance of the resonator structure. Here, the capacitance generated between two split gaps of the ELC resonator has a major influence on the structure's resonance frequency.

In this research, we have used a dielectric sensing material as the superstrate of the ELC resonator. The sensing material fills the gap between the two parallel strips of the capacitor. Hence, any change to relative permittivity, ε_r, of the material due to physical parameters results in a corresponding change in the value of capacitance generated between the two strips. This change in capacitance produces a corresponding resonant frequency shift. As shown in Figure 3.8, both real and imaginary parts of permittivity can cause resonant shift. In the case of real ε_r' change occurs in the resonant frequency, whereas for imaginary ε_r'' we expect drift in resonance power at a particular frequency. This attribute of smart-material-coated ELC has been used to sense the physical parameters of the surrounding environment. Detailed design of an ELC resonator for microwave sensing is presented in Section 3.4.5.

Here, instead of using one of the slot resonators, which are used for encoding tag identification data, a separate resonator is utilized for the following reasons:

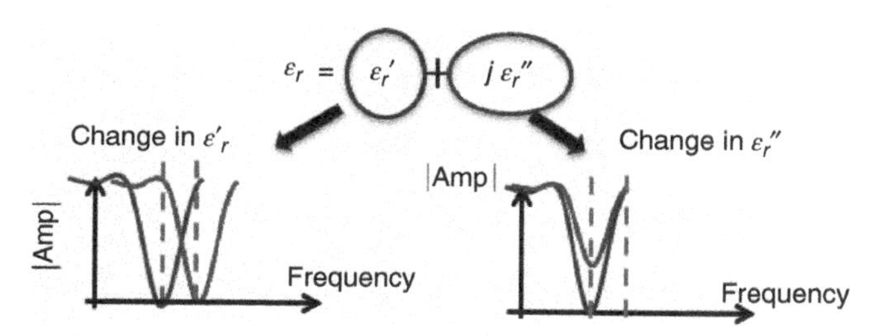

Figure 3.8 Dielectric sensing using superstrate

(i) The mechanism for humidity sensing used here is based on frequency shifts. For the closely located slot resonators, shifting or detuning of the resonance frequency of adjacent resonators also influence the location of the resonance frequency for a particular resonator due to mutual coupling. Therefore, the use of one of the slot resonators to carry out sensing operations might introduce ambiguity in measured sensing data, as any resulting frequency shifts may arise due to a change in environmental parameters, or a change in mutual coupling, or a combined effect of both.

(ii) The ELC resonator has a concentrated *E*-field within its parallel capacitor plates, which gives larger dielectric sensitivity compared with the distributed inductance of a slot resonator. This confirms that an ELC resonator has high dielectric sensitivity for analytes/materials placed between the parallel arms.

(iii) The use of a separate high Q resonator (i.e., ELC) allows multiple parameter sensing. For instance, more than one ELC resonator can be incorporated in our design together with multislot resonators, where each ELC resonator will carry particular sensing information (i.e., temperature, humidity, gas, pressure, and pH)

3.3.2 Integrated Chipless RFID Sensor

Based on the passive microwave components, we present two types of chipless RFID sensor in this research. These are (i) cascaded multiresonator-based sensor and (ii) backscatterer-based sensor. The following subsection describes the theory and general operation of these two sensors.

3.3.2.1 Cascaded Multiresonator-Based Sensor Cascaded multiresonator-based chipless RFID sensor consists of two modules: (i) an antenna R_x to capture UWB signal and (ii) a frequency modulation unit to produce certain frequency signature to the captured RF signal.

A general structure of cascaded multiresonator-based sensor is shown in Figure 3.9. The frequency-modulated signal can be directly sent to signal analyzer or retransmitted to a central server. In this study, we develop a PD sensor having a UWB patch antenna to capture PD signal. Also, a number of cascaded SIR as frequency-modulated unit. The modulated PD signal is sent to PD analyzer for further processing [12].

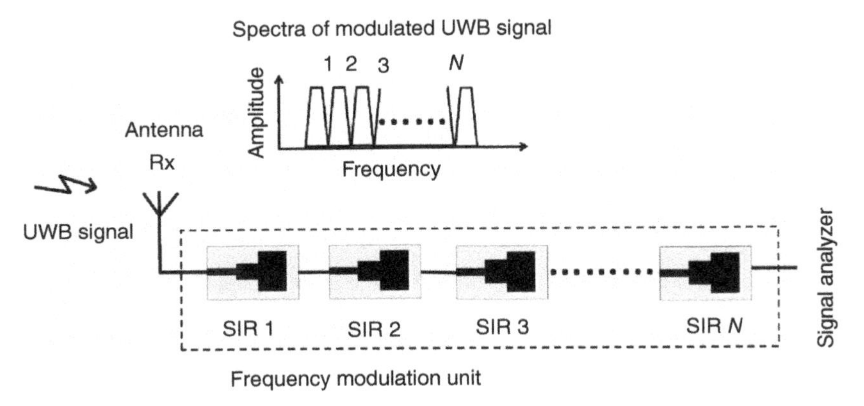

Figure 3.9 General structure of cascaded multiresonator-based chipless RFID sensor

We use the UWB nature of short-span PD pulses to encode data in frequency spectra. The resonators attenuate certain frequencies within the frequency band of the RF signal. A particular resonator's configuration embeds a distinct frequency signature to encode data as identification bits. The presence of a magnitude null resembles logic "0," whereas the absence of a magnitude null at a distinct frequency resembles logic "1." In addition, the combination of data bits when no filter is present is omitted in our proposed sensor system. Therefore, N number of cascaded resonators can produce $K = 2^N - 1$ different frequency signatures to detect K number of objects. Design of a cascaded multiresonator-based sensor is presented in Section 3.4.3. Also, detailed experimentation and measured results with high-voltage PD source are presented in Chapter 6.

3.3.2.2 Backscatterer-Based Chipless RFID Tag Sensor A backscatterer-based chipless RFID tag sensor is realized through integrating multislot patch resonator and ELC resonator in the same layout. Figure 3.10 shows an illustration of a multislot resonator cascaded with two ELC resonators. Here, slot resonators S_1–S_N are designated for data encoding. Moreover, ELC1 and ELC2 resonators are separated in the frequency spectrum and operate within band Δf_{ELC1} and Δf_{ELC2} to carry sensing information. Ideally, each resonator operates independently within their designated frequency band, and we can incorporate more than two sensing parameters, depending on the application. A multiple sensing chipless RFID tag sensor is designed in Section 3.4.6.

3.4 DESIGN

3.4.1 Tri-Step SIR

For the prototype of our cascaded multiresonator-based chipless RFID sensor, a frequency band of 100 MHz at 800 MHz center frequency is selected. Two CPW

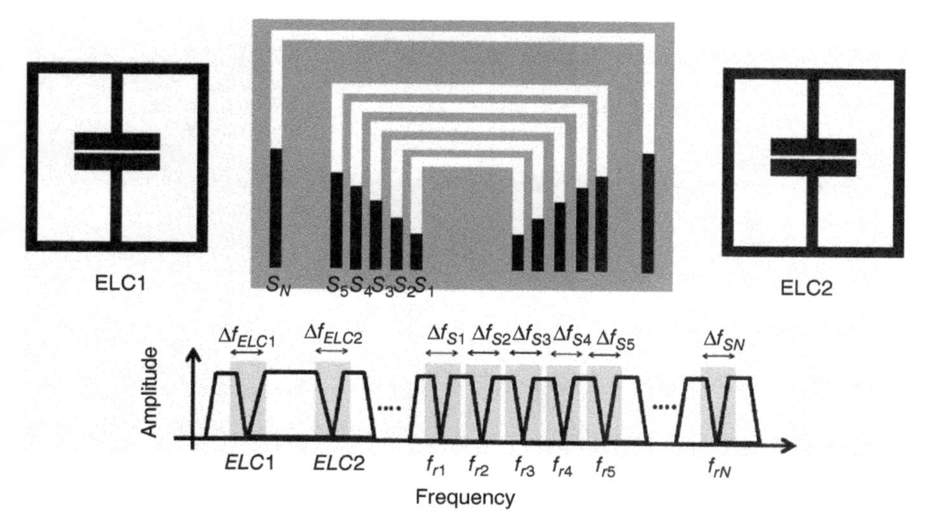

Figure 3.10 Illustration of chipless RFID tag with multiple parameter sensing

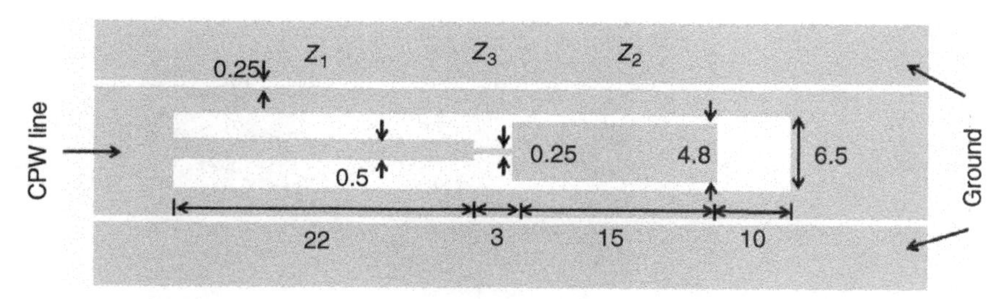

Figure 3.11 Layout of the tri-section SIR at 825 MHz. The simulation was performed on Taconic TLX_0 substrate with relative permittivity $\varepsilon_r = 2.45$ and tan $\delta = 0.0019$ and substrate thickness $h = 0.5$ mm. The CPW line was matched to 50 Ω. All the measurements are in millimeters

tri-section SIR filters are designed in CST Microwave Studio (MWS) operating at 775 and 825 MHz. To create an SIR filter in a CPW line, the microstrip SIR structure was cut away from the continuous 50-Ω line. The layout of SIR filter at 825 MHz is shown in Figure 3.11.

The resonance characteristics are explained using the surface current distribution in Figure 3.12. The surface current density is maximum along the edge of the CPW line at 1 GHz frequency (refer to Figure 3.12(a)). As this is outside the resonance frequency band, charge distribution along the SIR structure is negligible. The current passes through the edge of the CPW line as it has minimum impedance. However, at the resonant frequency (825 MHz), there is an intense current density along the SIR structure, which results in signal attenuation at the receiving end (Figure 3.12(b)).

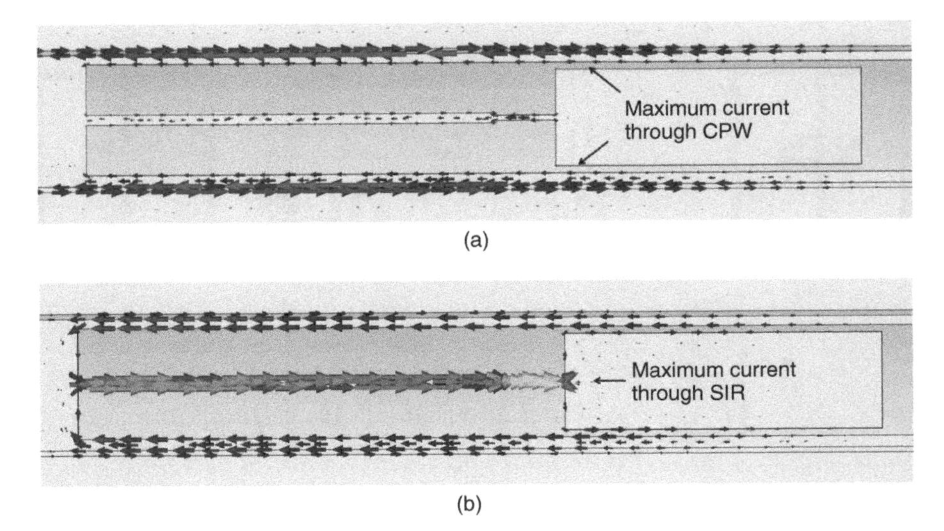

Figure 3.12 Surface current distribution of the SIR structure at (a) 1 GHz (outside resonant condition) and (b) 825 MHz

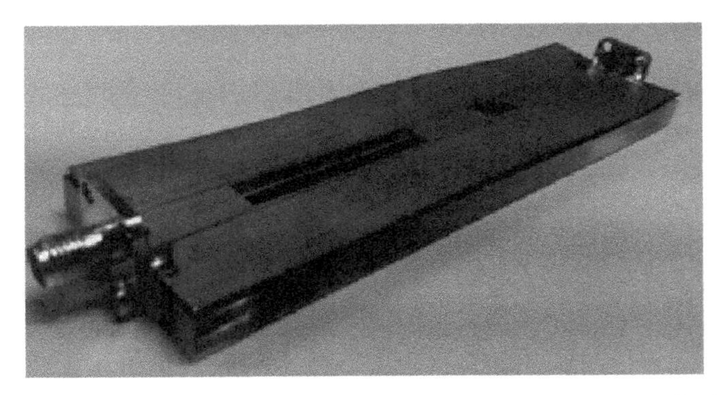

Figure 3.13 Photograph of fabricated SIR structure on Taconic TLX_0 substrate

Most of the power is reflected back to the incident port. Also, as the SIR filter acts as an open-ended quarter-wave transmission line, it has minimum current density at the open end and maximum density at the short end.

According to the design, the SIR structure is fabricated on Taconic TLX_0 substrate. Copper is etched out to create the microwave structure. A photograph of the fabricated SIR filter operating at 825 MHz is shown in Figure 3.13. The SIR operating at 775 MHz has the same layout but the length of tri-step impedances Z_1, Z_2, and Z_3 are varied to shift resonant frequency. Detailed simulation and measured results of the SIR filter are presented in Section 3.5.1.

3.4.2 Semicircular Patch Antenna

Fabricated semicircular patch antenna is shown in Figure 3.14. As shown in the top view, the antenna has an aluminum ground layer and *L*-slot semicircular brash plate as resonance patch. The antenna is energized through a feed line and it has air as dielectric medium. The antenna was designed in CST MWS and the optimum length is calculated using parametric sweep. The design goal for semicircular patch antenna was to achieve a bandwidth of 100 MHz at 800 MHz center frequency.

Table 3.1 shows the critical parameters of the antenna.

To increase the bandwidth of the antenna, second-order matching was used at the feed probe. In this method, extra reactance is introduced by placing a piece of coaxial cable in place of the feed probe. Depending on the *LC* of coax cable, the bandwidth could be adjusted. The measured return loss, radiation pattern, and gain are presented in Section 3.5.2.

3.4.3 Cascaded Multiresonator-Based Chipless RFID Sensor

Integrating the UWB patch antenna and SIRs, a cascaded multiresonator-based chipless RFID sensor is designed. In our chipless sensor, two SIR filter operating at 775 and 825 MHz are used to designate $K=2^2-1$ or 3 distinct frequency signature

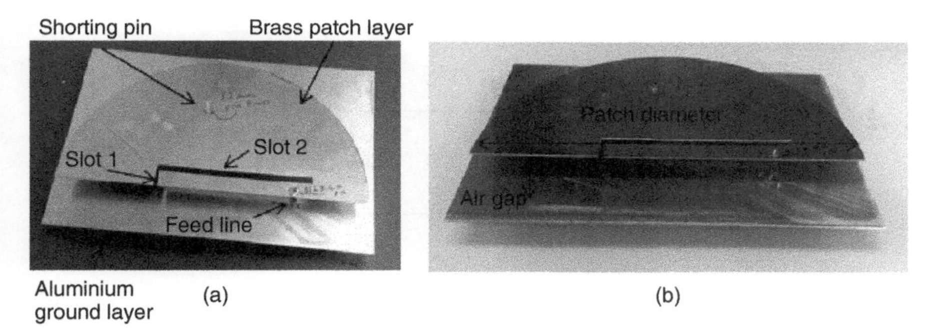

Figure 3.14 Fabricated semicircular patch antenna: (a) top view; (b) perspective view

TABLE 3.1 Critical Parameters of L-slot Semicircular Patch Antenna

Patch Antenna Design Parameters	Value (mm)
Length of slot 1	11
Length of slot 2	101.55
Width of slot 1	2
Width of slot 2	4
Patch diameter	202
Air gap layer thickness	20

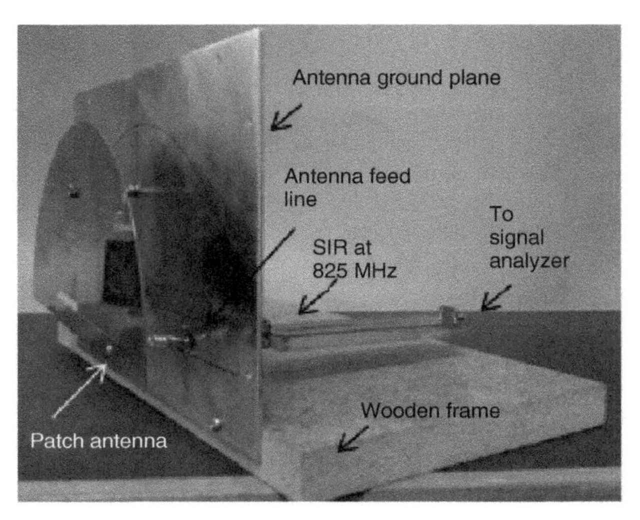

Figure 3.15 Photograph of UWB patch antenna connected to SIR filter for chipless RFID sensor operation

TABLE 3.2 Data Bit Represented by Cascaded Multiresonator-based Chipless RFID Sensor

Resonant Frequency		Data Bit	
775 MHz	825 MHz		
✓	×	0	1
×	✓	1	0
✓	✓	0	0

(Figure 3.15). These frequency signatures can be used to designate data bits "01," "10,", and "00," respectively, as shown in Table 3.2. Detailed results of cascaded multiresonator-based sensor are given in Section 3.5.3.

3.4.4 Multislot Patch Resonator

In this section, we present design guideline for multislot backscatterer for chipless RFID tag development. First a three-bit RCS scatterer is designed to encode data in binary configuration. Next, frequency shifting technique is implemented to encode high data density in compact footprint area. The operating frequency for the tag is 7–10 GHz.

3.4.4.1 Binary Encoding We simulated a three-bit tag to show binary data encoding by using a multislot RCS scatterer. In Figure 3.16, the three slots S_1, S_2, and S_3 are

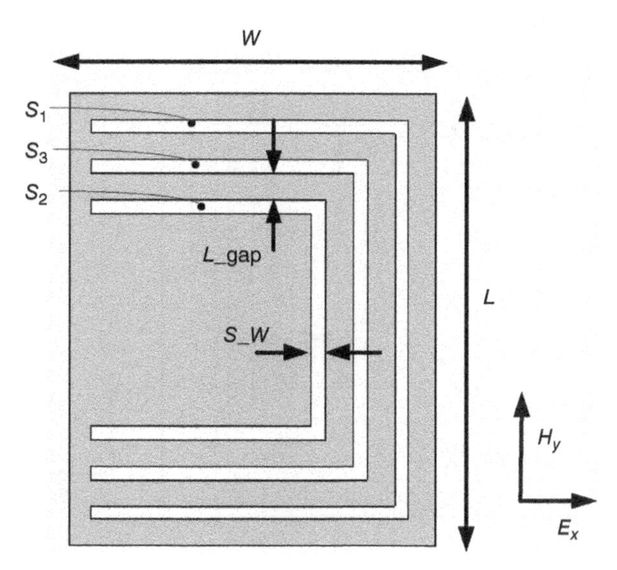

Figure 3.16 Layout of slot-loaded patch utilizing binary encoding. The dimensions are $W = 5.5$ mm; $L = 6.8$ mm; $S\_w = 0.2$ mm; $L\_gap = 0.4$ mm. Substrate Taconic TLX_0; height, $h = 0.5$ mm; $\varepsilon_r = 2.45$; tan $\delta = 0.0019$ (not drawn to scale)

tuned to operate within 7.0–10.0 GHz. Each slot resonates at a particular frequency, depending on its structural parameters. This resonance is observed as a magnitude dip in the backscattered RCS spectrum when the tag is illuminated by a plane wave. In contrast, if a slot is shorted at one corner, that changes its resonant condition. This eliminates the magnitude dip from the tag's RCS response [23]. Hence, three slots can encode three bits of data independently. Here, the tag is designed on Taconic TLX_0 substrate and a horizontal plane wave is used for exciting the tag and an RCS probe with identical polarization is placed 50 mm away from the tag to receive the far-field backscattered signal. The minimum gap of two slots is $S\_w = 0.2$ mm, and the overall footprint area is 6.8 mm × 5.5 mm.

3.4.4.2 Frequency Shifting Technique for Data Encoding Here, a rectangular patch loaded with three U-shaped slots is designed to designate bits for tag ID using the frequency shifting technique. Three slots, S_1, S_2, and S_3, shown in Figure 3.17 are tuned to operate within the 7.0–10.0 GHz frequency band. The tag is designed in CST MWS with Taconic TLX_0 as substrate. The slots are of equal widths (S_w) and gaps (L_gap). To allow changing the length of slot resonators and thus to facilitate shifting of resonance frequencies arising from three slots, three frequency shifting parameters, namely, x_1, x_2, and x_3, are, respectively, affiliated with slots S_1, S_2, and S_3.

Figure 3.18 shows the allocated frequency band in a graphical illustration of different slot lengths. In this design, we have set *the maximum frequency shift band* $\Delta f_{SN} = 1$ GHz for each slot, and the frequency shifting parameters are set to have

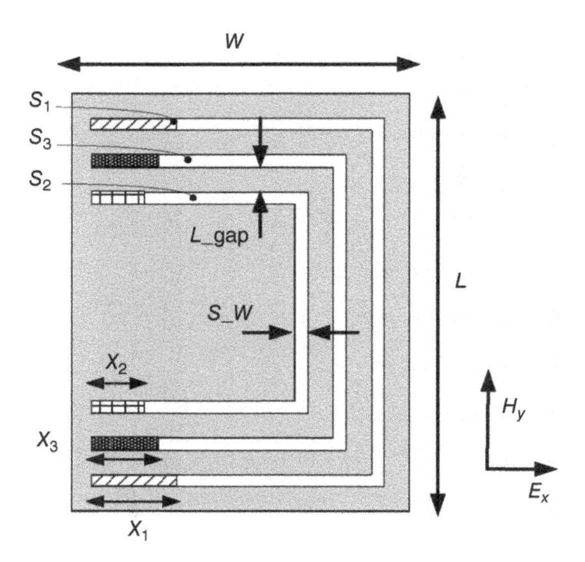

Figure 3.17 Layout of slot-loaded rectangular patch utilizing frequency shifting technique for data encoding. The simulation is performed in CST MWS with Taconic TLX_0 as substrate with a substrate height of 0.5 mm; $\varepsilon_r = 2.45$ and tan δ of 0.0019. The dimensions are $W = 7.3$ mm; $L = 6.8$ mm; $L\_gap = 0.3$ mm; $S\_w = 0.2$ mm

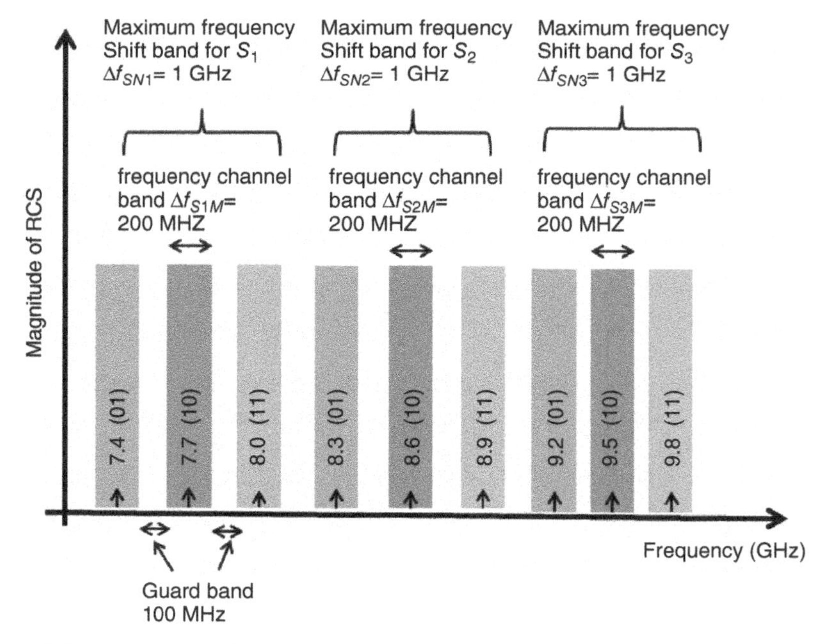

Figure 3.18 Illustration of frequency shifting technique for data encoding. Here, a particular slot can encode data bits 01, 10, and 11 depending on its resonant frequency

$M = 3$ frequency channels with bandwidths of $\Delta f_{SN1} = \Delta f_{SN2} = \Delta f_{SN3} = 200$ MHz. The presence of resonances in these frequency bands can represent data bits 01, 10, or 11, respectively, whereas no resonance detected at any of these three bands will imply that the resonance due to particular slot is detuned and thus can represent data bits 00. Specific allocated frequency bands for slots S_1, S_2, and S_3 along with the corresponding values of frequency shifting parameters x_1, x_2, and x_3 are shown in Table 3.3.

Here, we have kept a guard band of 100 MHz between adjacent frequency channels Δf_{SN1}, Δf_{SN2}, and Δf_{SN3}. As the compact multiple slots are prone to mutual coupling and fabrication error, introducing a guard band will reduce unwanted frequency overlapping.

The designed $N = 3$ slot tag will effectively encode a total of $(3 \times \log_2 (4)) = 6$ data bits. In contrast, with three slots we can have only three bits if data is encoded using binary data encoding. Moreover, the frequency shifting technique can be extended to have high data capacity. Two critical parameters for increasing bit capacity are (i) *maximum frequency shift band* Δf_{SN} and (ii) *frequency channel band* Δf_{SNM}. For the larger frequency shift band Δf_{SN}, we can accommodate more channels M. It also needs to be noted that the bit capacity largely depends on Δf_{SNM}. In this study, $\Delta f_{SNM} = 200$ MHz is taken, depending on reader accuracy and frequency selectivity. Therefore, for a high-sensitive reader, the channel band will be less. Here, a practical example of applying the frequency shifting technique for data encoding is provided, encountering physical limitations and challenges.

Three tags were designed and fabricated to show the different data combinations encoded by the proposed frequency shifting-based encoding technique. The designed tags are one sided, which means no metallic ground plane is required on the other side

TABLE 3.3 Value of Frequency Shifting Parameters and Corresponding Frequency Band

Slot name, S_N	Frequency shifting parameter, X_N	Frequency shifting parameter, X_N (mm)	Frequency band, Δf_{SNM} (GHz)	Corresponding bit combination
S_1	X_1	1.8	$\Delta f_{S11} = 7.3 - 7.5$	(01)
		2.1	$\Delta f_{S12} = 7.6 - 7.8$	(10)
		2.4	$\Delta f_{S13} = 7.9 - 8.1$	(11)
S_2	X_2	0.2	$\Delta f_{S21} = 8.2 - 8.4$	(01)
		0.4	$\Delta f_{S22} = 8.5 - 8.7$	(10)
		0.6	$\Delta f_{S23} = 8.8 - 9.0$	(11)
S_3	X_3	2.1	$\Delta f_{S31} = 9.1 - 9.3$	(01)
		2.3	$\Delta f_{S32} = 9.4 - 9.6$	(10)
		2.5	$\Delta f_{S33} = 9.7 - 9.9$	(11)

of the rectangular patch. Section 3.5.4 shows detailed simulation and measured result for the fabricated tags.

3.4.5 ELC Resonator for RF Sensing

According to the ELC resonator theory discussed in Section 3.3.1.5, we designed an ELC resonator at 6.96 GHz. The parametric values of the ELC resonator are shown

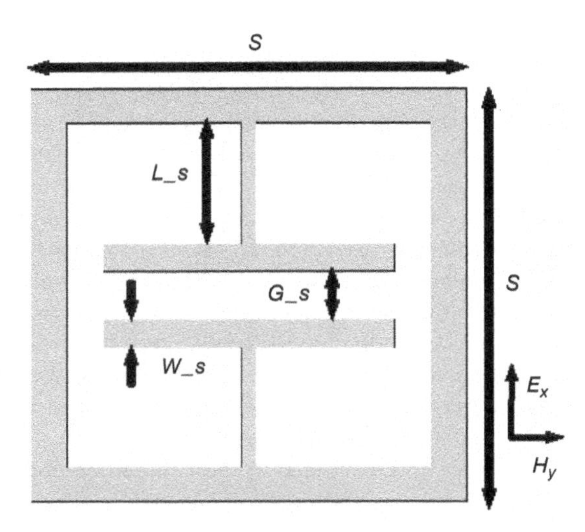

Figure 3.19 Layout of ELC resonator. The dimensions are $S = 6$ mm; $L\_s = 1.75$ mm; $G\_s = 0.7$ mm; $W\_s = 0.4$ mm. Substrate Taconic TLX_0; height, $h = 0.5$ mm; $\varepsilon_r = 2.45$; tan $\delta = 0.0019$

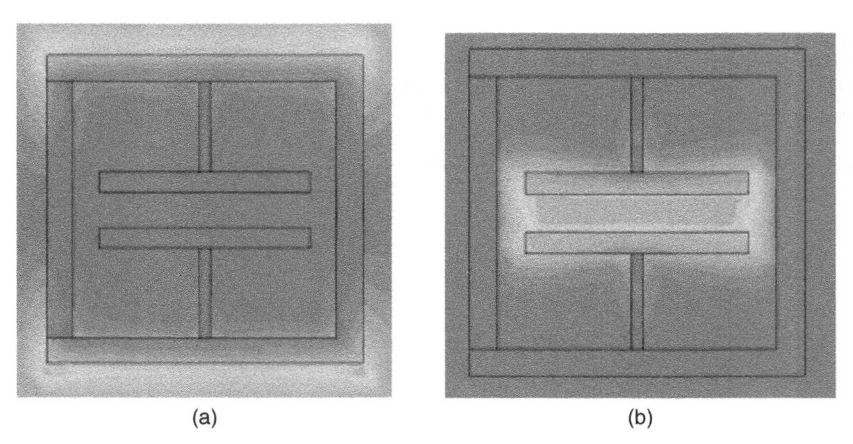

Figure 3.20 (a) Simulated E-field concentration at a frequency outside resonance (6 GHz). (b) E-field concentration at resonant frequency (6.96 GHz)

in Figure 3.19. In Figure 3.20(a), simulated electric field concentration is shown outside resonance frequency (6 GHz). It shows there is no E-field confinement between the parallel-plate capacitors. Moreover, Figure 3.20(b) shows the simulated E-field concentration of the ELC resonator at resonance. At this frequency, a prominent E-field is developed between the capacitive plates. This is due to the resonance of distributive LC element. The coupled LC produces a closed current loop within the circuit and attenuates interrogated EM signal. This indicates that ELC resonators are high Q bandstop filters suitable for dielectric RF sensing. Detailed simulation and measured results for the ELC resonator is shown in Section 3.5.5.

The next section presents integrated chipless RFID tag sensor using two prime components (i) multislot patch resonator and (ii) ELC resonator.

3.4.6 Backscatterer-Based Chipless RFID Tag Sensor

This section presents development of backscatterer-based chipless RFID sensor for multidimensional application. As discussed in the theory section, the tag sensor utilizes multislot resonator for carrying tag ID and ELC resonator for sensing.

An overall design method for chipless RFID sensor is shown in Figure 3.21. The first step of sensor design is to set the specifications for tag ID and sensor. The critical parameters for frequency shifting technique-based tag are (i) frequency band of operation, (ii) number of bits, (iii) maximum frequency shift band Δf_{SN}, and (iv) frequency channel band Δf_{SNM}. Besides, desired specifications for ELC resonator are (i) resonant frequency and (ii) frequency band of operation. After setting the specifications, individual passive components are designed to achieve desirable results. Finally, the two microwave components are integrated in a single layout to develop a chipless RFID tag sensor.

In this study, we explore three different RCS scatterer-based sensors. These are

- Chipless RFID tag for single parameter sensing
- Chipless RFID tag for multiple parameter sensing
- Highly compact ELC-coupled chipless RFID tag.

Detailed design of each sensor is presented in the subsequent sections.

3.4.6.1 Chipless RFID Tag for Single Parameter Sensing Here, an ELC resonator and multislot loaded patch is integrated in a single footprint area to develop an RFID tag sensor. The sensor has three slot resonators for data encoding and one ELC resonator for sensing. The overall bandwidth of the sensor is 6–10 GHz. However, the multislot patch resonators operate within 7–10 GHz and ELC resonator operates with 6–7 GHz. Figure 3.22 shows the layout of the overall sensor tag. The total dimensions are 15 mm × 6.8 mm. This design is more compact than the chipless tags reported to date [24].

The current density profile of the overall tag sensor is presented in Figure 3.23. The ELC resonator has resonant frequency at 6.96 GHz, hence at this frequency the current forms closed loops circulating through the two symmetric distributive inductances (Figure 3.23(a)). At this frequency, there is minimum current density circulating the slot resonators. At 7.9 GHz slot S_1 resonates, producing a null at RCS spectrum (Figure 3.23(b)). We observe high current density on the edges of this slot, confirming the resonant condition. Similarly, S_2 and S_3 slots have correspondingly high current densities at their resonant frequencies (Figure 3.23(c) and (d)). The tag is fabricated using chemical etching technique on Taconic TLX_0 substrate. The simulated and measured results are presented in Section 3.5.6.

3.4.6.2 Chipless RFID Tag for Multiple Parameter Sensing According to the theory summarized in Figure 3.10, a chipless RFID tag for multiple parameter sensing was designed and is shown in Figure 3.24. Here, two ELC resonators, ELC1 and ELC2, are coupled with a multislot resonator. The ELC resonators are located spatially apart to have less mutual coupling and cross-interference of sensing materials. Both ELC resonators have similar dimensions, and only the gaps between the parallel-plate capacitors are altered to have a resonant frequency variation in frequency spectra. The overall tag sensor operates from 6.5 to 10.6 GHz. Two ELC

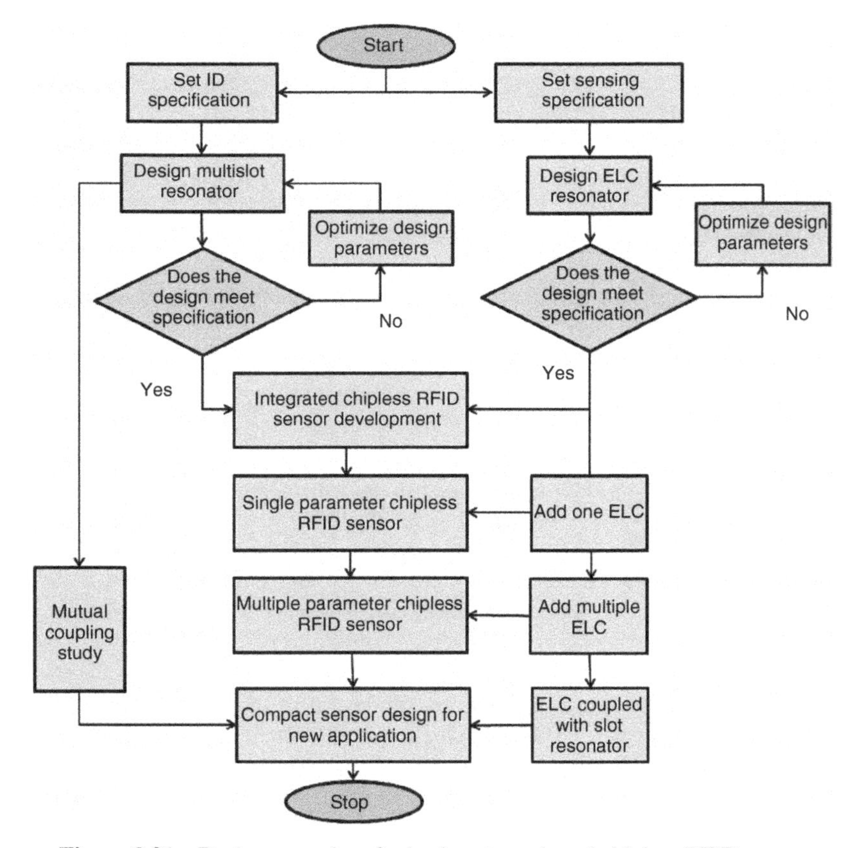

Figure 3.21 Design procedure for backscatterer-based chipless RFID sensor

resonators are dedicated to operate within 6.5–7.5 GHz and 7.5–8.2 GHz for sensing two distinct parameters. Also, the maximum frequency shift band for this design is $\Delta f_{SN} = 800$ MHz. Simulated and measured results for multiple parameter sensing tag are presented in Section 3.5.6.

3.4.6.3 Highly Compact ELC-Coupled Chipless RFID Sensor For mass deployment of chipless RFID sensor, the footprint area should be as small as possible. In this section, we present the design of a highly compact chipless RFID tag sensor, in which the fundamental sensing element is an ELC resonator. However, the ELC resonator is coupled with multislot resonators to create an extremely compact structure. The designed tag sensor operates between 6 and 10 GHz. Allocated frequency band for sensing is 6–7 GHz and for data encoding is 7–10 GHz.

The tag presented in Figure 3.22 was further downsized by analyzing the mutual coupling between the ELC resonator and multiple slots. By tuning the distance between the ELC and slot resonator (G), a new layout of the tag sensor is proposed in Figure 3.25. Here, the ELC resonator is embedded in rectangular slots to reduce the overall footprint area to about 60% of the previous design. The optimized gap between the ELC resonator and S_2 is $Gn = 0.6$ mm.

In this design, mutual coupling between adjacent slots at resonant frequency is reduced by increasing spatial distance. This is achieved by selecting slot lengths such that slots with maximum length variation are in close proximity. Figure 3.26 explains this effect of reduced coupling between adjacent slots at resonance using current density profile. Figure 3.26(a) shows current density profile at the resonance frequency of ELC resonator. At this frequency, the ELC resonator shows strong current-induced

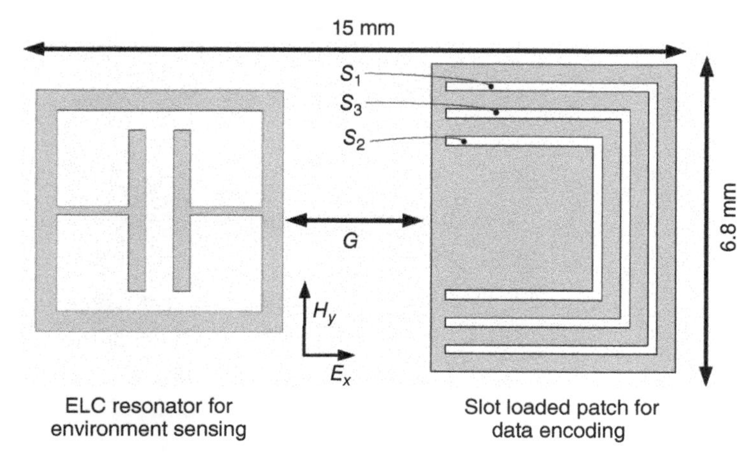

Figure 3.22 Layout of chipless RFID tag sensor. The simulation was performed in CST MWS with Taconic TLX_0 as substrate with substrate height of 0.5 mm; $\varepsilon_r = 2.45$, and $\tan\delta = 0.0019$. The overall dimensions are 15 mm × 6.8 mm. Gap between two resonators $G = 3$ mm

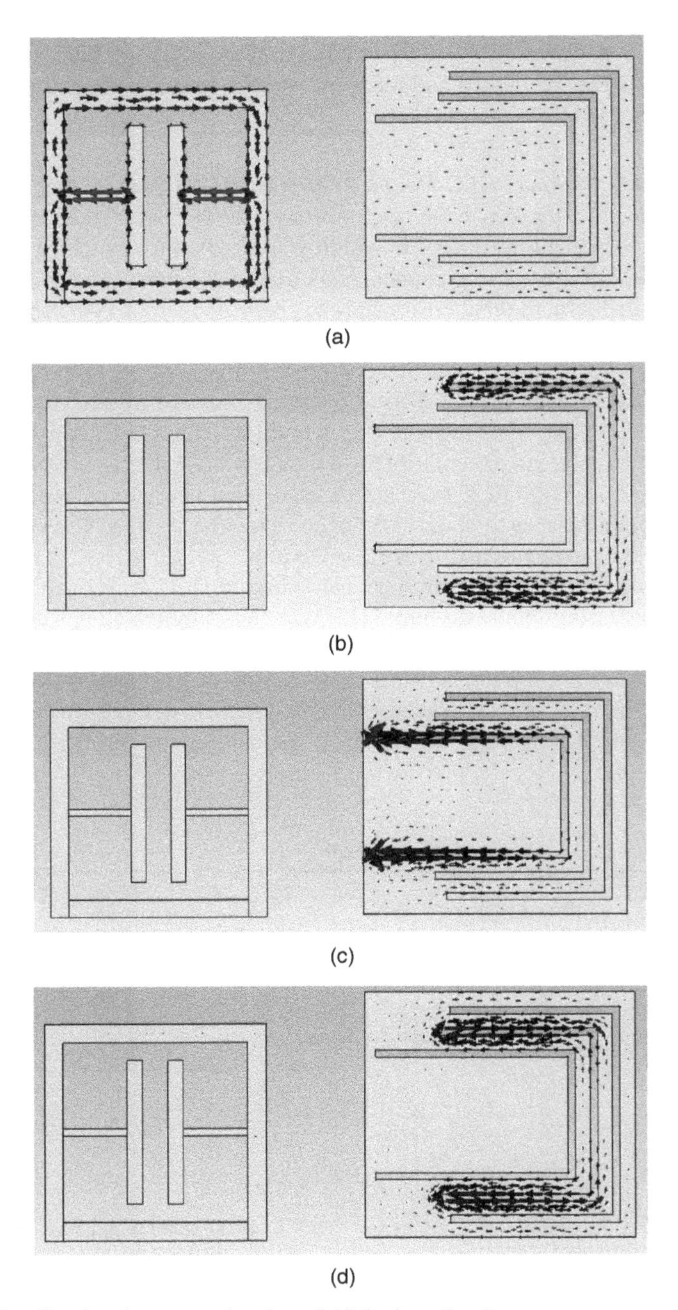

Figure 3.23 Simulated current density of high data density, compact tag at frequency, (a) ELC resonator, (b) slot S_1, (c) slot S_2, and (d) slot S_3

circulation through the structure. Although slots S_1 and S_2 show induced current at their edge, it is not sufficiently prominent to produce a resonant dip.

Furthermore, Figure 3.26(b) shows the current density profile at slot S_1 resonance frequency. Here, we selected the length of S_1 such that it resonates at a frequency closest to the ELC resonator but spatially positioned furthest away. This minimizes the mutual coupling between the resonators at their respective resonant frequencies. A similar approach was taken for slots S_2 and S_3 to reduce mutual coupling at resonance (Figure 3.26(c) and (d)). Hence, each slot resonates independently at their designed frequency with least interference from adjacent slots. Detailed simulation results and measured results for fabricated tag are presented in Section 3.5.6.

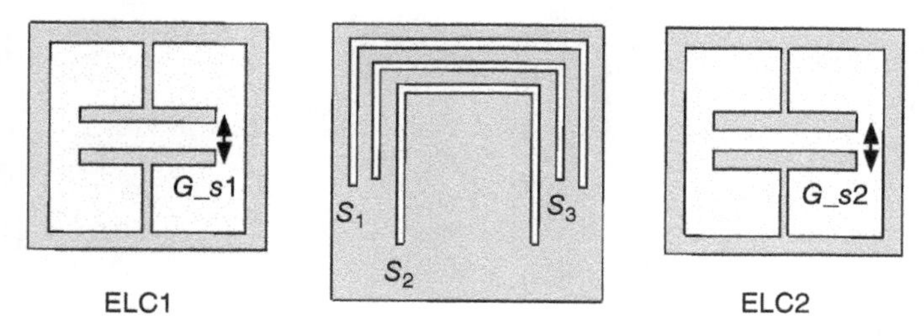

Figure 3.24 Layout of chipless RFID tag sensor for multiple parameter sensing. The simulation was performed in CST MWS with Taconic TLX_0 as substrate with substrate height of 0.5 mm, $\epsilon_r = 2.45$, and tan $\delta = 0.0019$. The overall dimensions are 25 mm × 8 mm. The gap between resonators is 3 mm

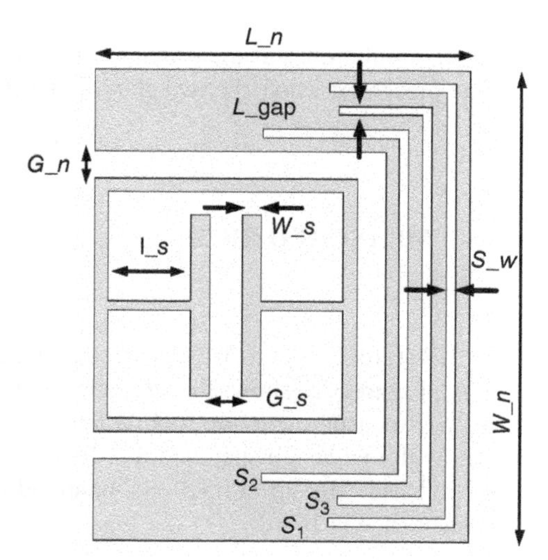

Figure 3.25 Layout of compact, high data density tag sensor. $W\_n = 10$ mm; $L\_n = 8$ mm; $S\_w = 0.3$ mm; $L\_gap = 0.2$ mm; $G\_n = 0.6$ mm; $l\_s = 1.75$ mm; $W\_s = 0.4$ mm; $G\_s = 0.6$ mm

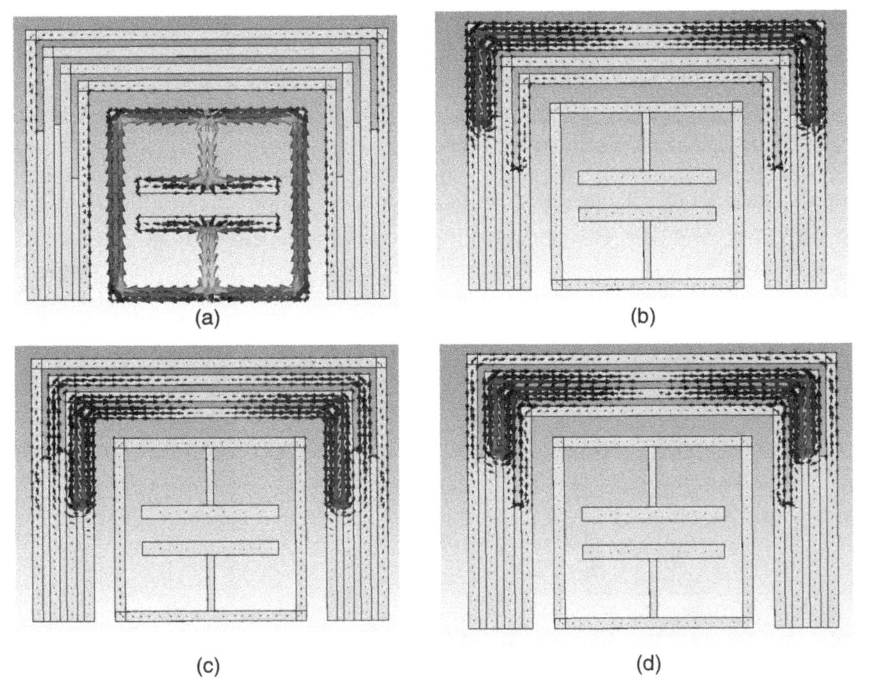

Figure 3.26 Simulated current density of high data density, compact tag at frequency, (a) ELC resonator, (b) slot S_1, (c) S_2, and (d) S_3

In the following section, simulation and measured results of individual passive microwave components are presented. Also, the two chipless RFID sensor responses are detailed.

3.5 SIMULATION AND MEASURED RESULTS

3.5.1 Tri-Step SIR

The insertion loss (S_{21}) of SIR (refer to Figure 3.13) is measured using a two-port vector network analyzer (VNA). Figure 3.27 shows the simulated and measured insertion loss for the insertion loss S_{21} measurement. Both simulation and measured results have good correspondence with the design specification. It shows SIR operates as a high Q bandstop filter at 825 MHz. Measured fractional bandwidth of SIR is 6% and maximum attenuation at resonance is 16 dB.

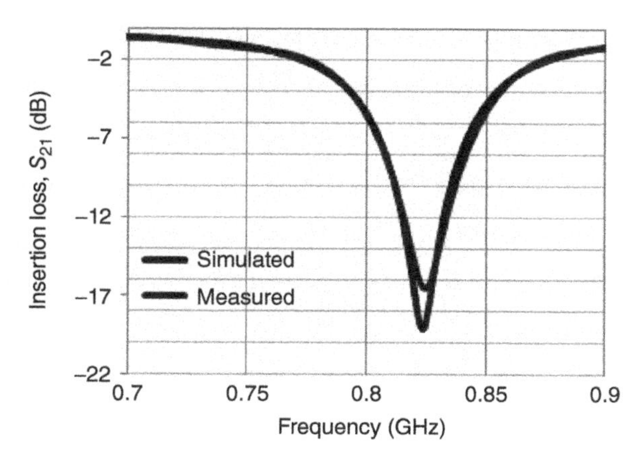

Figure 3.27 Simulated and measured insertion loss of 825 MHz SIR

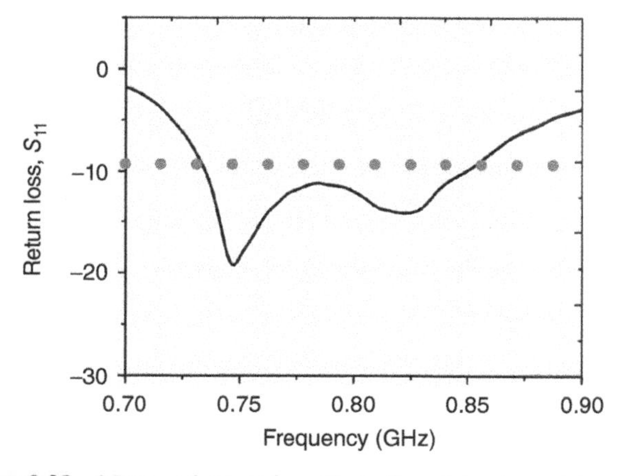

Figure 3.28 Measured return loss (S_{11}) of semicircular patch antenna

3.5.2 Semicircular Patch Antenna

The measured reflection loss (S_{11}) versus frequency of the antenna is given in Figure 3.28. The figure shows a bandwidth (less than −10 dB) from 740 to 855 MHz. This is in an acceptable range for our sensor operation. Also, the antenna has a gain of approximately 7.6 dBi at 800 MHz. This gain remains above 4 dBi for all frequencies within the 100 MHz bandwidth. Figure 3.29 shows measured radiation pattern for E plane and H plane.

The radiation pattern shows the antenna is directional and the pattern is slightly shifted in counterclockwise direction. However, the main lobe did not change direction as the frequency was varied from 750 to 850 MHz.

3.5.3 Cascaded Multiresonator-Based Chipless RFID Sensor

Integrated multiresonator-based chipless RFID sensor presented in Section 3.4.3 is tested for providing unique frequency signatures within a limited bandwidth. As shown in Table 3.2, two SIR filters operating at 775 and 825 MHz can produce three data bits "01," "10," and "00" depending on different combinations in the circuit. Figure 3.30 shows measured insertion loss of the three combinations for 775 and 825 MHz SIR filters. Each filter combination produces a distinct signature as expected. Hence, these high Q resonators are suitable for encoding data bits in frequency modulation-based chipless RFID tags. A practical application of SIR filters for chipless RFID sensor realization is presented in Chapter 6. The next section presents simulated and measured results for RCS scatterer-based chipless RFID sensor.

3.5.4 Multislot Patch Resonator

3.5.4.1 Binary Encoding Simulated RCS magnitude for multislot patch resonator is presented in Figure 3.31(a). The RCS response shows three distinct resonance nulls at 7.8, 8.4, and 9.0 GHz. Hence, the tag represents data bit "000." Also, by shorting out one of the slots, we can designate bit "1" in the frequency spectrum. Figure 3.31(b) and (d) shows the simulated RCS magnitude against frequency for the data bit combinations (b) "010," (c) "100," and (d) "001." In each of these three tags, one of the slots has been shorted to designate bit "1." Finally, the three-bit tag can represent 2^3 unique combinations with "0" and "1" and operates as a three-bit tag. The results verify there is a 1:1 correspondence with the number of slots (N) to number of data bits the tag can represent.

3.5.4.2 Frequency Shifting Technique Three tags were designed and fabricated to show the different data combinations encoded by the proposed frequency shifting-based encoding technique in Section 3.4.4. The designed tags are one sided, which means with no metal ground plane. Table 3.4 shows the encoded six-bit binary combinations for the three tags (column 2) and the corresponding values of the frequency shifting parameters (column 3).

In CST MWS, a horizontally polarized plane wave is used for exciting the tag and an RCS probe with identical polarization is placed 50 mm away from the tag to receive the far-field backscattered signal. Figure 3.32 shows the simulated RCS magnitude against frequency. As expected, the resonances are placed at the middle of 200 MHz band. Hence, the simulated tags do not have any frequency deviation from the design specifications shown in Table 3.4.

Figure 3.33 shows a photograph of the fabricated tags. The tags were fabricated using the chemical etching technique on Taconic TLX_0 substrate. The fabricated tags are read by putting two identical horn antennas on the same side of the tags. One of the antennas operates as a transmitter (right) and the other as a receiver (left), as shown in Figure 3.34. The antennas have a bandwidth (less than −10 dB) from 5.5 to 12 GHz. The distance between the horn antennas and tag is set to 30 cm.

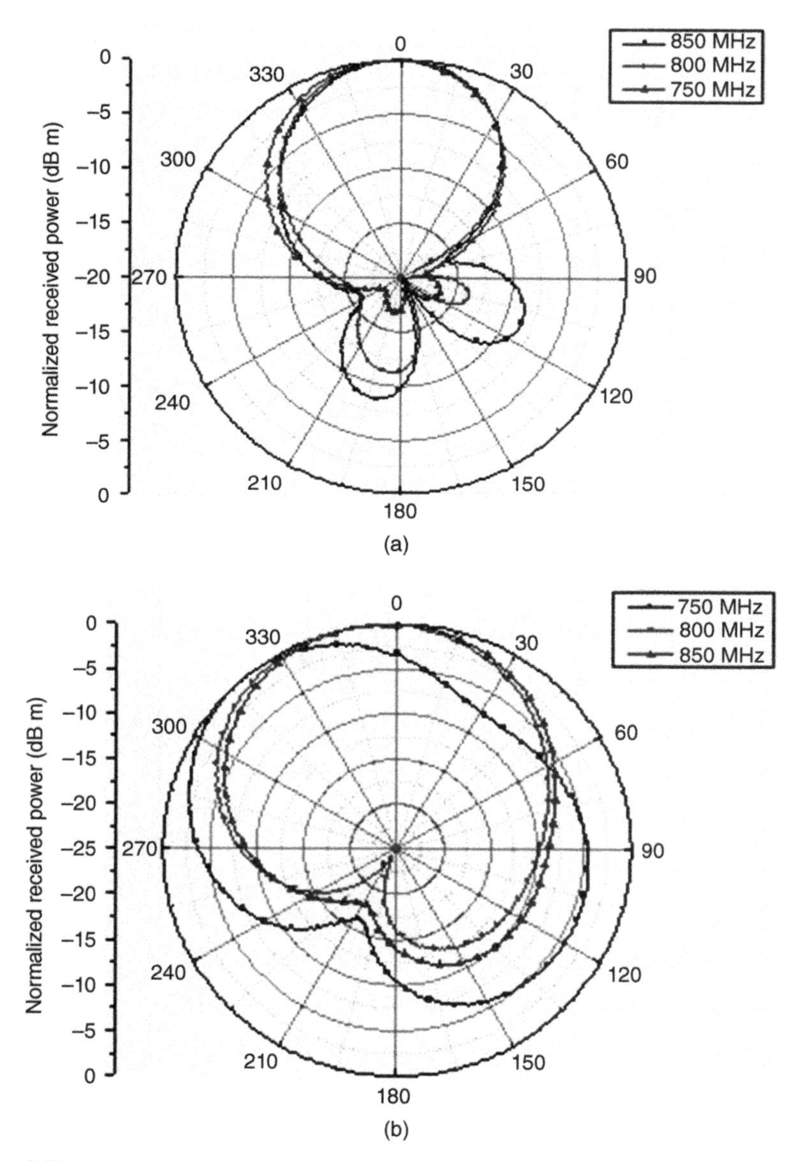

Figure 3.29 Measured radiation pattern of patch antenna on air gap layer with LC matching section at the feed probe (a) *E*-plane and (b) *H*-plane

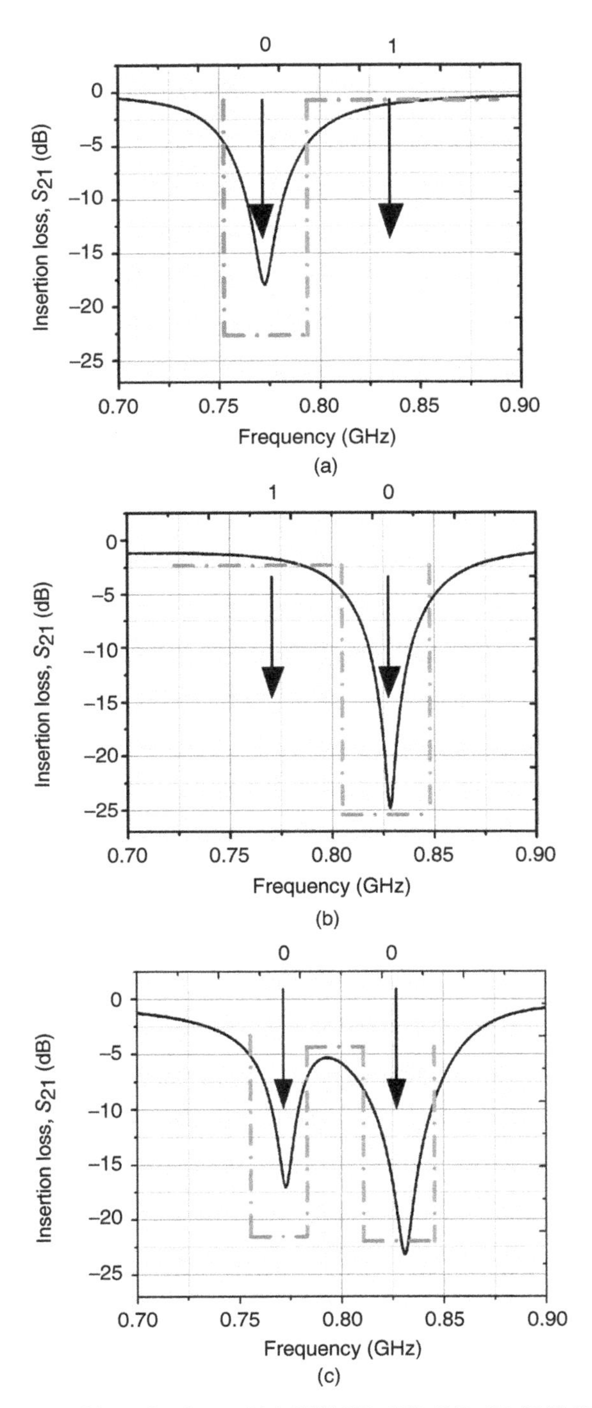

Figure 3.30 Measured insertion loss of (a) 775 MHz SIR (01), (b) 825 MHz SIR (10), and (c) cascaded SIR filters (00)

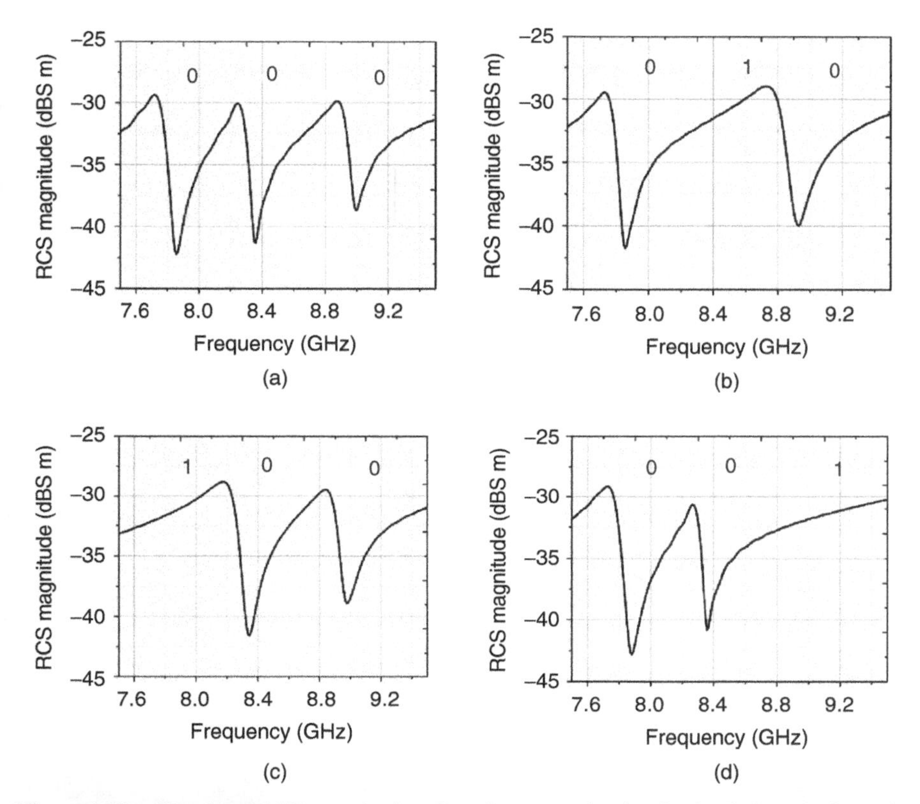

Figure 3.31 Simulated RCS magnitude versus frequency for the slot-loaded patch shown in Figure 3.16 for data bit combinations (a) "000," (b) "010," (c) "100" and (d) "001"

The two antennas are connected to a VNA for measuring the transmission coefficient (S_{21}). Using this setup, the RCS responses of fabricated tags are measured. Here, a calibration technique is used to nullify the background EM noise and path loss. The calibration is done using the "No tag" condition as reference. In the VNA,

TABLE 3.4 Simulated Frequency Shifting Parameter and Corresponding Resonant Frequency for Designed Tags

Tags	Encoded data (column 2)	Frequency shifting parameter value, x_N (mm) (column 3)			Measured resonant frequency, f_{rSN} (GHz) (column 4)		
		x_1	x_2	x_3	Slot S_1	Slot S_2	Slot S_3
Tag 1	101010	2.1	0.4	2.3	7.6	8.5	9.5
Tag 2	100101	2.1	0.2	2.1	7.6	8.2	9.2
Tag 3	011011	1.8	0.4	2.5	7.3	8.5	9.8

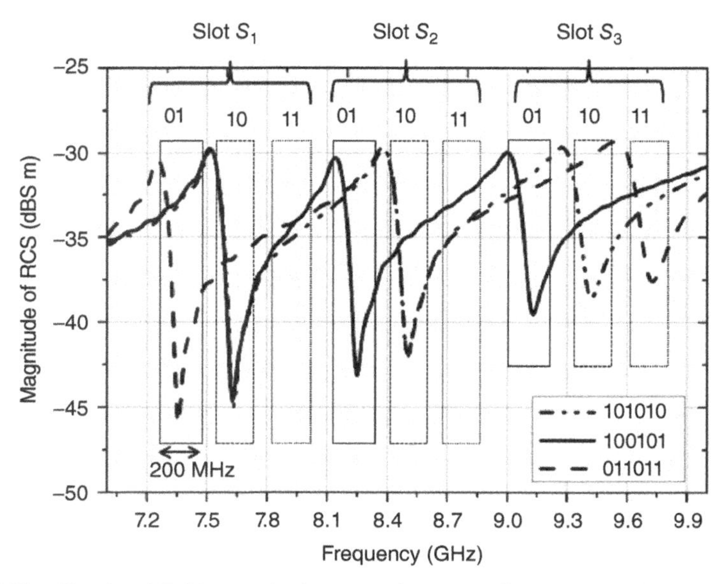

Figure 3.32 Simulated RCS magnitude versus frequency for tag 1, tag 2, and tag 3. The resonant frequencies of the tags are shown in Table 3.4 (column 4)

Figure 3.33 Photographs of fabricated tags

the "No tag" response (where the response is from the environment only) is measured and saved as "Memory data." Later, the normalized RCS response of each tag is measured and saved, which gives "Tag data"/"Memory data." Finally, the calibrated tag response is recalculated using the normalized data and "Memory data."

The calibrated transmission coefficient, $S_{21,}$ is measured and plotted against frequency in Figure 3.35. Due to fabrication error, none of the resonances is placed exactly at the center frequency. Table 3.4 (column 4) shows the measured resonant frequencies for three tags. Here, each of the slots resonances is within the ± 100 MHz

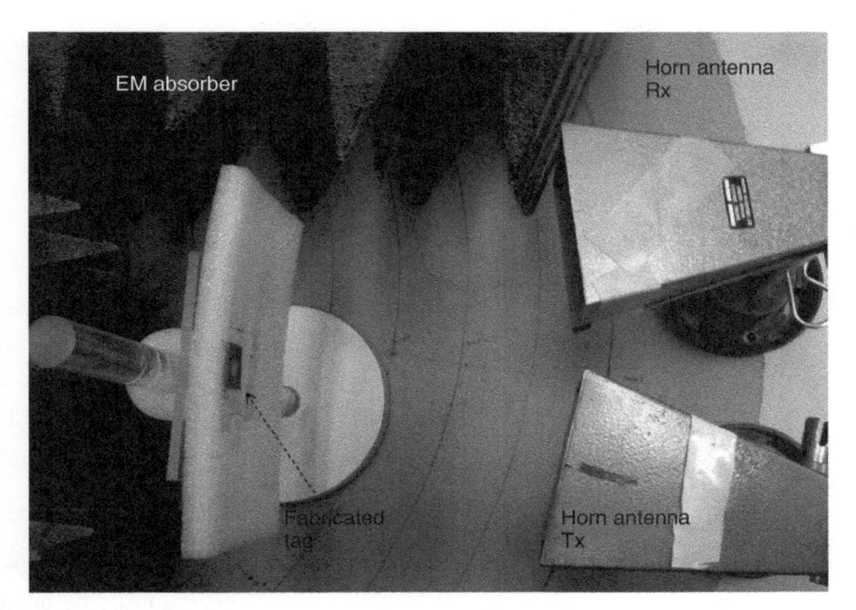

Figure 3.34 Experimental setup for tag measurement using two horn antennas. The right antenna is transmitting (Tx) and the left antenna is receiving (Rx)

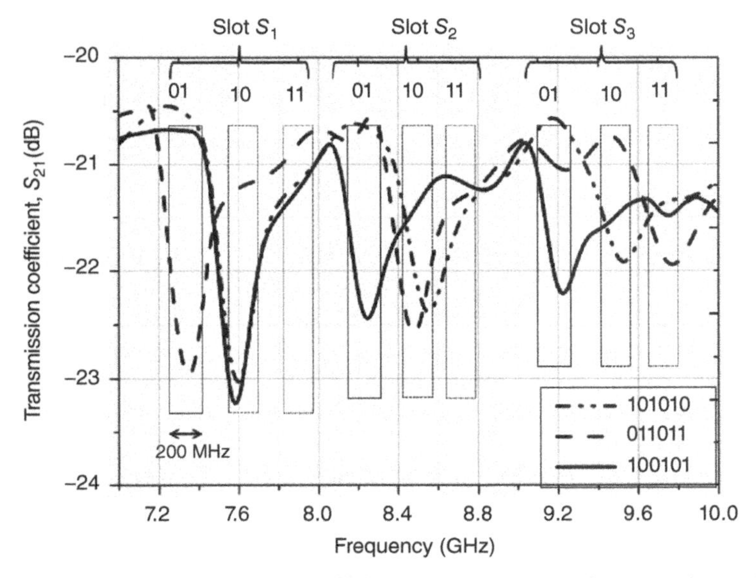

Figure 3.35 Measured transmission coefficient (calibrated) versus frequency for tag 1, tag 2, and tag 3. The resonant frequencies of the tags are shown in Table 3.4 (column 4)

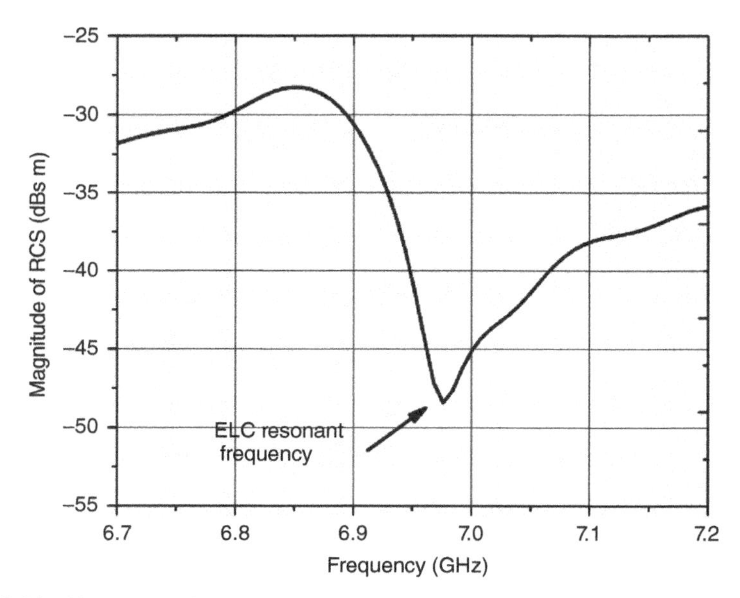

Figure 3.36 Simulated RCS magnitude versus frequency for the ELC resonator at 6.96 GHz

band around the designed frequency to represent designated data bits. Therefore, the binary IDs of all three tags can be extracted from the measured results.

So for the CST simulated and anechoic chamber measured results for data encoding element are presented. The following section presents the results for ELC resonator as the sensing element.

3.5.5 ELC Resonator

The simulated RCS response of ELC resonator presented in Figure 3.19 is shown in Figure 3.36. It shows that the ELC resonator has a 20 dB dip at resonant frequency at a narrow band around 6.96 GHz. Following the design, an ELC resonator is fabricated on Taconic TLX_0 substrate (see Figure 3.37(a)). The resonator is measured by using a two-horn antenna setup shown in Figure 3.34. The transmission coefficient S_{21} is plotted in Figure 3.37(b), which shows about 5 dB difference in the backscattered signal at resonant frequency.

The next section presents simulated and measured results for developed backscatterer-based chipless RFID sensor.

3.5.6 Backscatterer-Based Chipless RFID Tag Sensor

3.5.6.1 Chipless RFID Tag for Single Parameter Sensing Figure 3.38 shows the simulated RCS magnitude versus frequency for the integrated chipless RFID tag for single parameter sensing. The resonant frequency for ELC resonator is at 6.96 GHz

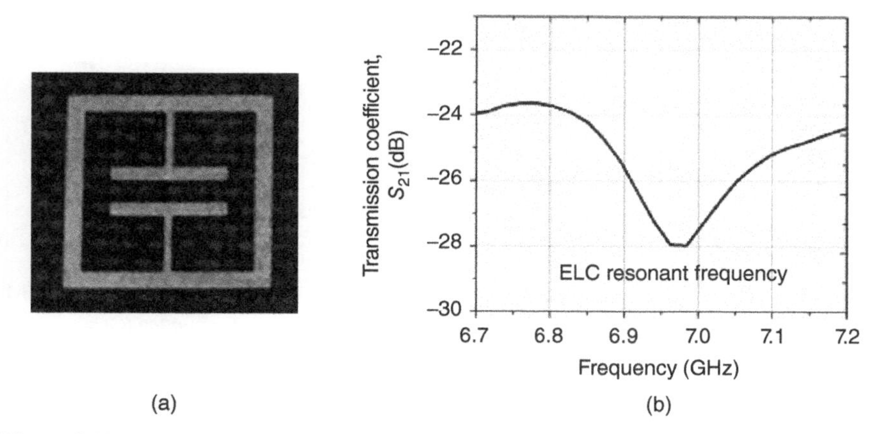

Figure 3.37 (a) Photograph of fabricated ELC resonator and (b) measured transmission coefficient of ELC resonator

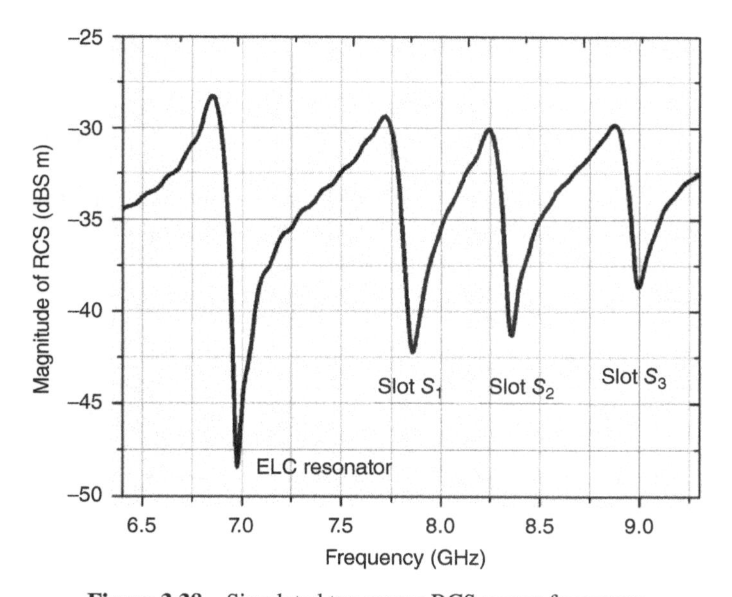

Figure 3.38 Simulated tag sensor RCS versus frequency

and the slots S_1, S_2, and S_3 resonate at 7.7, 8.3, and 9.0 GHz, respectively. Here, the resonant frequencies for data encoding and sensing can be clearly identified.

The tag sensor is fabricated by etching copper from the Taconic TLX_0 substrate (Figure 3.39(a)). The calibrated transmission coefficient, S_{21}, for the overall tag sensor in the room environment is measured and plotted against frequency in Figure 3.39(b). Here, the S_{21} result obtained for the "no tag condition" is taken as the calibration data.

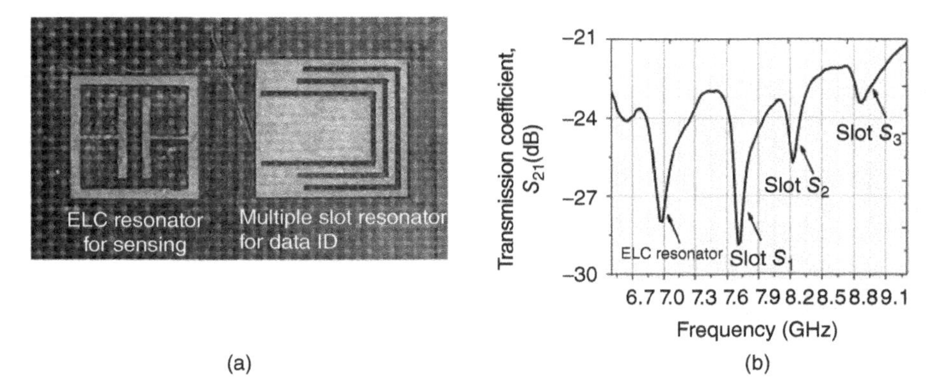

Figure 3.39 (a) Photographs of fabricated tag sensor and (b) measured transmission coefficient of overall tag sensor

The measured data correspond to the simulated RCS response for ELC resonator and slot S_1. However, there is minor frequency variation in the resonance dip of S_2 and S_3. This is due to fabrication inaccuracy.

Also it needs to be noted that the resonant frequency of ELC is independent of the slot resonators. Hence, change in ELC resonator does not affect the encoding of data bits. This makes multiple sensing possible, as described in the next section.

3.5.6.2 Chipless RFID Tag for Multiple Parameter Sensing Figure 3.40 shows simulated RCS spectrum of a chipless RFID tag for multiple parameter sensing. As mentioned in the design section, ELC1 and ELC2 are two resonators operating at 7.1 and 7.8 GHz to carry sensing information. Moreover, the slot resonators operate within their designated band to carry ID information.

The tag sensor is fabricated on Taconic TLX_0 substrate as shown in Figure 3.41(a). The overall footprint area is 25 mm × 8 mm. The measured transmission coefficient of the tag sensor (Figure 3.41(b)) corresponds to the simulated results. It needs to be noted that this tag sensor can be extended for more than two parameter sensing. Here, we present two parameter sensing for proof of concept. A practical example of a multiple parameter sensing tag is presented in Chapter 8.

3.5.6.3 Highly Compact ELC-Coupled Chipless RFID Tag Figure 3.42 shows the simulated RCS spectrum of the tag sensor. As expected, the ELC resonator operates independently without interfering with adjacent slot resonators. The resonant frequency of ELC resonator is 6.75 GHz and for slots S_1, S_2, and S_3 resonant frequencies are 7.7, 8.3, and 9.4 GHz, respectively. The measured transmission coefficient also corresponds to the simulated resonant behavior (see Figure 3.43(b)). Photo of the fabricated tag sensor (Figure 3.43(a)) shows it is highly compact compared with a single parameter sensing tag presented in Figure 3.39(a). This highly compact tag sensor is used in Chapter 8 for realizing a chipless RFID temperature threshold sensor.

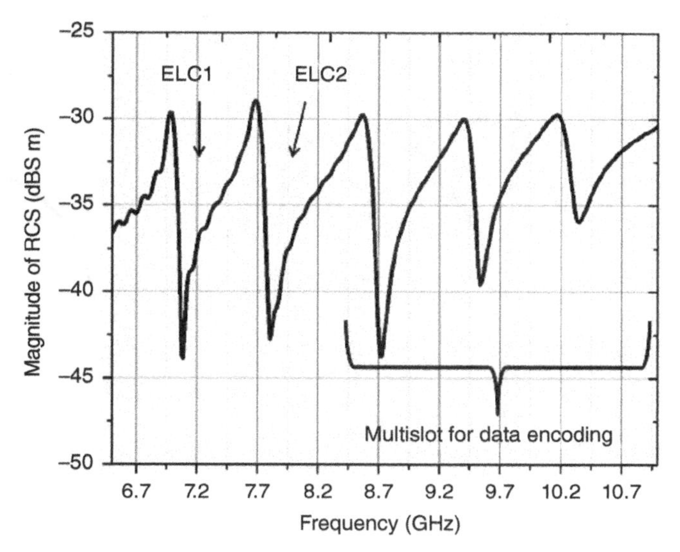

Figure 3.40 Simulated RCS spectrum of multiple parameter chipless RFID tag sensor

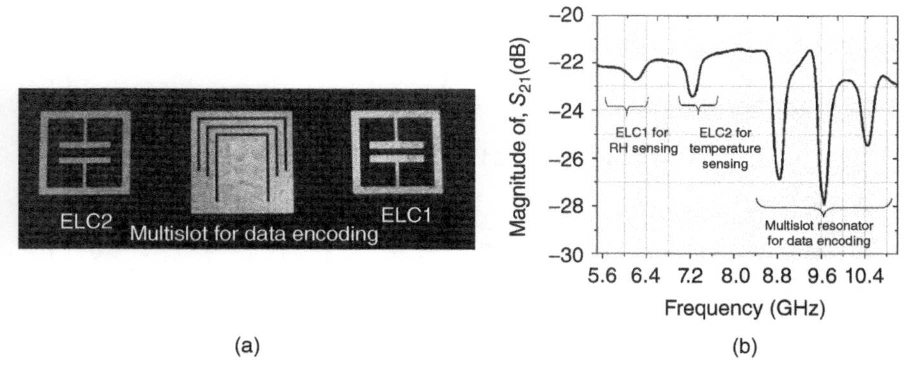

Figure 3.41 (a) Photographs of fabricated tag sensor and (b) measured transmission coefficient of overall tag sensor

3.6 CONCLUSION

In this chapter, detailed theories and designs of various passive microwave components have been presented. Next, two novel chipless RFID sensor platforms are presented integrating the passive microwave components. Important findings and conclusions are presented as follows.

Cascaded multiresonator-based sensor utilizes a patch antenna and a high Q SIR. Here, we present an L-shaped air gap patch antenna for capturing UWB signals. Moreover, the SIR is a high Q, planar CPW structure for designating frequency signature in UWB signal. Two tri-step SIRs are designed, fabricated, and measured to

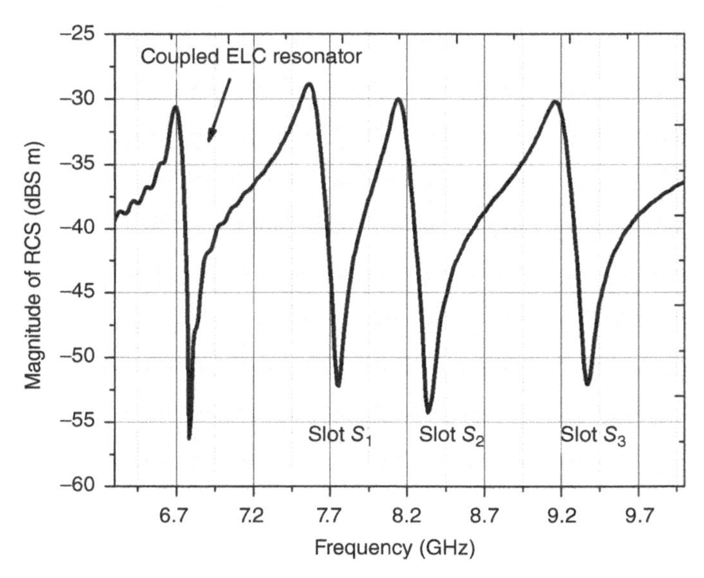

Figure 3.42 Simulated RCS spectrum of highly compact chipless RFID tag sensor

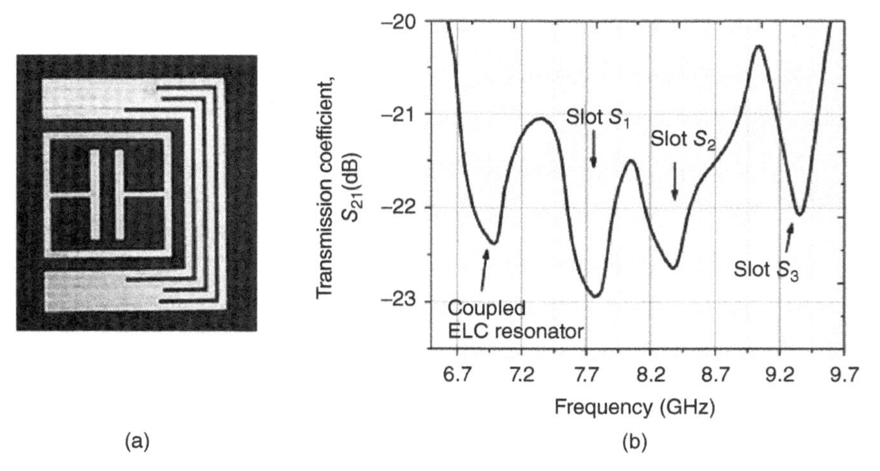

Figure 3.43 (a) Photograph of fabricated tag sensor and (b) measured transmission coefficient of the tag sensor

show data encoding in frequency spectra. The resonators are utilized in cascaded multiresonator-based chipless sensor for PD detection as described in Chapter 6. Also, SIR is used as a two-port microwave probe for characterizing smart materials presented in Chapter 7. Hence, SIR is one of the most significant passive components considered in this book.

The primary focus of backscatterer-based passive resonator design is to develop a high data density, compact, planar chipless RFID tag sensor. To achieve this goal, first a high data density multislot RCS scatter for data encoding has been presented. Two methods of data encoding are presented: (i) binary encoding and (ii) frequency shifting. The frequency shifting technique shows enhanced data density compared with binary encoding. This is because it utilizes unused bandwidth for each slot resonator. However, frequency shifting-based tags require highly sensitive readers and are prone to fabrication tolerances compared with binary encoding-based tags.

Next, a high Q ELC resonator for RF sensing has been designed and analyzed. Simulated results show high electric field concentration at a narrow region of the ELC resonator. This shows the suitability of the ELC resonator for dielectric sensing.

Finally, data encoding scatter and sensing metamaterial have been integrated to realize a high data density, compact tag sensor. The main features of the developed sensor are (i) data encoding is independent of sensing, (ii) multiple parameter sensing is possible, and (iii) there is minimum EM coupling and interference between adjacent resonators. To mitigate cross-coupling between highly compact slot resonators and ELC resonator, we carried out current density analysis. By carefully selecting the structural parameters and spacing between adjacent resonators, we came up with a highly compact tag sensor with enhanced data density and sensing capability. The developed tag sensor shows expected resonant characteristics both in simulated and measured results. A fully printable, planar sensor has tremendous potential in low-cost item tagging and sensing. In particular, our sensor has numerous applications in this era of IoT, where the primary goal is to collect and store status information of each and every object in cloud data. In the next chapter, smart materials for chipless RFID sensor applications are investigated.

REFERENCES

1. K.-S. Chin and C.-K. Lung, "Miniaturized Microstrip Dual-Band Bands-stop Filters Using Tri-section Stepped-Impedance Resonators," *Progress In Electromagnetics Research C*, vol. 10, p. 37, 2009.
2. S. J. K. Chang, "Compact microstrip Bandpass Filter Using Miniaturized Hairpin Resonator," *Progress in Electromagnetics Research Letters*, vol. 37, p. 65, 2013.
3. M. Sagawa, M. Makimoto, and S. Yamashita, "Geometrical Structures and Fundamental Characteristics of Microwave Stepped-Impedance Resonators," *IEEE Transactions on Microwave Theory and Techniques*, vol. 45, pp. 1078–1085, 1997.
4. Z. Hualiang and K. J. Chen, "Miniaturized Coplanar Waveguide Bandpass Filters Using Multisection Stepped-Impedance Resonators," *IEEE Transactions on Microwave Theory and Techniques*, vol. 54, pp. 1090–1095, 2006.
5. I. Jalaly and I. D. Robertson, "RF barcodes using multiple frequency bands," in *Microwave Symposium Digest, 2005 IEEE MTT-S International*, 2005, p. 4.
6. A. A. Deshmukh and G. Kumar, "Various Slot Loaded Broadband and Compact Circular Microstrip Antennas," *Microwave and Optical Technology Letters*, vol. 48(3), pp. 435–439, 2006.

7. K. P. Ray and D. D. Krishna, "Compact Dual Band Suspended Semicircular Microstrip Antenna with Half U-Slot," *Microwave & Optical Technology Letters*, vol. 48(10), pp. 2021–2024, 2006.
8. J. McVay, A. Hoorfar, and N. Engheta, "Space-filling curve RFID tags," in *IEEE Radio and Wireless Symposium, 2006*, San Diego, 2006, pp. 199–202.
9. I. Jalaly and I. D. Robertson, "Capacitively-tuned split microstrip resonators for RFID barcodes," in *European Microwave Conference, 2005* Paris, France, 2005, p. 4.
10. M. A. Islam and N. C. Karmakar, "A Novel Compact Printable Dual-Polarized Chipless RFID System," *IEEE Transactions on Microwave Theory and Techniques*, vol. 60, pp. 2142-2151, 2012.
11. T. Dissanayake and K. P. Esselle, "Prediction of the Notch Frequency of Slot Loaded Printed UWB Antennas," *IEEE Transactions on Antennas and Propagation*, vol. 55, pp. 3320–3325, 2007.
12. S. Preradovic, I. Balbin, N. C. Karmakar, and G. F. Swiegers, "Multiresonator-Based Chipless RFID System for Low-Cost Item Tracking," *IEEE Transactions on Microwave Theory and Techniques*, vol. 57, pp. 1411–1419, 2009.
13. E. M. Amin, N. Karmakar, and S. Preradovic, "Towards an intelligent EM barcode," in *Electrical & Computer Engineering (ICECE), 2012 7th International Conference on*, 2012, pp. 826–829.
14. E. M. Amin, S. Bhuiyan, N. Karmakar, and B. Winther-Jensen, "A novel EM barcode for humidity sensing," in *RFID (RFID), 2013 IEEE International Conference on*, 2013, pp. 82–87.
15. J. R. Terje, A. Skotheim, *Conjugated Polymers: Processing and Applications*, 3rd ed: CRC Press, 2007, p. 207.
16. M. Bellis. (2013). *Bar Codes*. Available: http://inventors.about.com/od/bstartinventions/a/Bar-Codes.htm.
17. D. P. Harrop and R. Das. (17 May). *Printed and Chipless RFID Forecasts,Technologies & Players 2009-2029*. Available: http://media2.idtechex.com/pdfs/en/R9034K8915.pdf.
18. S. Jaruwatanakilok, A. Ishimaru, and Y. Kuga, "Generalized surface plasmon resonance sensors using metamaterials and negative index materials," *Progress in Electromagnetics Research*, vol. 51, pp. 139–152, 2005.
19. M. R, O. Shamonin, C.J. Stevents, G. Faulkner, D.J. Edwards, O. Sydoruk, O. Zhuromskyy, E. Shamonina, and L. Solymar. "Waveguide and sensor systems comprising metamaterial element," in *DPG—Spring Meeting of the Division Condensed Matter*, Germany, March 2006.
20. M. Huang and J. Yang, "Microwave sensor using metamaterials," in *Wave Propagation*, A. Petrin, Ed.: Intech, 2011.
21. N. I. Zheludev, "The Road Ahead for Metamaterials," *Science*, vol. 328, pp. 582–583, 2010.
22. D. Schurig, J. J. Mock, and D. R. Smith, "Electric-field-coupled resonators for negative permittivity metamaterials," *Applied Physics Letters*, vol. 88, p. 041109, 2006.
23. M. A. Islam and N. C. Karmakar, "A Novel Compact Printable Dual-Polarized Chipless RFID System," *IEEE Transactions on Microwave Theory and Techniques, ,* vol. 60, pp. 2142–2151, 2012.
24. E. Perret, A. Vena, S. Tedjini, D. Kaddour, A. Potie, and T. Baron, "A compact chipless RFID tag with environment sensing capability," in *IEEE International Microwave Symposium*, Montreal, Quebec, Canada, 2012.

4

SMART MATERIALS FOR CHIPLESS RFID SENSORS

4.1 INTRODUCTION

In the preceding three chapters, our discussions were concentrated mainly on active and passive radio-frequency identification (RFID) sensors, the significance of chipless RFID tags and sensors for real-time low-cost item-level tagging and environmental monitoring, passive microwave design and metamaterials for integrated chipless RFID sensor development for wireless microwave measurement of physical parameters such as temperature, humidity, pressure, gas, pH, and light. Passive microwave resonant circuits provide identification and the very high-quality factor metamaterials such as electric inductive–capacitive (ELC) resonator loaded with smart materials provide real-time sensing data. Smart materials exhibit large and sharp physical and/or chemical changes in response to small physical or chemical stimuli in microwave frequency bands. This is a new field of investigation, where the characteristics of smart materials change its physical/chemical compositions with microwave stimuli and thus provide environmental data. These materials have tremendous potential in integrating with radio frequency (RF) devices for environment sensing. This chapter provides the list of smart materials and understanding of their functionality. The contents of this chapter and organization are shown in Figure 4.1.

Chipless RFID Sensors, First Edition. Nemai Chandra Karmakar, Emran Md Amin and Jhantu Kumar Saha.

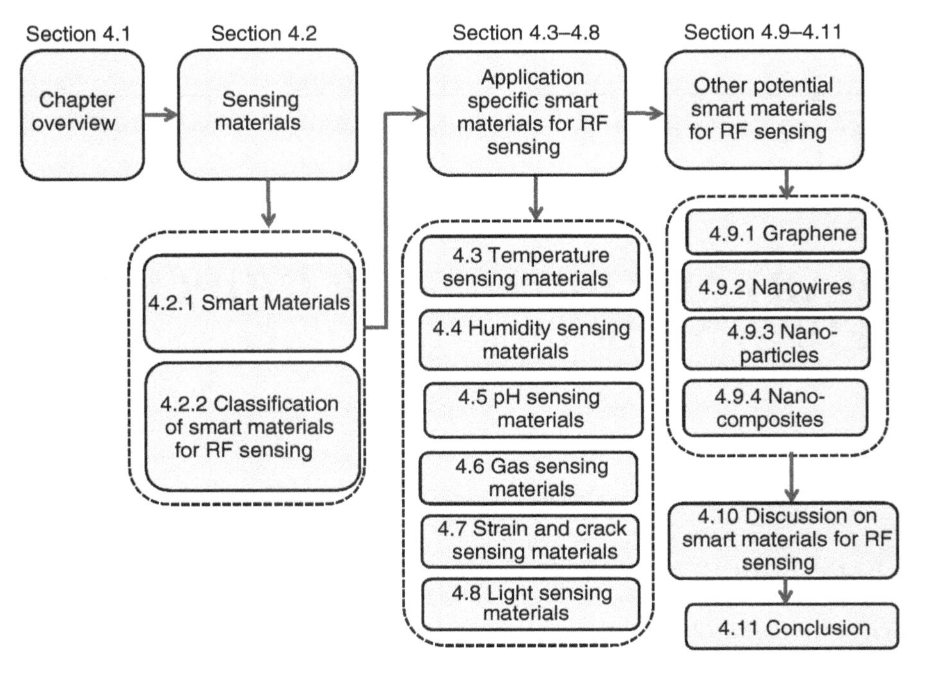

Figure 4.1 Flow diagram of content of the smart materials chapter

4.2 SENSING MATERIALS

The most useful characteristic of a smart material is its carrier mobility μ, defined as the proportionality constant between the applied electric field, E, and the corresponding average carrier drift velocity, v [1]. Usually, the carrier mobility of these materials is quite low, and they are not suitable for microwave circuit applications. However, they can be introduced as sensing materials that change transmission responses of microwave devices under the influence of varying physical parameters. Figure 4.2 shows a number of conductive, semiconductive, and nonconductive materials on the conductivity scale. It shows that pure conductors such as silver, copper, and aluminum can be used for microwave and millimeter-wave passive circuit design. However, these have low sensitivity to environment changes. On the other hand, some semiconductive materials such as indium tin oxide (ITO), silver flakes, and silver nanoparticles, conducting polymers such as poly-3,4-ethylene-dioxythiophene (PEDOT), polyaniline (PANI), and dielectric materials such as Kapton, polyvinyl chloride (PVC), and polyvinyl alcohol (PVA) exhibit much less conductivity than pure conductors. Nevertheless, these materials have noticeable dielectric and conductive property changes with certain physical parameters. Hence, these materials are investigated for RF sensing applications in the chipless RFID platform.

In this research, a number of smart materials have been reviewed for microwave sensing. These are classified according to their sensitivity to a particular physical

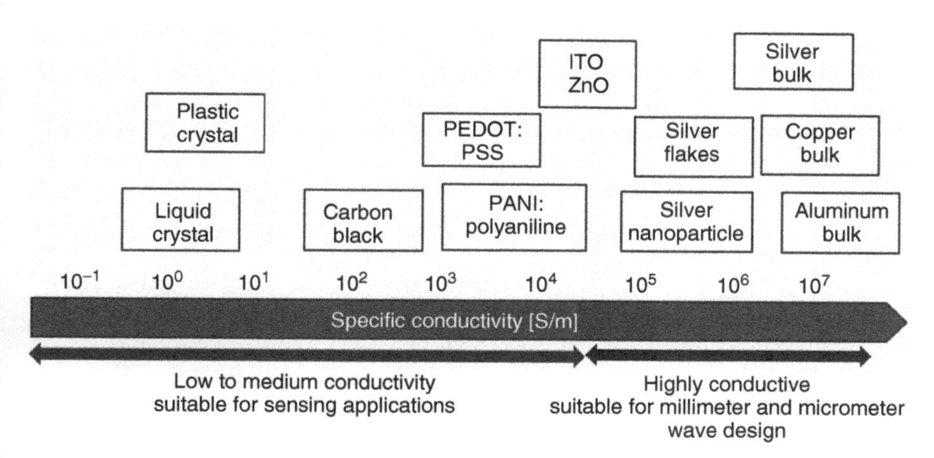

Figure 4.2 Classification of materials depending on the conductivity scale

parameter, as shown in Figure 4.2. As can be seen in Figure 4.2, Kapton and PVA are sensitive to humidity. These materials absorb water as ambient humidity increases, which change the dielectric property. Next, plastic crystals show multiphase glass transition at a low temperature range (−15 to +70 °C). Phenanthrene is a sublimate material that transforms from solid to gas without any liquid phase. There are metal oxides that show electric band gap variation with temperature change. Furthermore, PEDOT conducting polymer shows large conductivity change with pH. Cadmium sulfide (CdS) is a light-sensitive material investigated in this research. Finally, single-walled carbon nanotube (SW-CNT) and ZnO are listed as gas sensing materials.

In the following sections, detailed analysis of these smart materials for physical parameter sensing at microwave frequencies are discussed.

4.2.1 Smart Materials

The key element of the proposed chipless RFID sensor is "smart materials," also called nanostructured functional materials. These materials represent an important class of materials that has multifunctional properties.

These materials also have a wide range of applications in solar cells, fuel cells, sensors, and photoelectrochemical cells for water splitting. These materials exhibit large and sharp physical and/or chemical changes to external environmental stimuli, such as humidity, pressure, temperature, gas, and electric fields, and hence are suitable for sensing applications. These materials have tremendous potential in integrating with RF devices for environment sensing. However, the applications of these materials in microwave frequencies are relatively new. There is not much literature available in sensing applications in microwave frequencies in the field of investigation, where the characteristics of smart materials change its physical/chemical composition with microwave stimuli and provide environments data.

In this chapter, smart sensing materials for microwave sensing applications are explored. First, a comparative analysis of various nonconductive, semiconductive, and high conductive materials is discussed in the context of their microwave sensing properties. Next, various smart materials for microwave sensing applications are reviewed and their microwave characteristics in the influence of physical parameters are explored.

4.2.2 Classification of Smart Materials for RF Sensing

Some semiconductive materials and dielectric materials such as PEDOT:PSS (poly (3,4-ethylenedioxythiophene)-polystyrenesulfonicacid) [2, 3], phenanthrene [4], Kapton [5], PVA [6–8], PAni (polyaniline) [9], graphene [10], plastic crystals [11], hydrophilic polymer [12], SW-CNTs [13], metallic oxides[14], and nanoparticles [15] exhibit much less conductivity than that for the pure conductors. Usually, the carrier mobility of these materials is quite low and they are not suitable for microwave-/millimeter-wave passive design. However, they can be used for sensing applications. Nevertheless, these materials have noticeable dielectric and conductive property changes with certain physical parameters such as relative humidity, pressure, temperature, and electric field. Hence, these materials are investigated for microwave sensing materials in the chipless RFID platform. The study on smart sensing materials for physical RF sensing is reported in Ref. [16].

Figure 4.3 shows the classification of various smart materials for microwave sensing applications. These are classified according to particular physical parameters such as temperature, relative humidity, pH, strain, presence of gases, and light. The applicability of these materials is strictly dependent on the availability of optimized synthesis techniques that allow the processing and manipulation in a precise manner. Synthesis of these functional materials can be conducted through several novel fabrication and characterization techniques.

Various smart materials that exhibit noticeable responses against a particular temperature, relative humidity, pH, light, and presence of gases in gas sensing

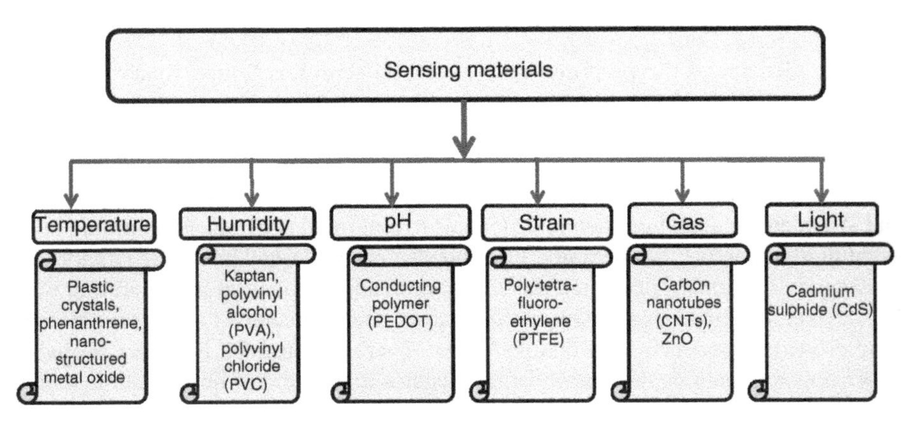

Figure 4.3 Classification of sensing materials

materials have been classified for microwave sensing as shown in Figure 4.3. As for example, Kapton, PVA are sensitive to changes in relative humidity in air. PEDOT is sensitive to pH level and CNT, ZnO are sensitive to the presence of noxious gases in air and finally CdS are sensitive to light. Therefore, these particular materials can be used to design the particular parameter sensing. In the following section, detailed analysis of these smart materials for physical parameter sensing at microwave frequencies are discussed.

4.3 TEMPERATURE SENSING MATERIALS

Temperature is imperative physical parameters in various fields such as chemistry, environment monitoring, shipping, pharmaceuticals, and health services. In the following section, various smart materials for temperature sensing are presented.

4.3.1 Phenanthrene

Phenanthrene is a sublimation substance from the polycyclic hydrocarbon group, which transforms directly from the solid to the gas phase without passing through an intermediate liquid phase. Sublimation is an endothermic phase transition that occurs at temperatures and pressures below a substance's triple point in its phase diagram. The enthalpy of phase transition for phenanthrene is 90.5 kJ mol^{-1} and the transition temperature (T_C) is around 72 °C. The dielectric behavior of phenanthrene is studied in Ref. [17]. The study shows that after the transition temperature, there is a drastic increase of dielectric constant, ε_r, which is permanent if the vapor is not desublimated (Figure 4.4). Therefore, this property can be used to realize a temperature threshold sensor for a chipless RFID tag that triggers at the transition temperature of phenanthrene.

This irreversible temperature sensing material is suitable for applications where a certain temperature violation record is crucial rather than real-time monitoring. It has a memory to store the event of temperature threshold violation. Phenanthrene is therefore one of the polycyclic hydrocarbons chosen for the present research. Other sublimate materials in this group can be used to obtain particular transition temperatures. For example, naphthalene (T_C = 298.15 K), benzene (T_C = 191 K), anthracene (T_C = 351 K) each has a different transition temperature and is suitable for integration in chipless RFID tag sensors [18]. Therefore, these materials can be used for versatile low-cost temperature threshold-sensing applications.

4.3.2 Ionic Plastic Crystal

Plastic crystal is a particular kind of crystal with weak interacting molecules that retain conformational degrees of freedom. Plastic crystals are regarded as an intermediate stage between real solids and real liquids and can be considered soft matter. *N*-methyl-*N*-alkylpyrrolidinium cation and PF_6 anion-based plastic crystals have been studied in Ref. [19]. These materials have a lower melting temperature (around

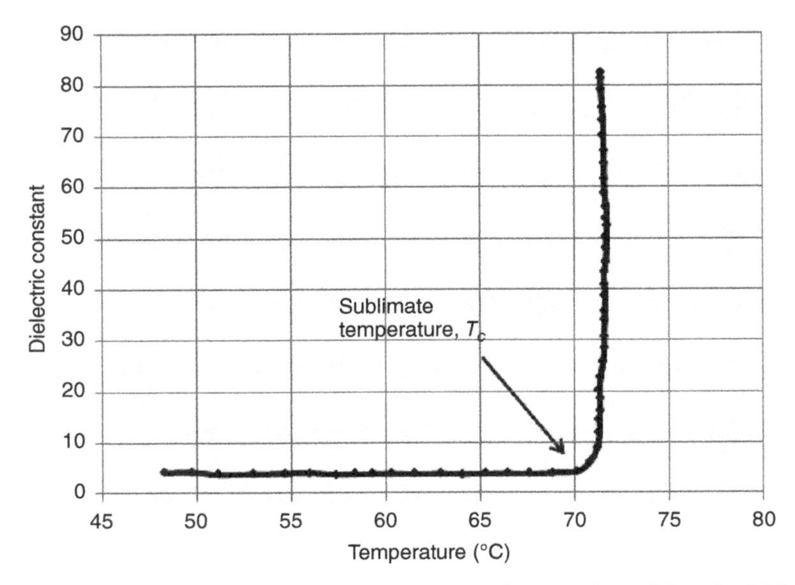

Figure 4.4 Dielectric constant change of phenanthrene during sublimation [17]

70 °C) and exhibit multiple thermal transitions during glass transition and melting. Golding *et al.* [19] measured three phase transition states for $P_{14}PF_6$ salt through thermogram analysis from −42 °C to its melting temperature. The salt shows intermediate glassy states at −15 , 14, and 42 °C before melting at 70 °C. At each glassy state, $P_{14}PF_6$ has particular polarization characteristics. This effect was confirmed by measuring the electrical conductivity of $P_{14}PF_6$ salt using electrochemical impedance spectroscopy (EIS).

EIS method is used to characterize the complex impedance of an electrode cell [20, 21]. During an impedance measurement, a frequency response analyzer imposes a small amplitude AC signal (typically 1–10 mV) to a test cell. The AC voltage and current response of the cell are analyzed to determine the resistive and reactive behavior of the impedance (Zs) for a particular frequency band. We performed EIS measurement on a test cell containing $P_{14}PF_6$ salt at Monash University Materials Engineering Department. The measured impedance values for 0.1–10 MHz are represented in a Nyquist plot (Figure 4.5). In the Nyquist plot, real and imaginary impedances (Z) are plotted on the horizontal and vertical axes. From an equivalent circuit model of an EIS test cell [22], the total resistance of the cell is found from point (A) of the Nyquist plot. Here, it is the real part of Z corresponding to a minimum value of imaginary Z. The measurement was repeated for a temperature range of −10 to +80 °C and values were taken for each 5 °C temperature increase.

Figure 4.5(a) and (b) show the Nyquist plots for temperatures −10 and +80 °C. The measured Nyquist plots show the resistance of the EIS test cell varies from 616,900 to 2699 Ω for total temperature variations from −10 to +80 °C. The conductivity can

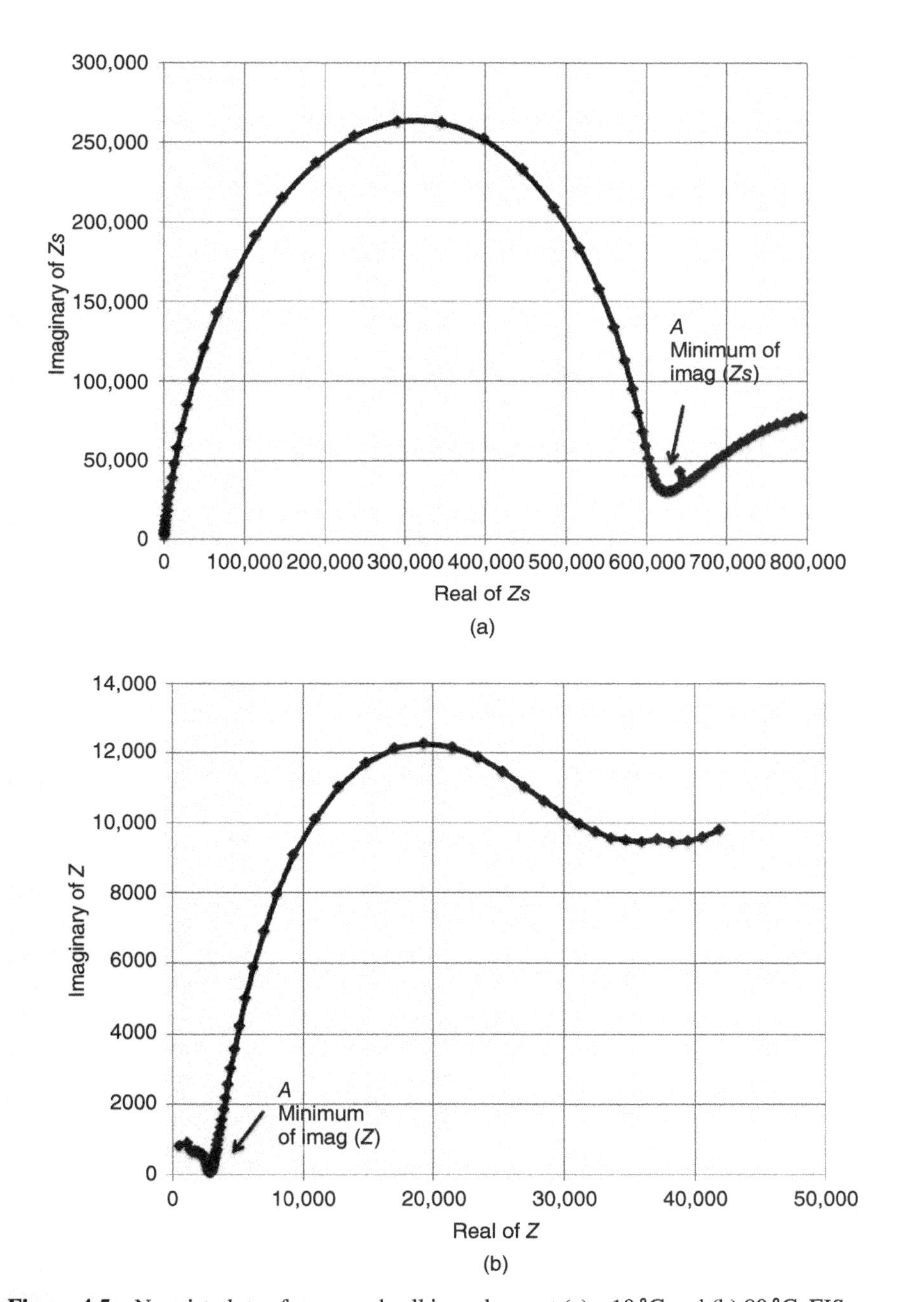

Figure 4.5 Nyquist plots of measured cell impedance at (a) −10 °C and (b) 80 °C. EIS measurements were conducted at Monash Materials Engineering Laboratory (*Source:* Authors acknowledge contribution from Assoc. Prof. Bjorn Jensen and Mega Kar)

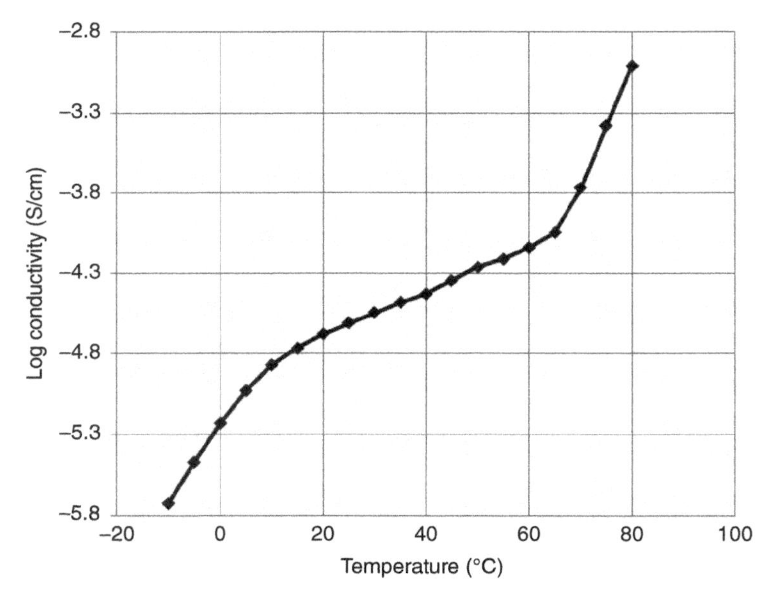

Figure 4.6 Measured conductivity of $P_{14}PF_6$ for various temperatures

be calculated using the cell resistance and cell constant (K) using Equation 4.2.

$$\sigma = \frac{K}{R_{\mathrm{cell}}} \tag{4.1}$$

In this experiment, K has a value of $1.157\,\mathrm{m}^{-1}$.

Hence, the conductivity at each temperature is measured using the cell coefficient and extracted resistance values. Figure 4.6 shows the conductivity of $P_{14}PF_6$ salt for temperatures ranging from −10 to +80 °C. The measured conductivity graph shows there is a prominent increase in electrical conductivity within the temperature range. However, change of conductivity is largely influenced by phase transition at 70 °C (solid to liquid). The intermediate phase transitions have least effect on electrical conductivity. $P_{14}PF_6$ salt has reversible temperature dependency and can be used to monitor real-time temperatures within −10 to +80 °C.

4.3.3 Nanostructured Metal Oxide

Semiconducting metal oxides such as ITO and zinc oxide (ZnO) are wide band gap materials that show promise for sensing. Both materials are very susceptible to external environmental changes, such as pressure, temperature, and electric field. ZnO also has a high melting point (2248 K) and good thermal stability. Due to the limited availability of Indium on the earth, investigations into the physical properties of ZnO films have indicated that the optical and electrical parameters of the ZnO films are similar to that of ITO. The optical absorption edge of the ZnO thin film has a regular redshift with temperature. This corresponds to a linear relation between the band gap energy of ZnO and temperature [23]. The optical band gap with direct transition can be calculated from the following relationship:

$$(\alpha h\upsilon) = B(h\upsilon - E_g)^{1/2} \tag{4.2}$$

where hv is photon energy, α is the absorption coefficient near the band edge, B is a constant between 10^7 and 10^8 m^{-1}, E_g is the optical ban gap. A detailed description regarding the measurement of the band gap of ZnO can be found in Ref. [24]. The results suggest that ZnO thin film is an excellent sensing material for temperature sensing. It is also robust enough to survive in real engineering environments.

4.4 HUMIDITY SENSING MATERIALS

4.4.1 Kapton

Kapton polyamide is a hydrophobic organic material and operates as a capacitive humidity sensor. It has a linear dielectric response while it absorbs water as the weight is proportional to RH. The structural formula is shown in Figure 4.7(a) and it absorbs water between the free spaces of adjacent polymeric molecules [25]. In Ref. [26], the linear dielectric behavior of Kapton film with RH is reported. Kapton film has a relative permittivity of 3.25 at 25% humidity and a room temperature of 23 °C. At low frequency (DC) Kapton's relative permittivity (ε_r) changes linearly with RH (%), which is given by Equation 4.3.

$$\varepsilon_r = 3.05 + 0.008 \times \mathrm{RH} \tag{4.3}$$

Also, the dissipation factor changes from 0.0015 at 0% RH to 0.0035 at 100% RH. During moisture absorption, Kapton goes through hydrolysis, which modifies the

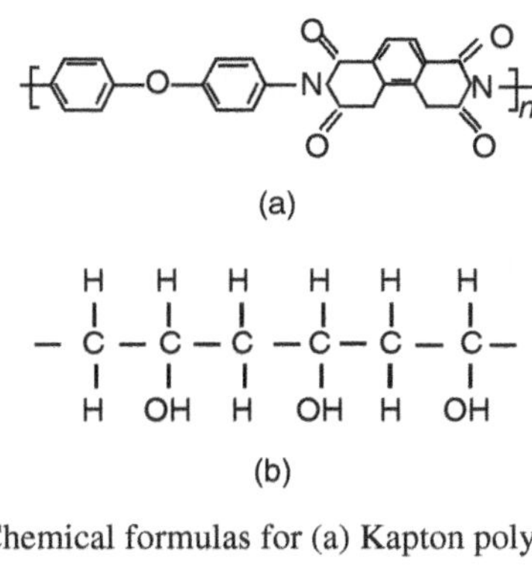

Figure 4.7 Chemical formulas for (a) Kapton polymer and (b) PVA

internal electrical polarization. The above properties suggest that Kapton can be introduced in ultra-high frequency (UHF) RFID humidity sensors [6]. Kapton 500 HN is a flexible, low loss, and extremely durable polyimide film dielectric. Kapton HN film was selected because of its special electrical properties: Kapton HN film's permittivity is dependent on the environmental humidity. This electrical property is the key factor in the sensor's functionality. Furthermore, Kapton HN is able to withstand high temperature levels. This property is very useful during sintering as a part of the inkjet printing process of the chipless RFID Tag sensor. In Chapter 7, we have performed detailed RF characterization of Kapton polymer to show RH sensitivity at UHF.

4.4.2 Polyvinyl Alcohol

PVA is a hygroscopic polymer material that absorbs water. The detailed chemical formula of the PVA polymer chain is shown in Figure 4.7(b). It has an OH group bonded to each carbon in the backbone chain $(–CH_2–CH_2–)_n$. It has high molecular weight and a glass transition temperature of about 70 °C. Standard products have 98–99 or 87–89 mol% of hydroxyl groups and can take up to 25% of water from humid ambient air [27, 28].

PVA is hydrophilic in nature and can therefore be used as a polyelectrolyte-based resistive sensor. In Refs. [29, 30], the microwave frequency characteristics of PVA in aqueous solution have been reported. In these studies, the dielectric behavior of water is investigated as the temperature and PVA concentration are changed. The results show that as the PVA concentration in water increases the real part of permittivity ε_r' decreases at any frequency (0.2–20 GHz). On the other hand, at high frequency (>5 GHz) the imaginary ε_r'' increases with the increase in water content [29]. This exhibits the twofold effect of humidity on the electric properties of PVA. With humidity change, PVA shows both dielectric and conductive sensitivities, which are useful for superior detectability.

Penza and Anisimkin have indicated the potential of a PVA film for humidity sensing [31]. They found negligible sensitivity of the PVA film to some gases such as NH_3, NO_2, CO and low hysteresis characteristics in sensing ambient humidity. This makes PVA a promising candidate as a chemically sensitive layer for sensing relative humidity.

In Chapter 7, we have carried out characterization of PVA polymer at UHF. Also, a chipless RFID humidity sensor is developed at microwave frequency incorporating PVA as sensing material.

4.5 pH SENSING MATERIALS

The amount of wasted food and perishable products throughout the globe is astonishing [32]. This in turn immensely reduces health hazards caused by having expired unhealthy items such as food and medicine. Every year, food spoilage leads to wastage and millions of food poisoning cases worldwide. The absence of adequate food monitoring technology may compromise the quality of food products during transit. The ability to wirelessly monitor the quality of food products reduces costs for producers and ensures consumer health and safety, hence making food quality measurement an indispensable requirement. Food spoilage is the process in which food deteriorates over time. Generally, food spoilage encourages the growth of microbes and yeast. The growth of these substances changes the chemical composition of the food, making them more acidic or alkaline depending on the food. Due to change in the acidity of the food and the general occurrence of microbial activities in food as spoilage occurs, pH can be used to monitor food spoilage. Bacteria growth is a major source of milk spoilage. As bacteria grow during spoilage, the pH of milk changes. Therefore, there is a huge economic aspect of perishable product's condition monitoring and storage. Timely sensing and alerting product conditions not only reduces wastage but also improves their in-time intake by the consumer.

A number of materials including polymer show sensitivity to pH variation of products [33, 34]. A number of conducting polymers also show noticeable changes in conductivity with pH. The advantage of organic semiconductors (conducting polymer) is their process ability. However, the instability toward environmental influences and the difficulty to achieve device-level performance impose challenges. Among the different conducting polymers used in practical applications, PEDOT is known as a particularly robust and well conductive material. These materials can be fabricated directly on flexible substrates to produce passive tags [20][84][95][97][101][103][105][111]. The conductivity of PEDOT depends apparently on the pH level, with the highest conductivities at low pH, but the change is not dramatic except for pH value exceeding 11. These changes are reversible over a wide pH range. The above result indicates that PEDOT may be used as a pH sensing material in passive chipless circuits.

4.6 GAS SENSING MATERIALS

Today, there is a major interest in the remote monitoring of environments such as sensing the air for the presence of hazardous gases in industrial environments, households, and corporate offices. In the supply chain management of perishable products such as vegetables, fish, milk, and meat, the monitoring of certain gases, such as oxygen, hydrogen, carbon monoxide and ammonia, is critical. In such applications, a batteryless, maintenance-free RFID sensor on flexible polymer would have a major impact. A number of nanostructures and organic materials show sensitivity to particular gases. For example, single- and multiwalled carbon nanotube films change electrical responses in the presence of ammonia gases [35]. Balachandran *et al.* [36] reported ethylene gas detection by integrating tin oxide in a capacitive RF sensor. In Ref. [37], carbon nanotubes and polymethylmethacrylate were incorporated in copolymer thin films to develop a wireless sensor for hazardous biological materials and vapors. A hydrogen gas sensor based on zinc oxide (ZnO) nanorods has been reported in Ref. [38]. Wang *et al.* [39] presented a hydrogen gas sensor comprising ZnO nanorod arrays grown on ZnO thin film. In addition, carbon dioxide, oxygen, and ammonia gas sensors based on carbon nanotube–silicon dioxide composites are presented in Ref. [40].

One major advantage of dispersible SW-CNTs is that they can be sucked into porous filter paper by using a common suction filtration method. They can also be assembled and distributed using paper, affording robust sensing arrays with the same thickness as the paper. Integrating these types of active nanomaterials into paper matrices opens up opportunities for flexible sensors on RFID platforms.

By using a common suction filtration method, water-dispersible single-walled carbon nanotubes (SW-CNTs) can be sucked into porous filter paper. It can be assembled and well distributed through the paper. Thus, it affords robust sensing arrays with the same thickness as the paper. Integrating these types of active nanomaterials into paper matrices opens up opportunities in flexible sensors and optoelectronic devices [13]. The paper sensors can be integrated into a common electric circuit, performing a similar function than traditional analytical equipment that is huge and hard to carry. And by using paper instead of plastic chips, there will be less environmental impact from nonbiodegradable waste. Estimating the cost of their paper-based gas-sensing chips, including all materials and equipment usage, energy and labor cost, the total price of below $1 per sensor chip is feasible. It is possible for paper chips to act as an antenna to recognize the target gas, deliver the information, and then catch, even eliminate, the toxic gas. This work will open the door to new, exciting applications of paper chips.

4.7 STRAIN AND CRACK SENSING MATERIALS

Various materials such as glass microfiber-reinforced polytetrafluoroethylene (PTFE) composite, polyester-based stretchable fabric, Nickel–Titanium (Nitinol) alloy can be used as strain and crack sensors for structural health monitoring [41–44]. PTFE-based sensor has a relatively large resonance frequency change due to its large dielectric constant variation under temperature fluctuation. While PTFE has very good electrical properties, other properties need to be well understood for several considerations.

4.8 LIGHT SENSING MATERIALS

Wireless pervasive light sensing has attracted great interest for agronomy research, safety and security, automatic indoor light intensity control, energy efficiency, and monitoring agricultural conditions. Sensors have been developed for light detection in UHF RFID platforms [45, 46]. These sensors essentially use a microcontroller or an independent light sensor and are integrated in CMOS platform. Therefore, they are of high cost, require a delicate manufacturing process, and need an onboard power supply.

Passive UHF RFID sensing has particular benefits over active sensing as passive sensors are of low cost, maintenance free, robust, and mass deployable. A number of photosensitive materials show noticeable changes in conductivity with light illumination. For example, lead sulfide, lead selenide, indium antimonide and most commonly cadmium sulfide and cadmium selenide are used in commercial photoresistors operating at low frequency [47]. Moreover, negative photoresistant materials, such as SU-8

TABLE 4.1 Measured NSL-6112 Photoresistor Impedances for Varying Light Intensities at UHF

Light intensity (lux)	0	250	500	750	1000
Re (Z_{1ph}) (Ω)	2.45	9.6	21	28.9	35.82
Im (Z_{ph}) (Ω)	−50	−41.34	−26.7	4.32	12.8

and poly(methyl methacrylate) (PMMA), permanently change their dielectric properties at certain ultraviolet (UV) illuminations [48]. These materials can be used as light threshold sensors.

One of the most commonly used photoresistors or light-dependent resistors is CdS. CdS resistance decreases with incident light intensity. However, these cells are mostly used for DC or low-frequency applications. In this study, we characterized a commercially available CdS cell Silonex NSL-6112 at 0.6–0.7 GHz to study its light sensitivity for RF applications. The Silonex NSL-6112 photoresistor has DC dark resistance of 1.3 MΩ (at 0 lux) and bright resistance of 170 Ω (at 1000 lux). We used a vector network analyzer (VNA) to measure impedance at different light conditions (refer to Table 4.1). The measured impedances show that in the RF range a CdS cell acts as a series RC circuit. It also shows variation in both resistance and reactance.

4.8.1 SIR Loaded with CdS Photoresistor

To verify the light sensitivity of CdS cells at UHF, an experiment was performed at the AutoID Laboratory, Massachusetts Institute of Technology (MIT). This experiment was part of the present thesis and aimed to develop a low-cost light senor using off-the-shelf components.

The stepped impedance resonator (SIR) discussed in Chapter 3 was designed and fabricated for the light sensing experiments. The photoresistor was connected as dielectric loading at the open end of the SIR (refer to Figure 4.8).

Experiments were conducted an enclosed box to maintain appropriate light conditions. Light intensity was varied using two cross-polarized lenses in front of a bright

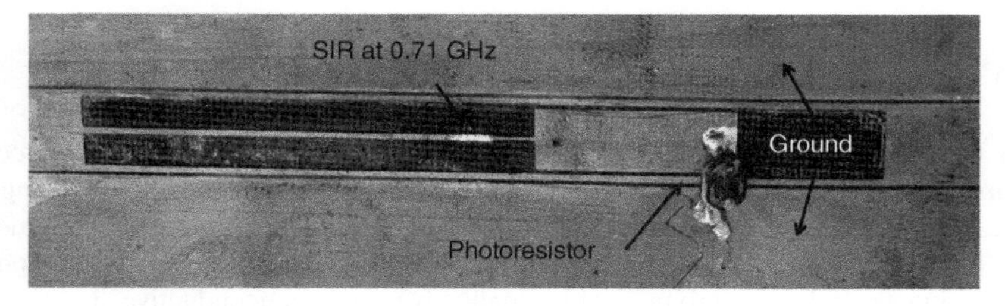

Figure 4.8 Photograph of SIR filter loaded with CdS photoresistor

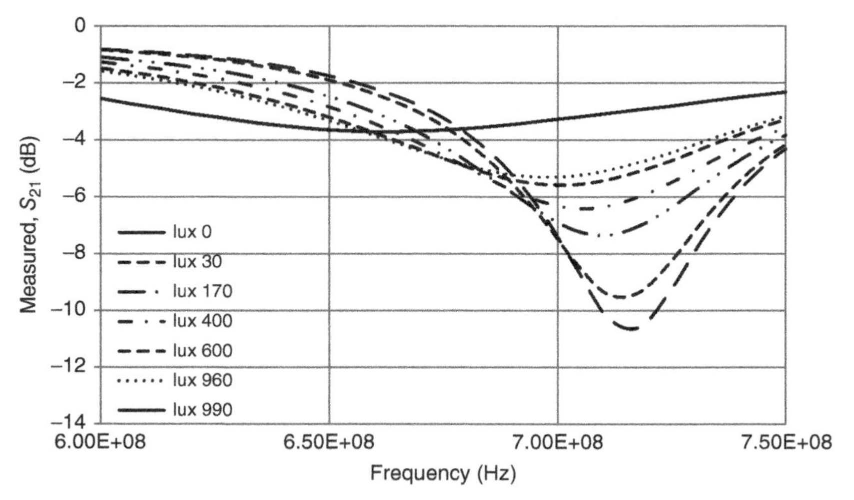

Figure 4.9 Measured $|S_{21}|$ versus frequency for photoresistor-loaded SIR for different light intensities

light source. The relative orientation of the two lenses produced varying lux intensity on our sensor. Moreover, we measured the different light conditions by using a standard TEMT6000X01 ambient light sensor.

Figure 4.9 shows measured insertion loss S_{21} versus frequency for the light sensor shown in Figure 4.8. Two port measurements were taken across the two ends of the SIR resonator loaded with photoresistor. As the light intensity varied from 0 to 1000 lux, we observed a noticeable shift in frequency as well as filter Q value. Figure 4.10 shows measured resonant frequencies and S_{21} at initial resonant frequency (710 MHz) for various light intensities. We obtained a maximum frequency shift of 70 MHz and a power difference of 7.5 dB for 0–1000 lux.

The results show that the CdS photoresistor has reduced dynamic range in light sensitivity in the RF range compared to DC. However, it can be integrated to passive RF circuits matched to ~50 Ω to provide light-dependent scattering parameters. Hence, CdS cells have the potential to be incorporated in chipless RFID tags for light sensing.

4.9 OTHER POTENTIALS SMART MATERIALS FOR RF SENSING

So far the materials that have been identified for the development of low-cost chipless RFID sensors are ionic plastic crystals, whose ionic conductivity changes due to organic molecules' defects and movements of crystals; conductive polymers (PEDOT), whose conductivity increases with frequency increases; composite/conjugate polymer mixing with conductive and nonconductive polymers; nanostructured metal oxides exhibit multifunctional properties, which are very

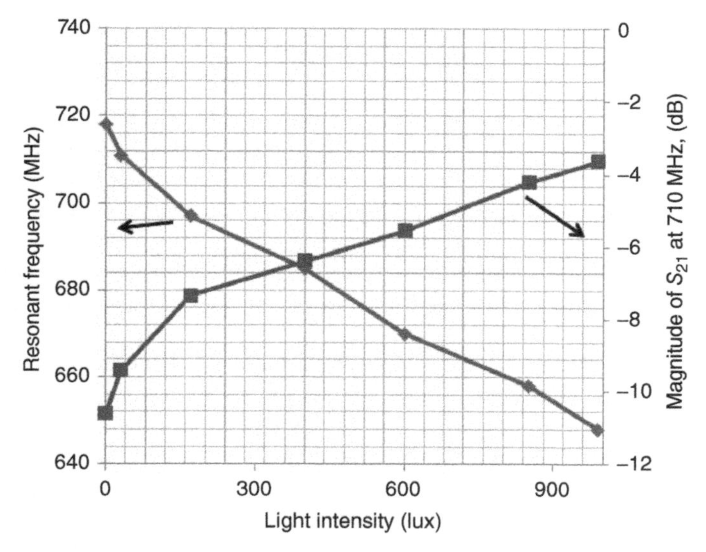

Figure 4.10 Plot of resonant frequencies and $|S_{21}|$ at initial resonance (710 MHz)

susceptible to external environmental changes such as external pressure, temperature, and electric field. In multifunctional materials, the sensing functionality arises from individual nanoscale components that can impart electrical conductivity, or responsiveness to light and other electromagnetic radiation. Candidate materials are percolating networks of nanoscale conductors such as SWNTs and doped semiconducting ceramic materials such as ZnO. However, both categories of materials, either carbon- or ceramic-based, suffer from drawbacks. For example, SWNT networks are flexible but lack the conductivity of the doped semiconductors, whereas doped ZnO, similar to most ceramic materials, is brittle and is prone to cracking when used in flexible electronics. It is also expected that such combinations of those materials will not only enhance the properties of thin films but also will impart additional functionality suitable for multiparameter sensing devices in a single sensor node. In the next section, other potential sensing materials such as graphene, nanocomposites, and plasmonic nanoparticles (NPs) are discussed.

4.9.1 Graphene

Graphene has attracted strong scientific and technological interests in recent years. It is a pure carbon in the form of a very thin, nearly transparent sheet, one atom thick. It is a flat monolayer of carbon atoms (of a diameter 0.34 nm) tightly packed into a two-dimensional honeycomb lattice. Graphene is a basic building block of graphite, quantum dots, and carbon nanotubes. It is remarkably strong for its very low weight (100 times stronger than steel) and it conducts heat and electricity very efficiently. There are currently several methods to produce graphene described in the scientific and commercial literature showing the intensity of development work globally. One is mechanical cleavage that is made of layers of multilayered graphite or by depositing one layer of carbon onto another material. The former is clearly how it was done using adhesive tape, but it is reported that the latter is more capable of making a monolayer with less defects. Graphene platelets can also be created by chemically cutting open carbon nanotubes, one method describing how the nanotubes are cut open in solution by the action of potassium permanganate and sulfuric acid. Chemical, solvent, or sonic exfoliation (separation) of graphene layers from graphite has also been developed. Plasma deposition techniques, reduction of graphene oxides (RGO), and other synthetic methodologies are being introduced as a route to scale up manufacture.

Electronic sensors based on single-atom-thick graphene are a particularly interesting material owing to its remarkable electrical, mechanical, and sensing properties. The growth of graphene films on supporting metallic films such as Ni or Cu using chemical vapor deposition methods, and then combined with postetching of the underlying metal, offers the ability to efficiently transfer graphene films to other substrates over large areas for biocompatible sensing and flexible electronics applications [49]. Figure 4.11 shows a biotransferable graphene wireless nanosensor. This is enabled by graphene's intrinsic strength as well as by the high interfacial adhesion exhibited by graphene to substrates. These properties render graphene an ideal active material for direct interfacing onto rugged surfaces.

Figure 4.11(a) shows graphene wireless nanosensor, where graphene is printed onto bioresorbable silk and converted into a coil antenna. Then, the wireless nanosensing architecture is obtained onto the surface of a tooth shown in Figure 4.11(b). Figure 4.11(c) shows the magnified schematic of the sensing element, illustrating wireless readout. Figure 4.11(d) shows the binding of pathogenic bacteria by peptides self-assembled on the graphene nanotransducer.

The usefulness of graphene oxides (GO) as a complement to wireless sensing technologies, highlighting their unique properties and ease of integration with existing wireless packaging technologies such as inkjet printing is also demonstrated in

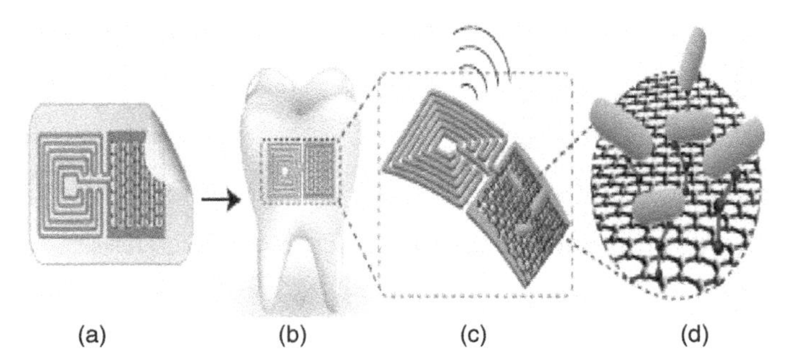

Figure 4.11 (a) Graphene wireless sensor, (b) sensor on tooth, (c) wireless readout of sensor, and (d) binding of pathogenic bacteria on graphene nanotransducer (*Source:* Nature Communications 2012. with permission)

Ref. [50]. These results can be improved upon by optimization of the deposition and curing techniques, and with enhancements to the output circuitry of the final chipless RFID sensor design.

4.9.2 Nanowires

Nanowires (NWs) have been defined as wires with at least one spatial dimension in the range of 1–100 nm. These new types of architectures exhibit variety of interesting and fascinating properties such as high conductivity and have been functioning as important potential candidates for the realization of the next generation of sensors. Among the nanostructures, NWs emerge as one of the best defined and controlled classes of the nanoscale building blocks in biosensing. These materials are attractive because they have very narrow diameters and provide a link between molecular and solid state physics. Figure 4.12 shows the surface charge state of the APTES-modified SiNW surface with pH. Due to quantum confinement effects and their high surface-area-to-volume ratios, NWs can be proposed as chemical and biological sensing element [52–55].

4.9.3 Nanoparticles

An emerging approach for robust real-world applications of flexible sensors relies on nanoparticles (NPs) with diameters ranging from 10 to 100 nm. Among the numerous reasons why exploiting (materials comprising) NPs for flexible sensors is promising. The first reason relates to the presumed ability to synthesize, nearly any type of NP. Several studies have shown the ability to control NPs, starting with cores

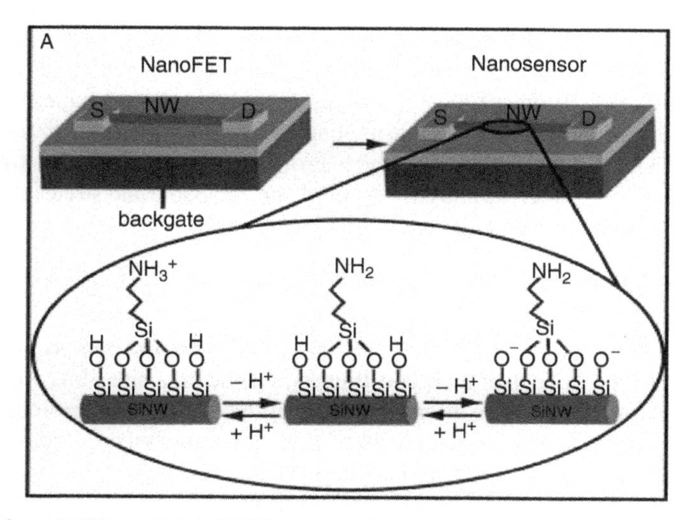

Figure 4.12 APTES-modified SiNW sensor with pH sensing (*Source:* Reproduced from Ref. [51] with permission)

made of pure metal (e.g., Au, Ag, Ni, Co, Pt, Pd, Cu, Al); metal alloys (e.g., Au/Ag, Au/Cu, Au/Ag/Cu, Au/Pt, Au/Pd, and Au/Ag/Cu/Pd, PtRh, NiCo, PtNiFe) metal oxides, and semiconducting materials (e.g., Si, Ge) [56–58]. The second reason is the ability to cap the NPs with a wide variety of molecular ligands, including alkylthiols, alkanethiolates, arenethiolate, alkyltrimethyloxysilane, dialkyldisulfides, xanthates, oligonucleotides, DNA, proteins, sugars, phospholipids, and enzymes. For sensing applications, this ability implies that one can obtain NPs with a hybrid combination of chemical and physical functions, which have a great effect on the sensitivity and selectivity of the sensors. The third reason is the ability to vary the NPs' size (1–100 nm) and shape (sphere, rectangle, hexagon, cube, triangle, and star, and branchlike outlines) and, consequently, the surface-to-volume ratio. For sensing applications, these features would allow deliberate control over the surface properties and the related interaction "quality" with the physical parameters such as pressure, temperature, and plasmon resonance. The fourth reason is attributed to the ability to prepare films of NPs with controllable porous properties. This attribute allows control over interparticle distance as well as controllable signal and noise levels, which, eventually, dominates the device sensitivity on exposure to either physical or chemical parameters. The fifth reason is the presumed ability of NPs to allow easier, faster, more cost-effective fabrication of flexible sensors compared with those currently in use, which mostly rely on complicated and multistep processes. For example, NP-based flexible smart sensors are extremely promising for a wide variety of applications in consumer electronics, robotics, prosthetics, health care, geriatric care, sports and fitness, safety equipment, environmental monitoring, homeland security, and space flight. (i) Flexible NP sensors can be applied in strain gauges, (ii) flexible multiparameter sensors with strain-tunable sensitivity toward different parameters, and (iii) sensors that are unaffected by mechanical deformation. To date, most flexible sensors are based on metal NPs, but the future of these sensors could be in semiconducting NPs with quantum-dot properties. Incorporating NPs into flexible sensors will soon become one of the most important applications of nanotechnology as shown in Figure 4.13 [59]. NP-based flexible sensors could help overcome the considerable technological challenges that hinder development of real-world applications as shown in Figure 4.13. These sensors are expected to provide a platform for future robust, simple, large-area, cost-effective, and easy-to-fabricate, bendable and stretchable sensing systems for a wide variety of applications.

4.9.4 Nanocomposites

Nanocomposites are a new frontier in materials science because composites can have very different properties than their constituents. Nanocomposite films are thin films formed by mixing two or more dissimilar materials having nanodimensional phase(s) in order to control and develop new and improved structures and properties. The properties of nanocomposite films depend not only on the individual components used but also on the morphology and the interfacial characteristics.

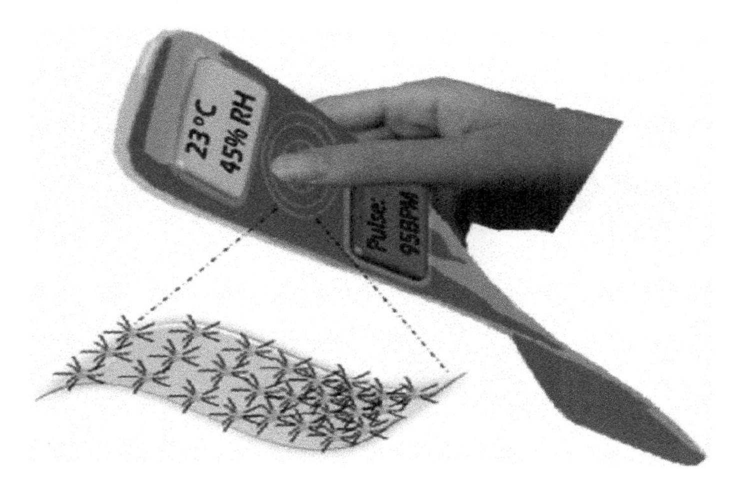

Figure 4.13 Flexible multiparameter sensors based on nanoparticle (*Source:* Reproduced from [59] with permission)

Nanocomposite films that combine materials with synergetic or complementary behaviors possess unique physical, chemical, optical, mechanical, magnetic, and electrical properties unavailable from that of the component materials and have attracted much attention for a wide range of device applications such as chipless RFID sensing application. Recently, various nanocomposite films consisting of either metal–metal oxide, mixed metal oxides, polymers mixed with metals or metal oxides, or carbon nanotubes mixed with polymers, metals or metal oxides have been synthesized and investigated for their application as active materials for sensors. Design of the nanocomposite films for gas sensor applications needs the considerations of many factors, for example, the surface area, interfacial characteristics, electrical conductivity, nanocrystallite size, surface and interfacial energy, stress and strain. All these factors depend significantly on the material selection, deposition methods, and deposition process parameters. Nanocomposite films consist of nanocrystalline or amorphous phase of at least two different materials. Depending on the nature of the component materials, micro/nanostructure and surface/interfacial characteristics, various unique gas sensing properties can be realized by using nanocomposite films as the active layer. Improvement in gas sensitivity, selectivity, stability and reduction in the response time and operating temperature have been demonstrated by various types of nanocomposite films [60]. Ultrasensitive-flexible-silver-nanoparticle-based nanocomposite-resistive sensor for ammonia detection is shown in Figure 4.14. Figure 4.14 shows ammonia sensing of GG/Ag film (a) Optical micrograph of the film on a flexible substrate. Optical micrograph of the film with electrical leads is shown in the inset. Figure 4.14(b) shows schematic representation of the experimental setup. The distance from the source to film is 1 cm.

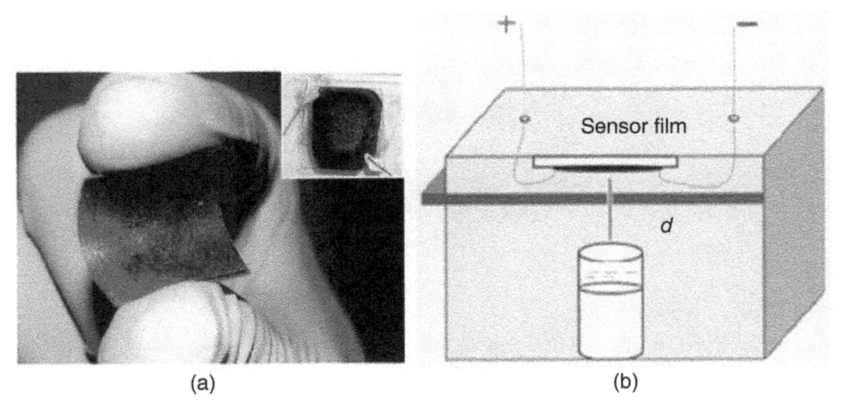

Figure 4.14 Ultrasensitive-flexible-silver-nanoparticle-based nanocomposite-resistive sensor for ammonia detection (*Source:* Reproduced from Ref. [61] with permission)

Recent developments in this new area of research including the fabrication methods are currently in use for preparing nanocomposite films including metal–metal oxides, mixed metal oxides, polymers mixed with metals or metal oxides, or carbon nanotubes mixed with polymers, metals, or metal oxides. Characterization and fabrication of these nanocomposites will be described in Chapters 5 and 9, respectively.

4.10 DISCUSSION

So far various materials such as temperature, humidity, pH, gas sensing materials and their basic properties are identified, analyzed, and classified. The identification of smart materials has been done based on dielectric and conductive properties analyzed for mm-wave and microwave RF sensing. The classification of materials for temperature, humidity, pH, gas, and strain sensing has also been presented based on physical parameter sensing. Moreover, these materials having multifunctional properties and having potential to sense more than one parameter are also analyzed in detail. In Table 4.2, the properties of these smart materials are summarized based on their attributes.

Table 4.3 presents a summary of smart materials discussed in this chapter and their microwave sensing capabilities. Also, it shows the critical sensing parameters for each material. Figure 4.15 shows the various smart materials versus physical parameters. As presented in the table, change in particular physical parameter can trigger chemical reaction, change of internal polarization, phase or band gap of a smart material. This results in microwave property variation that can be calibrated to measurable sensing data. Microwave sensing capabilities of smart materials is a novel field of study for passive RFID sensor development. This will lead to an interdisciplinary research area for high sensitive, passive sensor node development.

TABLE 4.2 List of Smart Materials with Their Advantages and Limitations

Smart Materials	Advantages	Limitations
Phenanthrene	Drastic increase of dielectric constant during sublimation after transition temperature 72 °C, which is permanent if the vapor is not desublimated It can be used as temperature threshold sensing. Irreversible temperature sensing material is suitable for applications where a certain temperature violation is crucial	Cannot be used in low temperature and real-time monitoring
Ionic plastic crystal	The ionic conductivity of plastic crystals changes due to organic molecule defects and the movement of crystals It has reversible temperature dependency and can be used to monitor real-time temperatures	As temperature is increased from −15 to +70 °C, it goes through three phase transitions starting from crystal state and ends up as liquid
Metal oxide	Doped semiconducting materials exhibit a remarkable combination of optical and electrical transport properties: (i) high electrical conductivity ($10^4\ \Omega^{-1}\ cm^{-1}$) and (ii) high optical transparency (>80%) in the visible range of the spectrum. ZnO thin film is an excellent sensing material for temperature sensing	Similar to most ceramic materials, they are brittle and prone to cracking when used in flexible electronics
Kapton	Significant advantages for both processing of circuitry and functionality of circuitry as well as laser ablatability. Processing advantages include the capability to fabricate from roll to roll, high mechanical strength, and unique distortional resistance to harsh environments such as high temperature bonding stations and corrosive aqueous etchants. It can be used as humidity sensing	The disadvantage of Kapton is its relatively high moisture absorption. The absorbed moisture can degrade Kapton's electrical properties
Polyvinyl alcohol (PVA)	Low price and low methanol permeability. Can be used as a humidity sensing material	Soluble in water

(continued)

TABLE 4.2 (*Continued*)

Smart materials	Advantages	Limitations
PEDOT	Can be fabricated on flexible substrates and has good mechanical properties. PEDOT may be used as a pH sensing material	Severe degradation at elevated temperature and inferior electrical and optical properties
Carbon nanotubes	Exhibit electrical conductivity as high as copper, have thermal conductivity as high as diamond, strength 100 times greater than steel at one-sixth the weight and high strain to failure. SWNT networks are flexible, doped semiconductor. It can be used for monitoring the toxic gas	Difficulties in manipulating individual NTs, difficult to control whether building blocks are semiconducting or metallic, stretching a nanotube can either increase or decrease its conductivity—depending on the type of nanotube, and so on
Microfiber-reinforced polytetrafluoroethylene (PTFE)	Lowest coefficient of friction of any polymer, broad working temperature range (500 °F/260 °C to −454 °F/−270 °C) Chemically resistant to all common solvents, acids, and bases, low extractable, excellent dielectric strength, biocompatible, precision extruded tolerances, flame resistant, limiting oxygen index (>95), easy to clean, mechanically resistant under severe conditions. It can be used as strain and crack sensing for structural health monitoring	High cost; low strength and stiffness; cannot be melt processed; poor radiation resistance
CdS	Resistance of CdS decreases with incident light intensity. It can be used as light sensing materials	Mostly used for DC or low-frequency applications
Graphene	It is the thinnest material known as well as the strongest. It is a superb conductor of both heat and electricity. Graphene-based wireless can be used for detection of bacteria; also it can be used for electrochemical sensors, biosensors, and so on. Due to extremely high electron mobility, it may be used for production of highly sensitive Hall effect sensors	Being a great conductor of electricity, although it does not have a band gap. Exhibits some toxic qualities

Nanowires	Exhibit electrical conductivity as high as copper, have thermal conductivity as high as diamond, strength 100 times greater than steel at one-sixth the weight and high strain to failure. NWs can be proposed as chemical and biological sensing element	Difficulties in manipulating individual nanowires and in controlling whether building blocks are semiconducting or metallic
Nanoparticles	Ability to control any shape of nanoparticles, a hybrid combination of chemical and physical functions, which would have a great effect on the sensitivity and selectivity of the sensors. Ability to vary the NPs' size (1–100 nm) and shapes (sphere-, rectangle-, hexagon-, cube-, triangle-, and star-, and branchlike outlines) and, consequently, the surface-to-volume ratio. For sensing applications, these features would allow deliberate control over the surface properties and the related interaction "quality" with the physical parameters such as pressure, temperature, plasmon resonance, and more Nanoparticles can be used for creating novel detection systems and for analyzing chemical and biological targets	Generally, more toxic than larger particles of the same composition. Nanoparticles no bigger than a few nanometers may reach well inside biomolecules, which is not possible for larger nanoparticles
Nanocomposites	Different properties than their constituents, electrical and thermal conductivities, optical properties, dielectric properties, heat resistance or mechanical properties such as stiffness, strength, and resistance to wear and damage can be tuned. It can be used in wide range of sensing applications by tuning the constituent materials	Dispersion difficulties, viscosity increase

TABLE 4.3 Summary of Smart Materials with Microwave Sensing Capabilities

Physical Parameters	Smart Materials	RF/Microwave Sensing Capabilities	Microwave Sensing Parameters
Relative humidity (RH)	Kapton	Absorbs water and changes internal electric polarization	Relative permittivity (ε_r)
	PVA	Creates hydrogen bond with water molecules	
Temperature	Plastic crystal ($P_{14}PF_6$)	Exhibits multiple thermal states during glass transition	Conductivity (σ)
	Phenanthrene	Evaporates directly from solid state at sublimation temperature	Relative permittivity (ε_r)
	Metal oxide	Band gap energy changes with temperature	Conductivity (σ)
pH	Conducting polymer: PEDOT	pH change results in reduction or oxidation reaction. This affects polymer sheet resistance	Conductivity (σ)
Light	CdS	Light illumination changes cell resistance	Equivalent impedance (Z)
Ammonia gas	SW-CNT	Presence of ammonia gas changes electrical property	Equivalent impedance (Z)

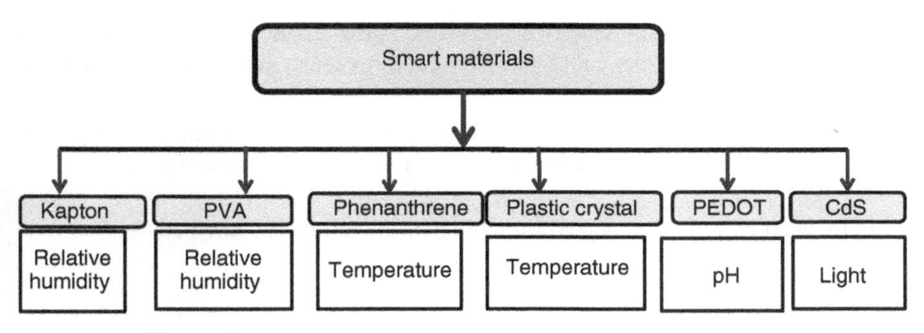

Figure 4.15 Various smart materials versus physical parameters

4.11 CONCLUSION

The aim of this chapter was to select and classify a number of smart materials according to sensitivity to particular physical parameters. This chapter has presented a comprehensive review of smart sensing materials for RF sensing and their applications. The identification and classification of smart materials for mm-wave and microwave/RF sensing of materials have been presented based on their physical parameter sensing characteristics. The materials exhibit high RF sensitivity of a particular physical parameter. Moreover, materials having multifunctional properties are investigated. These materials have potentials to sense more than one parameter. However, there remain challenges in characterizing them fully in high frequency and measuring the dielectric and conducting properties for a wide range of applications. From the above discussion, the following inferences are drawn:

(i) *Identification of smart materials*: Smart materials such as phenanthrene, ionic plastic crystals, metallic oxide, Kapton, PVA, PANI (polyaniline), cadmium sulfide, PEDOT:PSS(poly(3,4-ethylenedioxythiophene)-polystyrenesulfonicacid, hydrophilic polymer, SW-CNTs, metallic oxide, nanoparticles are identified for various physical parameter sensing based on their conductivity, relative humidity, surface impedance for mm-wave and microwave sensing.

(ii) *Classification of smart materials*: The smart materials are classified for various sensing capabilities such as humidity, temperature, pH, gas, strain, crack, and light. These particular classes of smart materials exhibit changes in their structural/chemical compositions when specific physical parameters in the environment change.

(iii) *Analysis of smart materials*: The properties of smart materials used for mm-wave and microwave sensing parameters are analyzed in detail. The analysis is based on their attributes.

(iv) *Emerging sensing materials*: The emerging sensing materials such as graphene, nanowires, nanocomposites, and nanoparticles have also been discussed in detail. Moreover, there are some smart materials having multi-functional properties and having potential to sense more than one parameter are also analyzed.

(v) *Discussion on sensing materials*: Finally, the properties of those sensing materials are summarized based on their attributes and salient features.

In the next chapter, characterization of smart materials is presented.

REFERENCES

1. D. B. Mitzi, "Solution-Processed Inorganic Semiconductors," *Journal of Materials Chemistry,* vol. 14, pp. 2355–2365, 2004.
2. L. Groenendaal, F. Jonas, D. Freitag, H. Pielartzik, and J. R. Reynolds, "Poly(3,4-ethylenedioxythiophene) and Its Derivatives: Past, Present, and Future," *Advanced Materials,* vol. 12, pp. 481–494, 2000.
3. T. Hayashi, T. Ino, K. Ueno, and H. Shirai, "Atmospheric-Pressure Argon Plasma Etching of Spin-Coated Polyethylene Dioxythiophene: Polystyrene Sulfonic Acid (PEDOT:PSS) Films for Copper Phthalocyanine (CuPc)/C-60 Heterojunction Thin-Film Solar Cells," *Thin Solid Films,* vol. 519, pp. 6834–6839, 2011.
4. J. Kroupa, J. Fousek, N. R. Ivanovb , B. Březina, and V. Lhotská, "Dielectric Study of the Phase Transition in Phenanthrene," *Ferroelectrics,* vol. 79, pp. 189–192, 1988.
5. J. Virtanen, L. Ukkonen, T. Bjorninen, A. Z. Elsherbeni, and L. Sydänheimo, "Inkjet-Printed Humidity Sensor for Passive UHF RFID Systems," *IEEE Transactions on Instrumentation and Measurement,* vol. 60, pp. 2768–2777, 2011.
6. M. R. Yang and K. S. Chen, "Humidity Sensors Using Polyvinyl Alcohol Mixed with Electrolytes," *Sensors and Actuators B: Chemical,* vol. 49, pp. 240–247, 1998.
7. N. C. Karmakar, E. M. Amin, and B. W. Jensen, "Polyvinyl-Alcohol (PVA)-based RF Humidity Sensor in Microwave Frequency" *Progress In Electromagnetics Research Journal B*, 54, pp. 149–166, 2013.
8. N. C. Karmakar and E. M. Amin, "Development of a low cost printable humidity sensor for chipless RFID technology", *2012 IEEE International Conference on RFID-Technologies and Applications (RFID-TA)*, pp. 165–170, 2012.
9. J. Stejskal and R. G. Gilbert, "Polyaniline. Preparation of a Conducting Polymer," *Pure and Applied Chemistry,* vol. 74, pp. 857–867, 2002.
10. X. Wang, L. Zhi, and K. Müllen, "Transparent Conductive Graphene Electrodes for Dye-Sensitized Solar Cells" *Nano Letters,* vol. 8, pp. 323–327, 2008.
11. J. Golding, N. Hamid, D. R. MacFarlane, C. Forsyth, C. Collins, and J. Huang, "N-Methyl-N-alkylpyrrolidinium Hexafluorophosphate Salts: Novel Molten Salts and Plastic Crystal Phases," *Chemistry of Materials*, vol. 13, p. 6; pp. 558–564, 2001.
12. H. Hatakeyama and T. Hatakeyama, "Interaction Between Water and Hydrophilic Polymer" *Thermochimica Acta*, vol. 308, pp. 3–22, 1998.
13. Z. Wu, X. Chen, X. Du, J. M. Logan, J. Sippel, M. Nikolou, K. Kamaras, J. R. Reynolds, D. B. Tanner, A. F. Hebard, and A. G. Rinzler, "Transparent, Conductive Carbon Nanotube Films," *Science,* vol. 305, pp. 1273–1276, 2004.

14. S. Chenghua, X. Juan, W. Helin, X. Tianning, Y. Bo, and L. Yuling, "Optical temperature sensor based on ZnO thin film's temperature-dependent optical properties," *Review of Scientific Instruments,* vol. 82, 084901, 2011.

15. X. Chen, B. Jia, J. K. Saha, B. Cai, N. Stokes, Q. Qiao, Y. Wang, Z. Shi, and M. Gu, "Broadband Enhancement in Thin-Film Amorphous Silicon Solar Cells Enabled by Nucleated Silver Nanoparticles," *Nano Letters,* vol. 12, pp. 2187–2192, 2012.

16. E. Amin, J. K. Saha, and N. C. Karmakar, "Smart Sensing Materials for Low-cost Chipless RFID Sensor," *IEEE Sensor Journal,* vol. 14, pp. 2198–2207, 2014.

17. J. Kroupa, J. Fousek, N. R. Ivanov, B. Březina, and V. Lhotská, "Dielectric Study of the Phase Transition in Phenanthrene," *Ferroelectrics*, vol. 79, pp. 189–192, 1988.

18. C. G. De Kruif, "Enthalpies of Sublimation and Vapour Pressure of 11 Polycyclic Hydrocarbons," *The Journal of Chemical Thermodynamics,* vol. 12, pp. 243–248, 1980.

19. J. Golding, N. Hamid, D. R. MacFarlane, M. Forsyth, C. Forsyth, C. Collins, *et al.*, "*N*-Methyl-*N*-alkylpyrrolidinium Hexafluorophosphate Salts: Novel Molten Salts and Plastic Crystal Phases," *Chemistry of Materials,* vol. 13, pp. 558–564, 2001.

20. E. Stavrinidou, M. Sessolo, B. Winther-Jensen, S. Sanaur, and G. G. Malliaras, "A Physical Interpretation of Impedance at Conducting Polymer/Electrolyte Junctions," *Journal of Applied Physics,* vol. 102, pp. 017127–0171276, 104111, 2007.

21. M. E. Orazem and B. Tribollet, "Methods for representing impedance," in *Electrochemical Impedance Spectroscopy*: John Wiley & Sons, Inc., 2008, pp. 307–331.

22. M. E. Orazem and B. Tribollet, "Electrical circuits," in *Electrochemical Impedance Spectroscopy*: John Wiley & Sons, Inc., 2008, pp. 61–72.

23. P. Kumar, H. K. Malik, A. Ghosh, R. Thangavel, and K. Asokan, "Bandgap Tuning in Highly c-axis Oriented Zn1−xMgxO Thin Films," *Applied Physics Letters,* vol. 102, pp. 2219031–2219035, 2013.

24. F. Urbach, "The Long-Wavelength Edge of Photographic Sensitivity and of the Electronic Absorption of Solids," *Physical Review,* vol. 92, pp. 1324–1324, 1953.

25. Z. Chen and C. Lu, "Humidity Sensors: A Review of Materials and Mechanisms," *Sensor Letters,* vol. 3, pp. 274–295, 2005.

26. *Kapton HN Polyimide Film Datasheet.* Available: http://www2.dupont.com/Kapton/en_US/.

27. Z. Chen and C. Lu, "Humidity Sensors: A Review of Materials and Mechanisms," *Sensor Letters,* vol. 3, 2005.

28. C. A. Finch, *Polyvinyl alcohol. Properties and Applications*: Wiley-Interscience 1973.

29. R. J. Sengwa and K. Kaur, "Dielectric Dispersion Studies of Poly(vinyl alcohol) in Aqueous Solutions," *Polymer International,* vol. 49, pp. 1314–1320, 2000.

30. Y. K. Yeow, Z. Abbas, K. Khalid, and M. Z. A. Rahman, "Improved Dielectric Model for Polyvinyl Alcohol-Water Hydrogel at Microwave Frequencies," *American Journal of Applied Sciences,* Vol. 7(2), pp. 270–276, 2010.

31. M. Penza and V. I. Anisimkin, "Surface Acoustic Wave Humidity Sensor Using Polyvinyl-alcohol Film," *Sensors and Actuators A: Physical,* vol. 76, pp. 162–166, 1999.

32. C. E. Clark. *Perfectly Good Food Is Thrown Away Daily While People Go Hungry In the U.S.* Available: http://aufait.hubpages.com/hub/Americans-throw-away-good-food-because-of-sell-by-dates.

33. S. Bhadra, G. E. Bridges, D. J. Thomson, and M. S. Freund, "A Wireless Passive Sensor for Temperature Compensated Remote pH Monitoring," *IEEE Sensors Journal,* vol. 13, pp. 2428–2436, 2014.

34. K. Z. Qingyun Cai, C. Ruan, T. A. Desai, and C. A. Grimes, "A Wireless, Remote Query Glucose Biosensor Based on a pH-Sensitive Polymer," *Analytical Chemistry,* vol. 76, pp. 4038–4043, 2004.

35. R. M. Sidek, F. A. M. Yusof, F. M. Yasin, R. Wagiran, and F. Ahmadun, "Electrical response of multi-walled carbon nanotubes to ammonia and carbon dioxide," in *Semiconductor Electronics (ICSE), 2010 IEEE International Conference on*, 2010, pp. 263–266.

36. M. D. Balachandran, S. Shrestha, M. Agarwal, Y. Lvov, and K. Varahramyan, "SnO_2 Capacitive Sensor Integrated with Microstrip Patch Antenna for Passive Wireless Detection of Ethylene Gas," *Electronics Letters,* vol. 44, pp. 464–466, 2008.

37. Y. Hargsoon, X. Jining, K. A. Jose, K. V. Vijay, and B. R. Paul, "Passive Wireless Sensors Using Electrical Transition of Carbon Nanotube Junctions in Polymer Matrix," *Smart Materials and Structures,* vol. 15, p. S14, 2006.

38. O. Lupan, G. Chai, and L. Chow, "Novel Hydrogen Gas Sensor based on Single ZnO Nanorod," *Microelectronic Engineering,* vol. 85, pp. 2220–2225, 2008.

39. J. X. Wang, X. W. Sun, Y. Yang, H. Huang, Y. C. Lee, O. K. Tan, *et al.*, "Hydrothermally Grown Oriented ZnO Nanorod Arrays for Gas Sensing Applications," *Nanotechnology,* vol. 17, p. 4995, 2006.

40. M. M. Tentzeris, "Inkjet-printed paper-based RFID and nanotechnology-based ultrasensitive sensors: The "Green" ultimate solution for an ever improving life quality and safety?," in *Radio and Wireless Symposium (RWS), 2010 IEEE*, 2010, pp. 120–123.

41. X. Yi, R. Vyas, C. Cho, C. Fang, J. Cooper, Y. Wang, R. T. Leon, and M. M. Tentzeris, "Thermal effects on a passive wireless antenna sensor for strain and crack sensing," in *Proceedings of the SPIE, Sensors and Smart Structures Technologies for Civil, Mechanical, and Aerospace Systems*, 2012, p. 11.

42. A. Vena, M. Hasani, L. Sydänheimo, L. Ukkonen, and M. M. Tentzeris, "Implementation of a Dual-Interrogation-Mode Embroidered RFID-Enabled Strain Sensor," *IEEE Antennas and Wireless Propagation Letters,* vol. 12, pp. 1272–1275, 2013.

43. C. Occhiuzzi, C. Paggi, and G. Marrocco, "Passive RFID Strain-Sensor Based on Meander-Line Antennas," *IEEE Transactions on Antennas and Propagation,* vol. 59, pp. 4836–4840, 2012.

44. R. Corporation. Properties of low dielectric constant laminates [Online]. Available: https://www.rogerscorp.com/documents/1798/acm/articles/Properties-of-Low-Dielectric-Constant-Laminates.pdf+&cd=1&hl=en&ct=clnk&gl=ca (accessed on 07 October 2015).

45. C. Namjun, S. Seong-Jun, K. Sunyoung, K. Shiho, and Y. Hoi-Jun, "A 5.1-μW UHF RFID tag chip integrated with sensors for wireless environmental monitoring," in *Solid-State Circuits Conference, 2005. ESSCIRC 2005. Proceedings of the 31st European*, 2005, pp. 279–282.

46. S. Folea and M. Ghercioiu, "Ultra-low power Wi-Fi tag for wireless sensing," in *Automation, Quality and Testing, Robotics, 2008. AQTR 2008. IEEE International Conference on*, 2008, pp. 247–252.

47. (2014). *Light Sensors*. Available: http://www.electronics-tutorials.ws/io/io_4.html.

48. J. M. Dewdney and W. Jing, "Characterization the microwave properties of SU-8 based on microstrip ring resonator," in *Wireless and Microwave Technology Conference, 2009. WAMICON '09. IEEE 10th Annual*, 2009, pp. 1–5.
49. M. S. Mannoor, H. Tao, J. D. Clayton, A. Sengupta, D. L. Kaplan, R. R. Naik, N. Verma, F. G. Omenetto, and M. C. McAlpine, "Graphene-based wireless bacteria detection on tooth enamel," *Nature Communications*, 2012.
50. L. Taoran, V. Lakafosis, L. Ziyin, C. P. Wong, and M. M. Tentzeris, "Inkjet-printed graphene-based wireless gas sensor modules," in *62nd IEEE Electronic Components and Technology Conference (ECTC)*, San Diego, CA, 2012, pp. 1003–1008.
51. Y. Cui, Q. Wei, H. Park, and C. M. Lieber, "Nanowire Nanosensors for Highly Sensitive and Selective Detection of Biological and Chemical Species," *Science,* vol. 293, pp. 1289–1292, 2001.
52. F. Patolsky, G. Zheng, and C. M. Lieber, "Nanowire Sensors for Medicine and the Life Sciences" *Nanomedicine,* vol. 1, pp. 51–56, 2006.
53. S. F. A. Rahman, N. A. Yusof, U. Hashim, and M. N. M. Nor, "Design and Fabrication of Silicon Nanowire based Sensor," *International Journal of Electrochemical Science,* vol. 8, pp. 10946–10960, 2013.
54. X. Chen, C. K. Y. Wong, C. A. Yuan, and G. Zhang, "Nanowire-based Gas Sensors," *Sensors and Actuators* vol. 177, pp. 178–195, 2013.
55. Y. Dan, S. Evoy, and A. T. C. Johnson, "*Chemical gas sensors based on nanowires*" in *Nanowire Research Progress*: Nova Science Publisher, 2008.
56. O. Masala and R. Seshadri, "Synthesis Routes for Large Volumes of Nanoparticles," *Annual Review of Materials Research,* vol. 34, pp. 41–81, 2004.
57. B. L. Cushing, V. L. Kolesnichenko, and C. J. O'Connor, "Recent Advances in the Liquid-Phase Syntheses of Inorganic Nanoparticles," *Chemical Reviews,* vol. 104, pp. 3893–3946, 2004.
58. C. J. Murphy, "Optical Sensing with Quantum Dots," *Analytical Chemistry,* vol. 74 issue 19, pp. 520A–526A, 2002.
59. M. S. Bar and H. Haick, "Flexible Sensors Based on Nanoparticles," *ACS Nano,* vol. 7, pp. 8366–8378, 2013.
60. D. Yang, *Nanocomposite Films for Gas Sensing*: InTech, 2011.
61. S. Pandey, G. K. Goswami, and K. K. Nanda, "Nanocomposite Based Flexible Ultrasensitive Resistive Gas Sensor for Chemical Reactions Studies," *Scientific Reports,* vol. 3, pp. 1–6, 2013.

5

CHARACTERIZATION OF SMART MATERIALS

5.1 INTRODUCTION

In the preceding chapter, various smart materials have been identified for RF/microwave/millimeter-wave sensing. In the beginning of the investigation, it has been also shown that the materials that exhibit low conductivities but susceptible to physical parameters are used for sensing devices, whereas those materials are of high conductivities and whose conductivity does not change with physical parameters are used for printing chipless identification circuits. Therefore, there are clear demarcation between the two classes of materials for identification and sensing. However, there are other certain classes of advanced materials that show very favorable properties that can be used for both identification and sensing. For example, graphene is used to print a patch antenna for communication, and at the same time, they can also be used for sensing. Based on the properties of these materials, they are classified for particular physical parameters sensing. For example, metal oxides and plastic crystals can be used for multifunctional sensing.

As stated earlier, characterization of these materials at microwave and millimeter-wave frequency bands are a new field. Therefore, it is imperative to find their physical properties and electrical properties at microwave and millimeter-wave frequency bands. In general sense, material characterization denotes a systematic analysis, measurement, testing, modeling, and simulation procedure that yields both qualitative and quantitative data of the specific attributed of specific purposes/applications.

Chipless RFID Sensors, First Edition. Nemai Chandra Karmakar, Emran Md Amin and Jhantu Kumar Saha.

As for example, for chipless radio-frequency identification (RFID) sensor, we are more interested in the constitutive parameter changes (ε_r, tan δ, б) at microwave and millimeter-wave frequencies with physical parameters such as temperature, pressure, relative humidity, pH, stress, and light.

Materials are broadly classified as structural, functional, and smart materials. While structural materials are engineered for specific mechanical and thermal properties, we are more interested in functional and smart materials that are used for sensing and actuating devices. The last two materials react in response to external loading and provide the opportunity to optimize their responses according to any specific requirements from the materials' performances.

This chapter presents various characterization procedures for smart materials for microwave and millimeter-wave sensing. Various novel analysis and characterization techniques including microstructural and surface morphology (X-ray diffraction, XRD; atomic force microscopy, AFM; scanning electron microscopy, SEM; transmission electron microscopy, TEM), optical (spectroscopic ellipsometry, ultraviolet visible infrared spectroscopy, UV–vis), electrical and thermal (DC conductivity, stability, etc.), microwave (scattering parameters i.e., complex permittivity, dielectric loss, and reflection loss) in the gigahertz range for sensing materials are also described. The contents of this chapter and organization are shown in Figure 5.1.

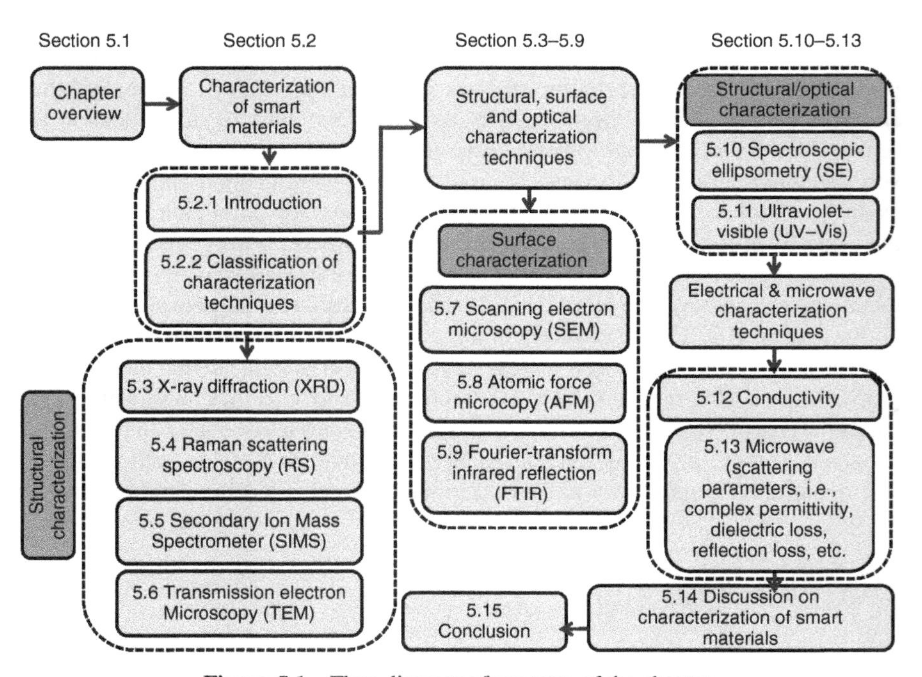

Figure 5.1 Flow diagram of contents of the chapter

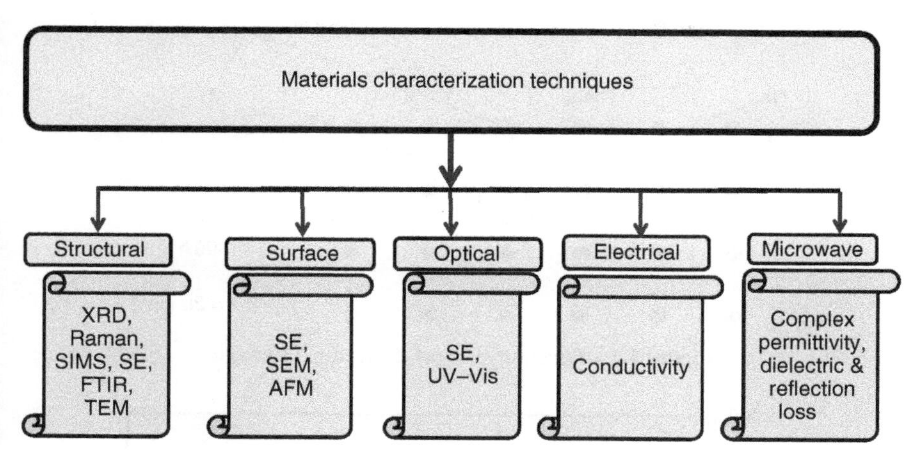

Figure 5.2 Smart materials characterization techniques

5.2 CHARACTERIZATION OF MATERIALS FOR MICROWAVE SENSING

The main objective of this chapter is to investigate the properties of low conductive materials for sensing applications. For the investigation, material properties are analyzed by diagnostic techniques to identify (a) the structural properties, (b) the elemental and chemical composition, (c) optical, (d) electrical and thermal parameters (DC conductivity, stability, etc.), (e) microwave (scattering parameters i.e., complex permittivity, attenuation, dielectric loss, and reflection loss) in the gigahertz range. Optical, scanning electron, and transmission electron microscopy techniques are used to observe the structure, and X-ray diffraction and transmission electron microscopy are used for analysis of crystal structure and defects.

The analysis and characterization techniques of smart materials include microstructural and surface morphology (XRD), Raman spectroscopy (RS), Atomic Force Microscopy (AFM), Scan Electron Microscopy (SEM), and Transmission Electron Microscopy (TEM), secondary ion mass spectroscopy (SIMS), spectroscopic ellipsometry (SE), Fourier transform infrared spectroscopy (FTIR), optical (UV–Vis), electrical and thermal parameters (DC conductivity, stability, etc.), microwave (scattering parameters i.e., complex permittivity, dielectric loss, and reflection loss) in the gigahertz range[1,2]. Figure 5.2 shows the characterization techniques of smart materials for microwave sensing applications. In the following sections, these techniques are described in detail.

5.3 X-RAY DIFFRACTION

XRD is a nondestructive technique for the qualitative and quantitative analysis of crystalline materials, form of powder or solid. Basically, XRD is obtained as reflection of an X-ray beam from a family of parallel and spaced atomic planes,

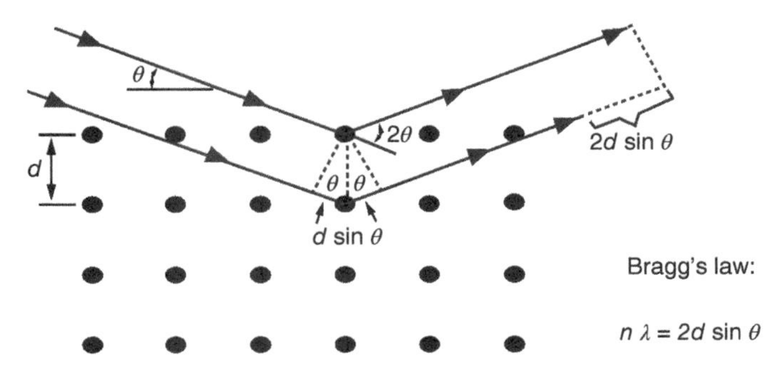

Figure 5.3 Principle of X-ray diffraction technique

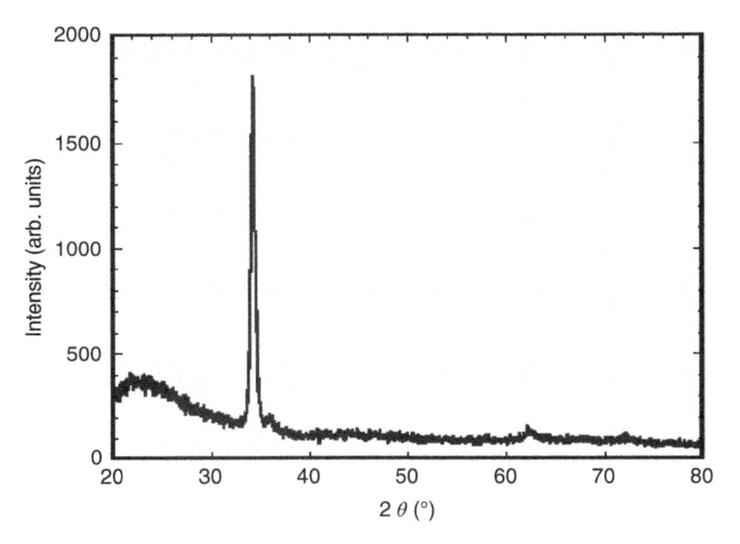

Figure 5.4 XRD picture of ZnO thin film

following Bragg's law: when a monochromatic X-ray beam with wavelength λ is incident on lattice planes with an angle θ, diffraction occurs if the path of rays reflected by successive planes (with distance d) is a multiple of the wavelength. Figure 5.3 shows the principle of X-ray diffraction technique.

The good crystalline quality with preferred orientation (0002) along the c-axis of ZnO with the sharp feature peak at around 35° can be observed along the c-axis perpendicular to the substrate surface, which is critical for piezoelectric applications for film is shown in Figure 5.4. X-ray diffraction provides most definitive structural information such as interatomic distance, bond angles, lattice parameters, phase identity, phase purity, crystallinity, crystal structure, and percent phase composition.

5.4 RAMAN SCATTERING SPECTROSCOPY

Raman scattering has been widely used in the fundamental spectroscopic study of excitations in solids, liquids, and gases and has also been extensively used in

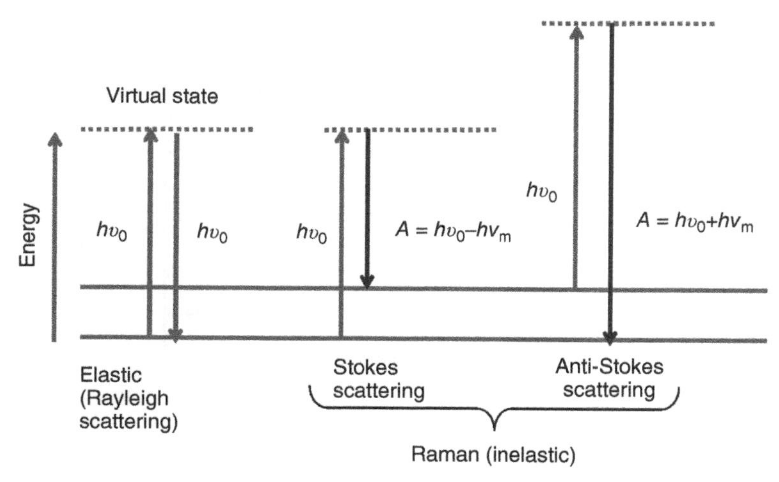

Figure 5.5 Energy level diagrams for Stokes and anti-Stokes inelastic scattering

material characterization. Raman spectra refer to inelastic scattering of quantized form of electromagnetic radiation. Photons enter a medium and lose energy to the sample, which is referred to as Stokes scattering, or lost energy, which is anti-Stokes scattering as shown in Figure 5.5 [3]. We can see that this results from the incident photon exciting the molecule into a virtual energy state. As a result, most Raman measurements are performed considering only the Stokes shifted light.

In comparison to other vibrational spectroscopy methods such as NIR, Raman has several major advantages such as Raman Effect manifests itself in the light scattered off a sample as opposed to the light absorbed by a sample. As a result, Raman spectroscopy requires little to no sample preparation and is insensitive to aqueous absorption bands. This property of Raman facilitates the measurement of solids, liquids, and gases not only directly, but also through transparent containers such as glass, quartz, and plastic.

Figure 5.6 shows typical Raman spectra of microcrystalline silicon film. The sharp feature peak at around $520\,cm^{-1}$ (c-Si phase) could be observed for films, and the broad band peaks attributed to an amorphous phase were observed at $480\,cm^{-1}$ region for microcrystalline silicon films. The nanocrystalline silicon film could be observed $510\,cm^{-1}$ region for microcrystalline silicon films. Thus, microcrystalline silicon network possesses a mixture of amorphous and crystalline Si phase. As discussed above, using Raman spectral libraries, it is easy to see how easily Raman spectra can be used for material identification and verification.

5.5 SECONDARY ION MASS SPECTROMETER

The technique of SIMS is the most sensitive of all the commonly employed surface analytical techniques because of the inherent sensitivity associated with mass spectrometric-based techniques. In SIMS, the surface of the sample is subjected to

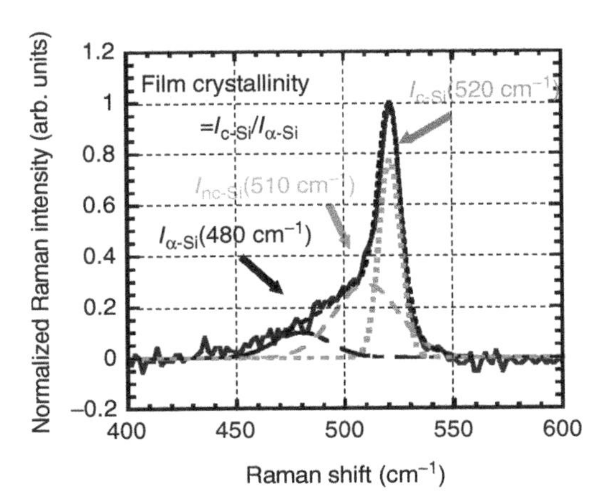

Figure 5.6 Raman spectra of microcrystalline silicon film (*Source*: Reproduced with permission Ref. [4])

bombardment by high-energy ions and this leads to the sputtering of both neutral and charged species from the surface. The ejected species may include atoms, clusters of atoms, and molecular fragments. The emitted ions are analyzed with a mass spectrometer, resulting in positive or negative mass spectra consisting of parent and fragment peaks characteristic of the surface.

The instrumentation for SIMS can be divided into two parts: (a) the primary ion source in which the primary ions are generated, transported, and focused toward the sample and (b) the mass analyzer in which sputtered secondary ions are extracted, mass separated, and detected, which is limited to explain in detail. The quantity of secondary ions is detected by very highly sensitive second electron multiplier tube. The SIMS measurement is performed to understand the residual atomic ion concentrations in the film.

5.6 TRANSMISSION ELECTRON MICROSCOPY

TEM is a technique used to observe modulations in chemical identity, crystal orientation, and electronic structure [5]. Figure 5.7 presents high-resolution micrographs of silver nanoparticles with diameter ranging from 2 to 12 nm [6]. Figure 5.7 shows aberration-corrected TEM images of silver nanoparticles synthesized free of stabilizing ligands. Particles with diameters of 2 nm (a), 3 nm (b), 4.5 nm (c), 6 nm (d), 7.5 nm (e) 9 nm (f), 10.5 nm (g), and 12 nm (h) are shown. Scale bars: (a–d), 2 nm; (e–h), 5 nm.

5.7 SCANNING ELECTRON MICROSCOPE

A scanning electron microscope (SEM) is an imaging technique of a sample. Field emission SEM (FE-SEM) produces clearer, less electrostatically distorted images

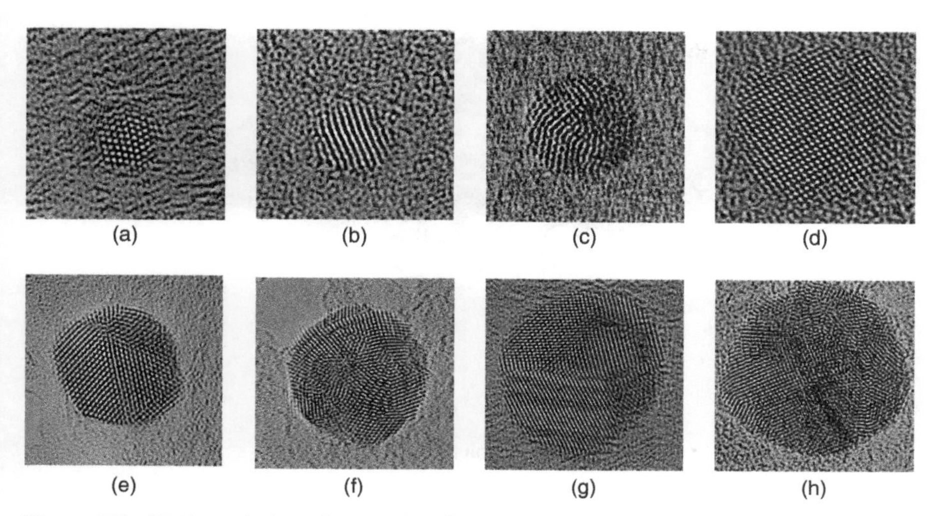

Figure 5.7 High-resolution micrographs of silver nanoparticles with diameters ranging from 2 to 12 nm (*Source*: Reproduced from [6] with permission)

Figure 5.8 FE-SEM images of flower-like ZnO nanorods with different magnifications synthesized by sol-hydrothermal process [7]

with spatial resolution down to 1.5 nm and it is 3–6 times better than conventional SEM. FE-SEM images of the flower-like nanorods synthesized by sol-hydrothermal process are shown in Figure 5.8; these nanoflowers were composed of shorter nanorods assembles because the orientation growth along the *c*-axis was inadequate in lacking the driving force of the growing units.

5.8 ATOMIC FORCE MICROSCOPY

AFM provides topographic information down to the Angstrom level [8]. The AFM picture of ZnO thin film is shown in Figure 5.9. The smooth surface morphology with the surface roughness of ZnO thin film is about 7.4 nm.

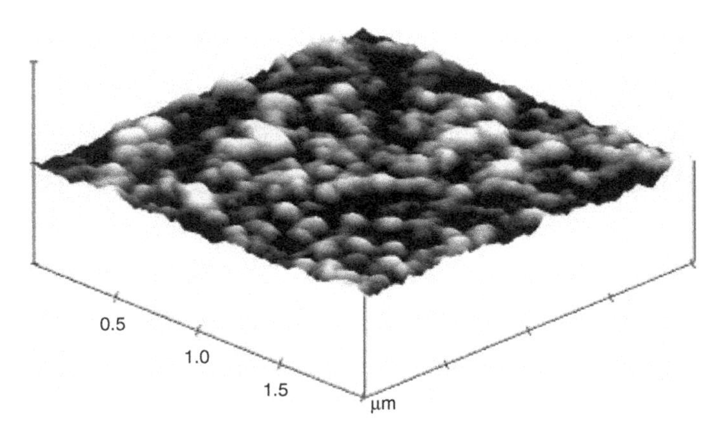

Figure 5.9 AFM picture of ZnO thin film

5.9 INFRARED SPECTROSCOPY (FOURIER TRANSFORM INFRARED REFLECTION)

Fourier transform spectroscopy is a measurement technique, whereby spectra are collected based on measurements of the temporal coherence of a radiative source, using time-domain measurements of the electromagnetic radiation or other types of radiation. Infrared (IR) spectroscopy is the subset of spectroscopy that deals with the infrared region of the electromagnetic spectrum. It covers a range of techniques, the most common being a form of absorption spectroscopy. As with all spectroscopic techniques, it can be used to identify compounds or investigate sample composition. The infrared portion of the electromagnetic spectrum is divided into three regions: the near-, mid-, and far-infrared, named for their relation to the visible spectrum. The mid-infrared, approximately 4000–400 cm^{-1} (30–1.4 μm), may be used to study the fundamental vibrations and associated rotational–vibrational structure [9].

This technique works almost exclusively on samples with covalent bonds. Simple spectra are obtained from samples with few IR active bonds and high levels of purity. More complex molecular structures lead to more absorption bands and more complex spectra. The technique has been used for characterization of very complex mixtures. Figure 5.10 shows typical FTIR spectra of microcrystalline silicon films from dichlorosilane at different T_sS. The typical FTIR spectrum for the SiH_x stretching absorption region in the microcrystalline silicon films from SiH_4 is also shown as a reference.

5.10 SPECTROSCOPIC ELLIPSOMETRY

Spectroscopic ellipsometry (SE) is a nondestructive diagnostic test of thin films. SE is essential for the determination of layer thickness, surface roughness, and the optical and electrical properties of the films. As stated earlier, all these attributes of materials are very significant for the perspective of microwave design. For thin films, the optical properties vary considerably depending on the microstructure and growth conditions.

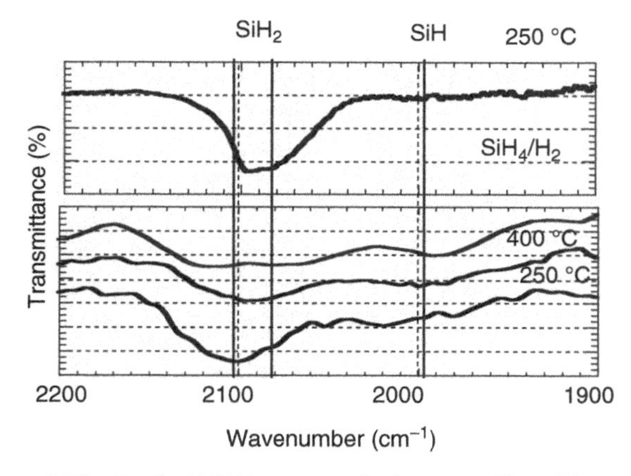

Figure 5.10 Typical FTIR spectra of microcrystalline silicon films

The principle of ellipsometry is based on the measurement of the change in polarization when light reflects on a surface, due to the difference in the reflection for the *p*- and *s*-direction. After reflection of the sample surface, a linearly polarized light beam is generally elliptically polarized. SE measures as a function of the photon energy (or wavelength). The complex ratio ρ between the *p*- and *s*-reflection coefficients is used to express this change of polarization

$$p = \frac{R_p}{R_s} \tag{5.1}$$

Both the phase (R_p) and amplitude ratio (R_s) from the *p*- and *s*-waves are of interest. Fortunately, the two reflection coefficients are independent. Therefore, two independent parameters of the surface can be determined, for instance, the real and imaginary parts of the refractive index, Equation 5.1 can be written as [10]

$$p = \frac{R_p}{R_s} = \tan \Psi ei\Delta \tag{5.2}$$

where Ψ and Δ are two ellipsometric angles that characterize the polarization effects of the surface. Ψ is the angle whose tangent gives the ratio of the amplitude attenuation for the *p* and *s* polarizations, while Δ gives the difference between the phase shifts experienced by the *p* and *s* polarizations. They depend on the angle of incidence, the wavelength, the optical constants, and the morphology of the film. Figure 5.11 shows the schematic of the SE measurement.

It consists of the measurement of the change in polarization state of a beam of light upon reflection from (or transmission through) the sample as shown in Figure 5.11 [11].

Two factors make ellipsometry particularly attractive:

1. It is essentially a nondestructive measurement procedure; hence, its suitability for *in situ* measurements is well recognized [12].
2. It is remarkably sensitive to minute interfacial effects such as the formation of sparsely distributed submonolayer of atoms or molecules [13].

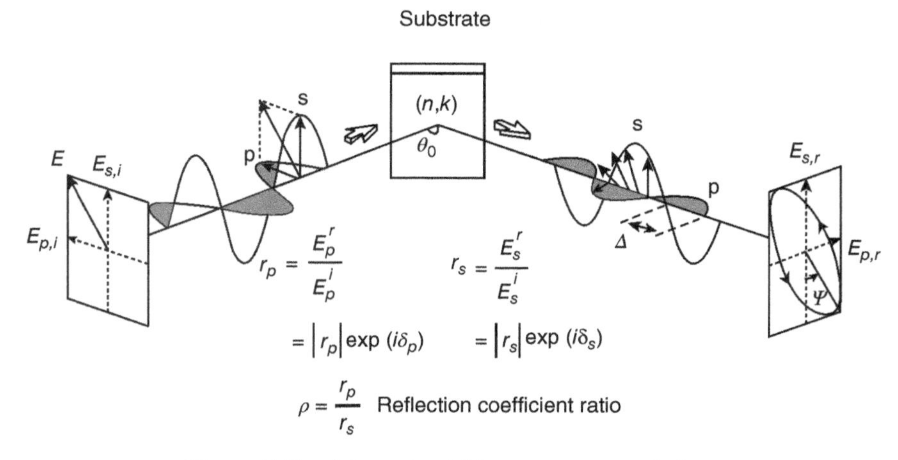

Figure 5.11 Schematic of SE measurement technique

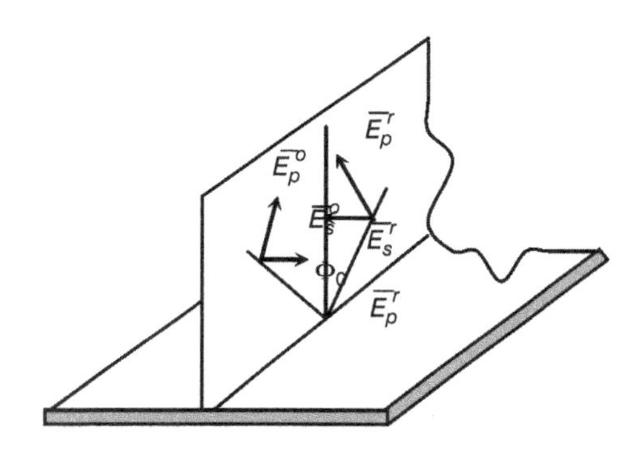

Figure 5.12 Linearly polarized light incidence at interface with a semi-infinite medium

The quantity $r_p(r_s)$ is the Fresnel reflection coefficient for light polarized parallel (perpendicular) to the plane of incidence and the angles Ψ, Δ are the traditional ellipsometry parameters. Ellipsometry measures the two values (Ψ, Δ) that express the amplitude ratio and phase difference between p- and s-polarizations, respectively as shown in Figure 5.11 and 5.12.

In the simplest case of an interface with a semi-infinite medium, the pseudodielectric function ε can be easily deduced from the ellipsometric ratio:

$$\overline{\varepsilon} = \varepsilon_1 + j\varepsilon_2 = \sin^2 \Phi_0 \left\{ 1 + \frac{\mathrm{tg}^2\, \Phi_0 \left(1 - \overline{\rho}\right)^2}{(1 + \overline{\rho})^2} \right\} \tag{5.3}$$

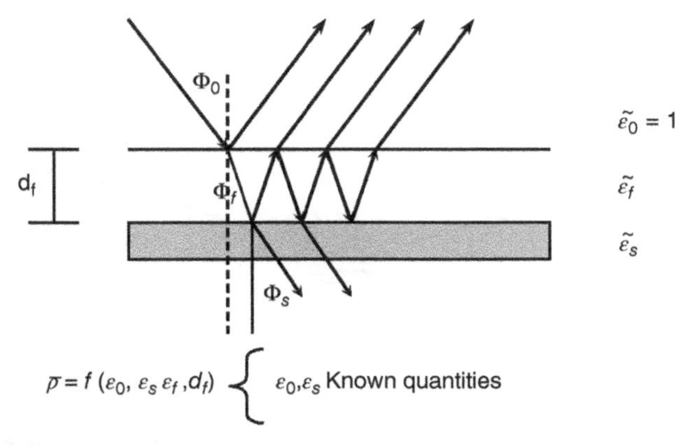

Figure 5.13 Schematic of light incidence into ambient-film-substrate

Here Φ_0 is the angle of incidence. tg Φ_0 is the tangent of angle Φ_0 of incident light. Dielectric function ε is related to the refractive index (n) and extinction coefficient (k) by the following relation:

$$\bar{\varepsilon} = \bar{n}^2 = (n^2 - k^2) + 2jnk \tag{5.4}$$

$$\left.\begin{array}{l} p : // \\ s : \perp \end{array}\right\} \text{Plane of incidence}$$

The accurate determination of dielectric function by SE allows, in particular, the optimization of devices or analysis of composite materials. In the case of ambient-film-substrate system, the scattering of light incidence is shown in Figure 5.13:

$$\bar{\rho} = \text{tg}\,\Psi \exp j\Delta = f(n_f, k_f, d_f) \tag{5.5}$$

where n_f, k_f, and d_f are the thickness, refractive index, and extinction coefficient of the film, respectively. By optical model of this system, we can extract n_f, k_f, and d_f, which are the main parameters of interest.

When used at a fixed wavelength in an *in situ* configuration:

$$\bar{\rho} = f[d(t)] \tag{5.6}$$

Here, $d(t)$ is thickness versus time, which is the situation during a growth/etching process. In this situation, we obtain a trajectory of the ellipsometric parameters, which represents the whole history of the process.

Figure 5.14 shows a typical ellipsometer used where the measurements were performed at an incident beam angle of 71°.

Generally, the measurement spectra include information of not only bulk but also the surface roughness. Therefore, the spectra analysis (fitting of this data) is required.

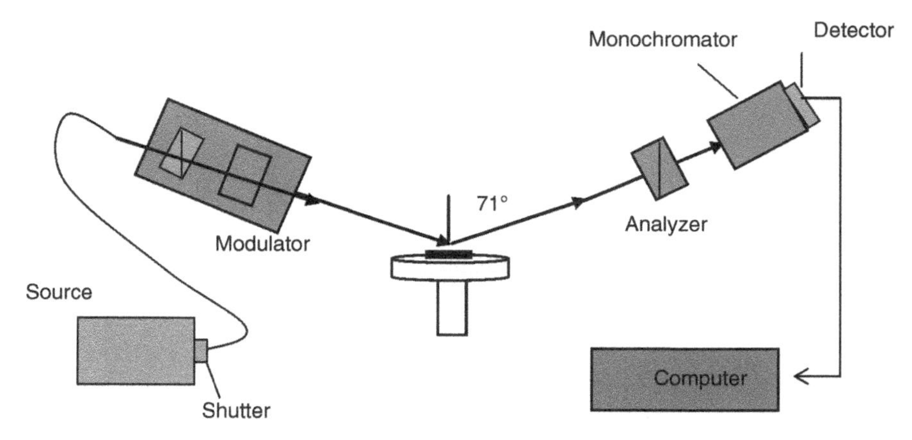

Figure 5.14 Measurement arrangement of spectroscopic phase modulated ellipsometer

In this study, the fitting procedure was performed combined with the Bruggeman effective medium approximation (BEMA). For example, Tauc–Lorentz model (for the best-fitted parameter sets (ε, A, E_0, C, and E_g) is used for the a-Si component

$$\sum_i f_i \frac{\langle \varepsilon_i \rangle}{\langle \varepsilon \rangle + 2\langle \varepsilon \rangle} = 0$$

$$\sum_i f_i \frac{\langle \varepsilon_i \rangle - \langle \varepsilon \rangle}{\langle \varepsilon_i \rangle + 2\langle \varepsilon \rangle} = 0, \quad \sum f_i = 1 \tag{5.7}$$

$$\begin{aligned} \varepsilon_2 &= \frac{AE_{n0}C\left(E_n - E_g\right)^2}{\left(E_n^2 - E_{n0}^2\right)^2 + C^2E_n^2} \frac{1}{E_n} \\ &= 0\left(E_n < E_g\right) \end{aligned} \tag{5.8}$$

where $<\varepsilon_i>$ and f_i indicate function and volume fraction of the ith component, respectively. A probable structure was determined by minimizing the mean square error, χ^2, between the measured and calculated ellipsometric error parameters using a linear regression method.

$$x^2 = \frac{1}{2N - M} \sum_{i=1}^{N} \left[\left(\tan \Psi_i^c - \tan \Psi_i^m\right)^2 - \left(\cos \Delta_i^c - \cos \Delta_i^m\right)^2 \right] \tag{5.9}$$

where the superscripts c and m represent the calculated and measured values, and N and M are the numbers of the measured and calculated wavelengths, respectively. For all the spectra analyzed here, χ^2 was confined to within 5 to ensure the analysis results reasonably fitted with the measured data.

5.10.1 Basic Steps for a Model-Based Analysis

Data analysis is a very important part of spectroscopic ellipsometry (SE). Without data analysis, SE measures only the ellipsometric parameters Psi (Ψ) and Delta (Δ) versus wavelength. To determine sample properties of interest, such as layer thicknesses and optical constants, a model-based analysis of the SE data must typically be performed.

The basic five steps of this approach are as follows:

1. SE data is measured on the sample.
2. A layered optical model is built, which represents the nominal structure of the sample.
3. Model fit parameters are defined and then automatically adjusted by the software to improve the agreement between the measured and model-generated SE data. This is known as "fitting" the data.
4. The results of the fit are evaluated. If the results are not acceptable, the optical model and/or defined fit parameters are modified and the data is fit again.
5. Keep in mind that while the basic SE data analysis approach is straightforward, "real-world" samples can often be difficult to analyze, but there is no substitute for experience when dealing with complex samples.

5.10.2 Layered Optical Model

For a multilayered substrate, exact layer information is vital for microwave design. This information is more critical for microwave sensing devices design.

The SE data analysis process for most samples begins by building a layered optical model, which corresponds to the nominal sample structure as shown in Figure 5.15.

Each layer is parameterized by thickness (d_1, d_2, etc.) and optical constants. Optical constants describe how light interacts with and propagates through the layer. Using the optical and standard textbook thin film equations (Snell's law, Fresnel equations, etc.), the software can calculate "generated" or "simulated" SE data. If the model is a good representation of the sample, the model-generated SE data will be in good agreement with the SE data measured on the sample. Sometimes, an ideal layered model does not adequately describe the optical behavior of the actual sample because of two common nonidealities: surface roughness and index gradients.

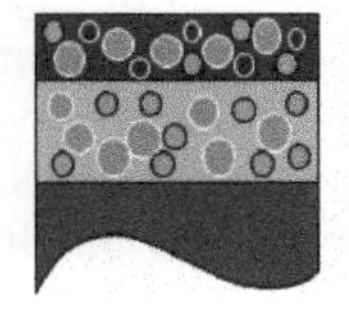

Figure 5.15 Nominal sample structure

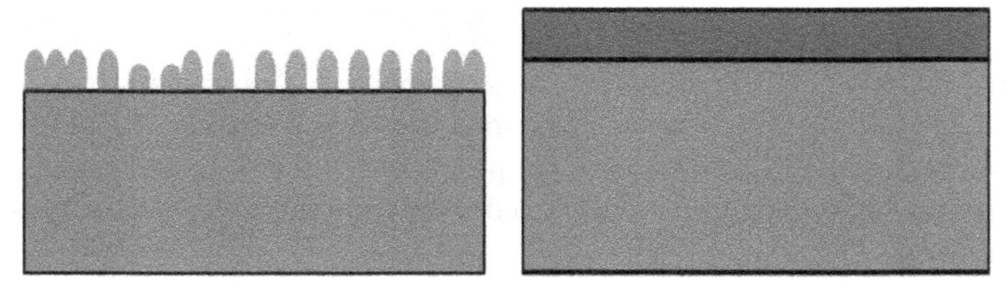

Figure 5.16 Optical model for surface roughness: (a) actual sample with nonabrupt "rough surface," (b) optical model with "effective" roughness layer

5.10.3 Optical Model for Surface Roughness

Knowing the surface roughness is very important for microwave design. The surface roughness data of a substrate is a critical parameter for microwave passive design.

Surface roughness is shown in Figure 5.16 to model the actual sample, which may have a nonabrupt "rough" surface; an "effective" roughness layer is added to the model. The optical constants of the "effective" roughness layer are derived by mixing the optical constants of the underlying material with the optical constants of "void" (which has optical constants of $n = 1$, $k = 0$). The BEMA is used to calculate the optical constants of this "mixed" layer assuming 50% void content. While the effective roughness layer approach is certainly an approximation to the actual sample, this approach works extremely well for modeling SE data when the size of the surface roughness is much less than the wavelength of light used to measure the sample. For most SE systems, this implies that surface roughness features must be less than 40 nm.

5.10.4 Approximation of Surface Roughness As an Oxide Layer

In the case where there is not a lot of roughness on the surface (<100 Å or so), it is often possible to model the roughness fairly well as an oxide film. This is particularly useful for polycrystalline and amorphous silicon films and metal films, where some degree of surface oxidation will exist in addition to the surface roughness. It is nearly impossible to separate the effects of the roughness from the presence of the thin oxide at the surface. This is beneficial as we do not actually wish to measure the roughness. Rather we want to model the surface effects properly in order to get accurate measurements of what lies underneath the surface.

5.10.5 Optical Model for Index Gradients

Sometimes, the optical "constants" of a layer are not constant throughout the layer. This may be caused by process variations during the film deposition. Figure 5.17 shows how to model a "graded" film: the layer is divided into sublayers with small thicknesses, and each sublayer has slightly different optical properties. A linear variation in the film index "n" is assumed for the graded layer. This simplified approach, which is automated in most SE software, works well for modeling many types of samples.

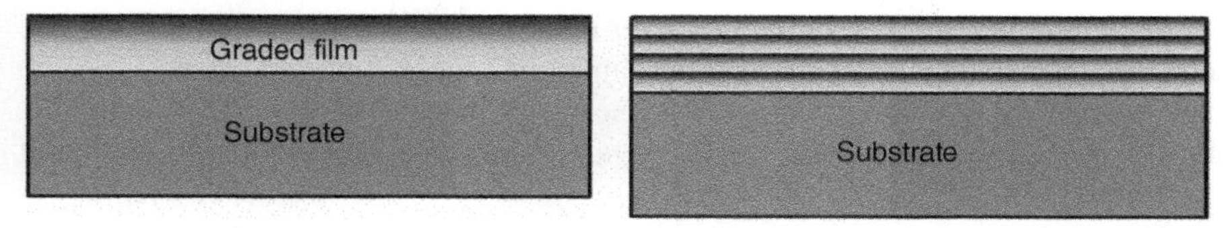

Figure 5.17 (a) Continuous variation in the optical properties of the "graded" film and (b) approximation of the graded film with discrete layers

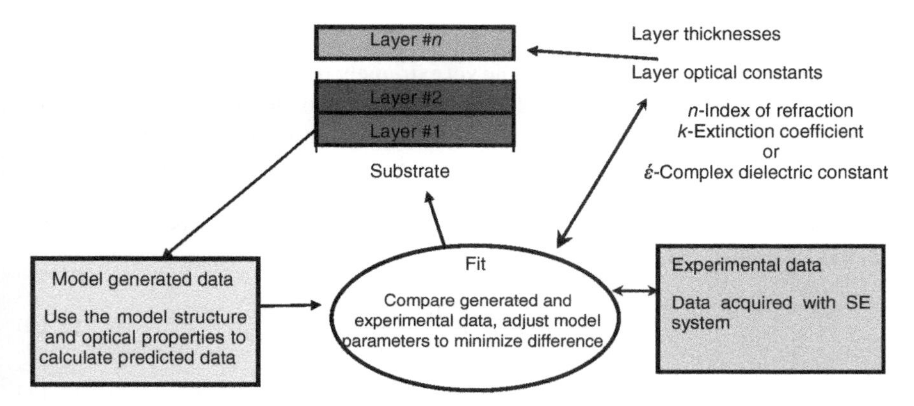

Figure 5.18 Flow chart of the procedure for an ellipsometric experiment

Ellipsometry is a model-dependent technique in which the measured quantities are the physical quantities we wish to determine, and a numerical analysis of the experimental data based on some mathematical model is required to obtain useful physical information about the sample under study.

5.10.6 Procedure for an Ellipsometric Modeling

Assume that we are studying a sample from which we have measured ellipsometric data as a function of wavelength and angle of incidence. We also have an optical model for our sample, consisting of any number of layers on a substrate and parameterized by the optical constants of the various materials and the thickness of the films on the sample. We now wish to vary some parameters in this model such that ellipsometric data calculated from the model matches our experimental data as closely as possible. A flowchart of this general procedure is shown in Figure 5.18:

5.10.7 Regression

The probable structure was determined by minimizing the mean-squared error, σ (or chi-square, χ^2) between the measured and the calculated ellipsometric parameters using a linear regression method.

The unknown parameters such as layer thickness, optical constants, and volume fractions of constituent phases can be numerically determined by minimizing the following mean-squared error using a linear regression algorithm, LRA program. In SE, software used a linear regression algorithm, that is, Levenberg–Marquardt algorithm [14].

The goodness of the fit is estimated by the unbiased estimator, σ. The SE analysis software adjusts to fit parameters to find the best match between model and experimental curves.

5.10.8 Dielectric Film

The Cauchy dispersion model is a simple yet extremely powerful dispersion model for describing the index of refraction of dielectric and semiconducting materials. It tends to work best when the material under study shows little or no optical absorption in the spectral region of interest. The Cauchy dispersion model describes the dependence of the index of refraction on wavelength as follows:

The UV term follows a Cauchy law on n and k, n and k parameters are calculated with the following expressions:

$$n(\lambda) = A + \frac{B}{\lambda^2} + \frac{C}{\lambda^4} \tag{5.10}$$

$$k(\lambda) = \frac{D}{\lambda} + \frac{E}{\lambda^2} + \frac{F}{\lambda^5} \tag{5.11}$$

where $n(\lambda)$ is the index of refraction, λ is the wavelength, and A, B, and C are parameters. Generally, the quantities to be determined are the thickness of the film and the values of A, B, and C.

For example, D, E, F parameters are kept to zero; we assume the extinction coefficient $k=0$ if the material is transparent.

5.10.9 Mixed or Composite Materials

Introducing several components in a layer by choosing the mixing type (Bruggeman) which is no more than generalization of the Clausius–Mossotti Formula. The model is then self-consistent and the two media play exactly the same role. For this reason, this model in SE is more frequently used. Indeed, the other models make some geometrical approximations on the mixing properties of the medium.

5.10.10 Accuracy and Precision of SE Experiments

Ellipsometric measurements are, by nature, very precise. Also, if the model used to fit the ellipsometric data is unique (i.e., does not contain any strongly correlated parameters), the results of the modeling of the ellipsometric data will be very precise as well. It is almost impossible to specify the accuracy of an ellipsometer to the level of precision. To do so we would require another measurement technique such as Rutherford backscattering, transmission electron microscopy, and atomic force

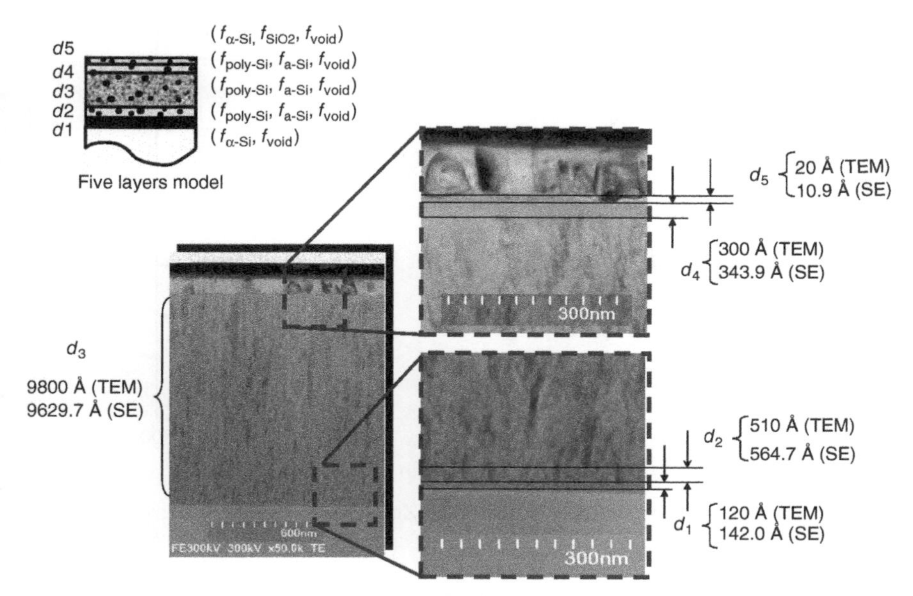

Figure 5.19 A typical cross-sectional TEM image and SE analysis (*Source*: Reproduced from [15] with permission)

microscopy with accuracy comparable to the extreme (submonolayer) precision of the ellipsometer. Figure 5.19 shows typical cross-sectional TEM image of microcrystalline silicon film fabricated by using the high-density microwave plasma source. The best-fitted thicknesses and $f_{c\text{-}Si}, f_{a\text{-}Si}, f_{void}, f_{SiO2}$ of layers 1–5 determined by SE analysis are also included [15].

5.11 UV–VISIBLE SPECTROPHOTOMETERS

Ultraviolet–visible (UV–vis) spectroscopy provides information about transmittance, reflection, and absorption coefficient in thin films. The transmittance spectra of Al-doped ZnO film is shown in Figure 5.20. From the spectra, band gap of materials can be derived. The detailed description regarding the measurement of the band gap of materials can be found in Ref. [16]. Based on these spectra of the materials, absorption edge and band gap energies of the sensing materials have been calculated and compared, which can be used to design the sensor.

5.12 ELECTRICAL CONDUCTIVITY MEASUREMENT

The electrical conductivity of a dielectric/semiconductor represents the propagation loss of the signal impinged on the medium. At microwave frequency design, this conductivity is quantified and loss tangent (tan δ) conductivity, and dielectric constant are interrelated.

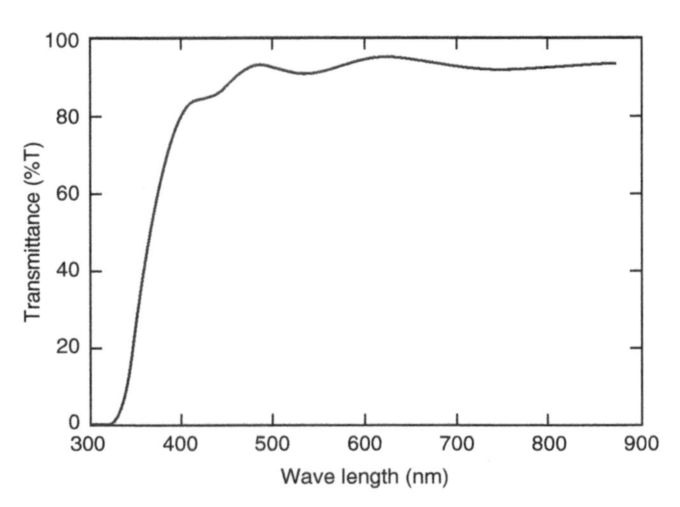

Figure 5.20 Transmittance spectra of Al-doped ZnO film measured by ultraviolet–visible (UV–vis) spectroscopy

The dark conductivity of a semiconductor [17] is

$$\sigma_d = e(n_e\mu_e + n_p\mu_p) \tag{5.12}$$

where e is electron charge, n_e and n_p are electron and hole densities, μ_e and μ_p are their motilities.

The majority-carrier-caused dark conductivity in Si film is

$$\sigma_d = en_e\mu_e \tag{5.13}$$

Carrier concentration in Si conduction band is

$$n = \int g_c(E)f_e(E)dE = N_c \exp\left[\frac{-(E_C - E_F)}{kT}\right] \tag{5.14}$$

where $g_c(E)$, N_c, $f_c(E)$, E_C, E_F, k, T, are state density, effective state density, Fermi–Dirac distribution function, conduction band energy, Fermi level energy, Planck constant, and temperature, respectively. The dark conductivity in conductive band is

$$\sigma_d = eN_c \exp\left[\frac{-(E_C - E_F)}{kT}\right]\mu_e = \sigma_0 \exp\left(\frac{-\Delta E}{kT}\right)\mu_e \tag{5.15}$$

The activation energy is determined from temperature dependence conductivity.

The photo conductivity is described as

$$\Delta\sigma_p = e\Delta n\mu \tag{5.16}$$

Δn: concentration of light-induced carrier

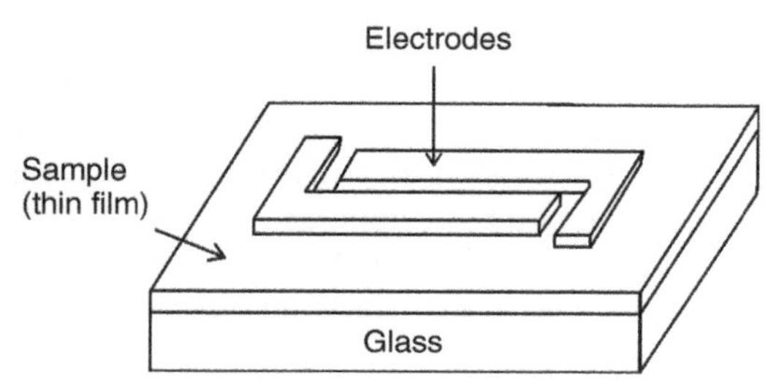

Figure 5.21 Schematic of photoconductivity measurement (*Source:* Reproduced from [14] with permission)

The concentrations of light-induced carrier were determined on life time τ and carrier generation rate F.

$$\Delta n = F\tau \tag{5.17}$$

Here, under light exposure condition, F is described as

$$F = \eta N_0 (1 - R)\{1 - \exp(-\alpha D)\} \tag{5.18}$$

η: quantum efficiency (number of carriers generated per photon), N_0: incident photon number, R: reflection coefficient, α: absorption coefficient, D: film thickness, τ: carrier lifetime.

$$\sigma_p = e\eta\mu\tau N_0 (1 - R)\{1 - \exp(-\alpha D)\} \tag{5.19}$$

The schematic of measurement for the conventional measurement of photoconductivity with coplanar electrodes is shown in Figure 5.21. Measurement results include the contribution of both electron and hole.

So far different aspects of material characterization techniques have been described in detail. The outcomes of these characterization techniques are known composition, surface morphology/roughness, refractive index, surface, conductivity of the materials at DC/low frequency. In the next section, the microwave characterization of sensing materials is presented.

5.13 MICROWAVE CHARACTERIZATION (SCATTERING PARAMETERS—COMPLEX PERMITTIVITY, DIELECTRIC LOSS, AND REFLECTION LOSS) FOR SENSING MATERIALS

Microwave measurements of the dielectric properties of materials are finding increasing applications in new electrotechnology. The interest in dielectric properties of materials has historically been associated with the design of electrical equipment, where various dielectrics are used for insulating conductors and other components of electric equipment. Measurement of the bulk dielectric properties (dielectric constant, dielectric loss factor) is not only testing. Rather, these properties are an intermediary vehicle for understanding, explaining, and empirically relating certain physicochemical properties of the test material. Therefore, in this section, an attempt is made to fully explore the existing knowledge of dielectric properties (complex permittivity), their role, and importance in various sectors, and the concept of various measurement methodologies and their development. An extensive review of the literature on measurement techniques and the comparison and potential applications of dielectric properties is reported.

The EM wave absorption characteristics of materials depend on its dielectric properties (complex permittivity), magnetic properties (complex permeability), thickness, and the frequency of operation. This section deals with characterization of the materials at microwave frequency bands. There are various characterization procedures for dielectric, loss tangent, and conductivity (attenuation) measurement for unknown dielectric and conductive materials and can be found in Refs. [18, 19].

5.13.1 Basic Microwave-Material Interaction Aspects

When microwaves are directed toward a material, part of the energy is reflected, part is transmitted through the surface, and of this latter quantity, a part is absorbed. The proportions of energy, which fall into these three categories, have been defined in terms of the dielectric properties. In microwave electronics, the fundamental electrical property through which the interactions are described is the complex relative permittivity of the material, ε. It is mathematically expressed as

$$\varepsilon_r = \frac{\varepsilon}{\varepsilon_0} = \frac{\varepsilon' - j\varepsilon''}{\varepsilon_0} = \varepsilon_r' - j\varepsilon_r'' = \varepsilon'(1 - j\tan\delta_e) \tag{5.20}$$

where ε is the complex permittivity; ε_r, relative complex permittivity; ε_0, 8.854×10^{-12} F/m is the permittivity of free space; ε', the real part of relative complex permittivity; ε'', the imaginary part of relative complex permittivity; $\tan\delta_e$, dielectric loss tangent; and δ_e, dielectric angle. The absolute permittivity of vacuum or air, ε_0, is determined by

$$\varepsilon_0 = \frac{1}{\mu_0 C^2} \tag{5.21}$$

where C is the speed of light in vacuum and μ_0, the permeability constant, and the value of $\mu_0 = 4\pi \times 10^{-7}$ H/m.

In solid, liquid, and gaseous media, the permittivity has higher values and is usually expressed relative to the value in vacuum [20]. The relative permittivity of a material, ε_r, is equal to $\varepsilon_{abs}/\varepsilon_0$, where ε_{abs} is the absolute permittivity of the material, also called dielectric constant of a material. Materials that do not contain magnetic components respond only to the electric field. The penetration depth or skin depth, d_p, is usually defined as the depth into a sample where the microwave power has dropped to $1/e$ or 36.8% of its transmitted value. Sometimes, d_p is defined as the distance at which the microwave power has been attenuated to 50% of transmitted power (P_{trans}). The penetration depth is a function of e' and e'':

$$d_p = \frac{\sqrt{\varepsilon'}}{2\pi\varepsilon''} \tag{5.22}$$

where λ_0 is the free space microwave wavelength for example at 2.45 GHz, $\lambda_0 = 1.22$ cm.

5.13.2 Methods of Measurement of Dielectric Properties

The measurement of dielectric properties has gained importance because it can be used for nondestructive monitoring of specific properties of materials undergoing physical or chemical changes. There are several techniques to measure the dielectric properties of materials. The particular method used depends on the frequency range of interest and the type of target material. The choices of measurement equipment and sample holder design depend on the dielectric materials to be measured, the extent of the research, available equipment, and resources for the studies. A vector network analyzer (VNA) is an expensive equipment but is very versatile and useful if studies are extensive. Scalar network analyzers and impedance analyzers are relatively less expensive but still too expensive for many programs. For limited studies, more commonly available RF and microwave laboratory measurement equipment can suffice if suitable sample holders are constructed. Nyfors and Vainikainen [20] gave four groups of measurement methods, namely, lumped circuit, resonator, transmission line, and free-space methods. The lumped circuit techniques are no longer used to any great extent since they are only suitable for low frequencies and high loss materials. The last three and the open-ended coaxial probe developed by Hewlett Packard (HP 1992) use impedance, spectrum, or network analyzers.

At low- and medium-frequency ranges, bridge and resonant circuits have often been used for characterizing dielectric materials. At higher frequencies, however, transmission line, resonant cavity, and free-space methods are commonly used. In general, dielectric measurement techniques can be categorized as reflection or transmission type, using resonant or nonresonant systems, with open or closed structures [21]. To determine the effective dielectric constant, loss tangent, and attenuation of the material, transmission measurement techniques are used. Figure 5.22 shows various measurement techniques. For liquid, solid, and powder material measurement, techniques shown in Figure 5.22(a–c) are usually used.

In-house-developed codes using advanced microwave waveguide modal expansion methods and diffraction theories of multilayered unbounded and bounded media (materials under tests) are envisaged to prefect the prediction of materials characterization methods. For liquid and solid samples, the input impedance reflection coefficient (S_{11} vs frequency) is measured, as shown in Figure 5.22(a). More accurate measurement is achieved with full 2-port calibration and recording the forward transmission coefficient (S_{21}) versus frequency of material-filled jigs (containers, waveguides, and resonators) as shown in Figure 5.22(b).The third method is etching integer

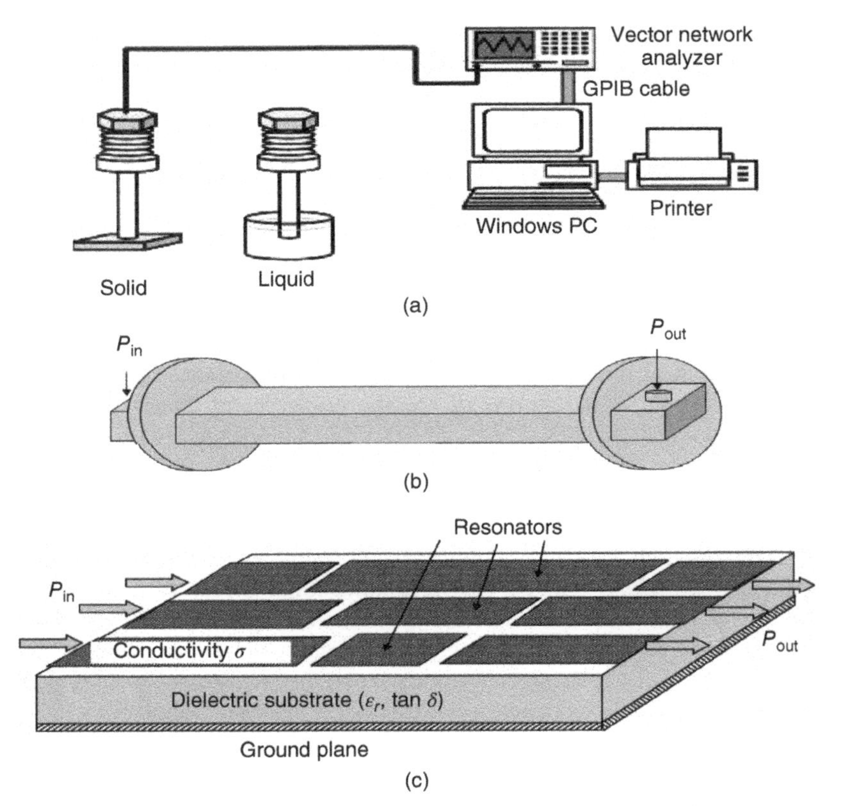

Figure 5.22 Dielectric measurement kits: (a) 1-port solid/liquid; 2-port (b) waveguide and (c) line resonators

multiple half-wavelength resonators on the material under test and testing S_{21} versus frequency. The measured S_{21} resonant frequency and Q (1/BW) are used to determine the electrical properties of the unknown materials at the test frequency. A four-point probe is used to measure the DC conductivity in ($\Omega/\square$). For these data, the ac conductivity is calculated for a particular frequency [22]. These parameters are used to design microwave and mm-wave passive circuits. Figure 5.23 shows the measured electric permittivity versus frequency for PVA water solution. Figure 5.23 shows the dielectric constant change of phenanthrene during sublimation.

5.14 DISCUSSION ON CHARACTERIZATION OF SMART MATERIALS

Table 5.1 presents the summary of characterization techniques for sensing materials that have been discussed so far. Table 5.1 shows the critical material properties for each characterization technique. This results in microwave property variation that can

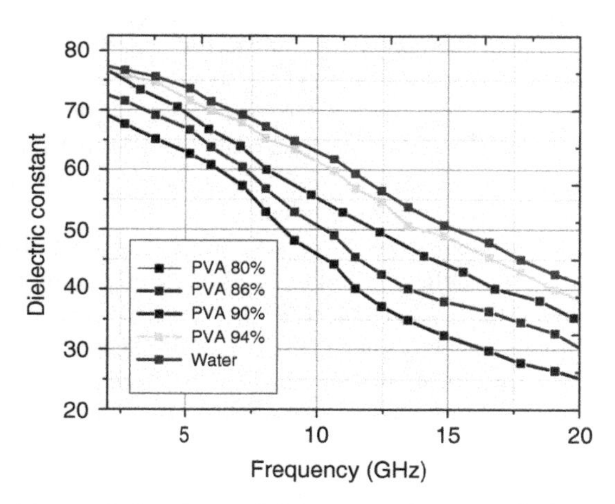

Figure 5.23 Measured electric permittivity versus frequency for PVA water solution

TABLE 5.1 A Summary of Characterization Techniques for Sensing Materials

Name of Techniques	Type of Techniques	Materials Properties
XRD	Structural	Crystalline quality
Raman	Structural	Crystalline quality
SE	Structural, surface, optical, electronic	Layer thickness, surface roughness, and dielectric constant, refractive index
UV–vis	Optical, electronic	Transmittance, reflection, and absorption coefficient
SIMS	Structural	Residual atomic ion concentrations
AFM	Surface	Surface roughness, morphology
TEM	Structural	Examine fine detail—even as small as a single column of atom
FTIR	Surface	Surface bonding
Conductivity	Electrical	Conductivity
Microwave characterization	Optical, electrical, electronic	Dielectric properties

be calibrated to measurable sensing data. Microwave sensing capabilities of smart materials is a novel field of study for passive RFID sensor development. This will lead to an interdisciplinary research field of highly sensitive, passive sensor node development.

5.15 CONCLUSION

Microwave characterization of smart materials for chipless RFID sensors is a crucial step to develop appropriate materials for specific sensing purposes.

The characterization provides accurate constitutive parameters such as relative permittivity (dielectric constant), conductive dielectric losses (loss tangent) at microwave frequencies. These parameters are used in developing microwave modeling of chipless RFID sensors. Besides, the constitutive parameters such as extraction, surface roughness, stability of the materials, material structures, and so on are determined at microwave frequency bands.

This chapter has presented various novel analysis and characterization techniques for smart materials. These include microstructural and surface morphology, optical, electrical, and thermal (DC conductivity, stability, etc.), microwave scattering parameters, that is, complex permittivity, dielectric loss, and reflection loss in the gigahertz range for sensing materials. From the above discussion, the following inferences are drawn:

(i) Structural characterization of smart materials for microwave sensing: The XRD, Raman spectroscopy, SIMS, SE, TEM, and FTIR techniques are used for structural characterization of smart materials. This structural characterization of smart materials is required for material identification, verification, and to understand the definitive structural information in order to develop various smart materials for microwave sensing.

(ii) Surface characterization of smart materials for microwave sensing: The SE, SEM, and AFM techniques are used for surface characterization of smart materials. These surface characterizations are used to understand the topographic information in order to develop various smart materials for microwave sensing.

(iii) Optical characterization of smart materials for microwave sensing: The SE, UV–vis techniques are used for optical characterization of smart materials for microwave sensing. The SE with an optical model is used to extract the complex refractive index, dielectric constant, layer thickness, surface roughness, and so on of each layer of smart materials. The UV–vis provides information about the transmittance, reflection, absorption coefficient, band gap, and so on in thin films.

(iv) Electrical and microwave characterization of smart materials for microwave sensing: The electrical conductivity measurement is used for a dielectric/semiconductor material for microwave sensing. The microwave characterization is used for measurement of scattering parameters, that is, complex permittivity, dielectric, reflection loss of materials at microwave and mm-wave frequencies. The results in microwave property variation that can lead to research for highly sensitive passive sensor node development.

So far various sensing materials for temperature, relative humidity, pH, impact, presence of noxious gases and light has been identified in Chapter 4 and their characterization techniques have been described in Chapter 5. In the following three Chapters 6–8, various sensors are presented. The first sensor is a passive noninvasive partial discharge (PD) detection sensor utilizing SIR passive circuits and antenna.

The second sets of sensors are described with smart-material-loaded ELC resonator RH sensor. The third set of sensors is described for temperature and multiparameter sensing.

The next chapter presents one of the three applications of chipless RFID sensor considered in this book, that is, noninvasive PD detection and localization using frequency modulation-based chipless RFID sensor.

REFERENCES

1. Y. Fan, H. Yang, M. Li, and G. Zou, "Evaluation of the Microwave Absorption Property of Flake Graphite," *Materials Chemistry and Physics* vol. 115, pp. 696–698, 2009.
2. A. M. Nicolson and G. F. Ross, "Measurement of the Intrinsic Properties of Materials by Time-Domain Techniques" *IEEE Transactions on Instrumentation and Measurement* vol. 19, pp. 377–382, 1970.
3. I. P. Herman, *Optical Diagnostics for Thin Film Processing*, 1996.
4. J. K. Saha, "Fast Growth of Microcrystalline Silicon Thin-Films from Dichlorosilane by the High-Density Microwave Plasma Source using the Spoke Antenna for Si Thin-Film Solar Cells", Ph.D. Thesis, Saitama University, Japan, p. 46, 2008.
5. K. Raja, "On the UHF partial discharge measurement in transformers," in *Electrical Insulation and Dielectric Phenomena, 2003. Annual Report. Conference on*, 2003, pp. 349–352.
6. J. A. Scholl, A. L. Koh, and J. A. Dionne, "Quantum Plasmon Resonances of Individual Metallic Nanoparticles," *Nature,* vol. 483, pp. 421–427, 2012.
7. B. Shouli, L. Chen, D. Q. Li, W. S. Yang, P. C. Yang, Z. Y. Liu, A. Chen, and C. C. Liu, "Different Morphologies of ZnO Nanorods and Their Sensing Property," *Sensors and Actuators* vol. B146, pp. 129–137, 2010.
8. Available: http://en.wikipedia.org/wiki/Thermal_printing (accessed on 11 October 2015).
9. P. C. Baker, M. D. Judd, and S. D. J. McArthur, "A Frequency-Based RF Partial Discharge Detector for Low-Power Wireless Sensing," *IEEE Transactions on Dielectrics and Electrical Insulation,* vol. 17, pp. 133–140, 2010.
10. R. M. A. Azzam and N. M. Bashara, *Ellipsometry and Polarized Light*: North-Holland Publishing Company, Amsterdam, 1977.
11. H. Fujiwara, *Spectroscopic Ellipsometry: Principles and Applications*: Wiley 2003.
12. M. Kondo, H. Fujiwara, and A. Matsuda, "Interface-Layer Formation in Microcrystalline Si:H Growth on ZnO Substrates Studied by Real-Time Spectroscopic Ellipsometry and Infrared Spectroscopy," *Journal of Applied Physics*, vol. 93, pp. 2400–2409, 2003.
13. F. A. Modine and G. E. Jellison, "Parameterization of the Optical Functions of Amorphous Materials in the Interband Region," *Applied Physics Letters*, vol. 62, p. 3, 1993.
14. W. H. Press, S. A. Teukolsky, W. T. Vetterling, and B. P. Flannery, *Numerical Recipes: The Art of Scientific Computing*, 3rd ed.: Cambridge University Press, 2007.
15. J. K. Saha, N. Ohse, K. Hamada, K. Haruta, T. Kobayashi, T. Ishikawa, Y-Ichiro Takemura, and H. Shirai, "Synthesis of Microcrystalline Silicon Films Using High-Density Microwave Plasma Source from Dichlorosilane," *Japanese Journal of Applied Physics,* vol. 46, p. 3, 2007.

16. S. Chengdua, X. Juan, W. Helin, X. Tianning, Y. Bo, and L. Yuling, "Optical Temperature Sensor Based on ZnO Thin Film's Temperature-Dependent Optical Properties," *Review of Scientific Instruments*, vol. 82, pp. 084901 (1–3) 2011.
17. J. Singh, *Semiconductor Devices*: John Wiley & Sons, 2001.
18. Y. Li, M. Tentzeris, and A. Rida, *RFID-Enabled Sensor Design and Applications*: Artech House, 2010.
19. M. S. Venkatesh and G. S. V. Raghavan, "An Overview of Dielectric Properties Measuring Techniques," *Canadian Biosystems Engineering*, vol. 47, pp. 7.15–7.30, 2005.
20. N. Nyfors and P. Vainikainen, "*Industrial Microwave Sensors*," Norwood, Boston, MA: Artech House, 1989.
21. K. L. Wong, "Application of Very-high-Frequency (VHF) Method to Ceramic Insulators," *IEEE Transactions on Dielectrics and Electrical Insulation*, vol. 11, pp. 1057–1064, 2004.
22. S. M. Wentworth, *Fundamentals of Electromagnetics with Engineering Applications*: John Wiley & Sons, 2005.

6

CHIPLESS RFID SENSOR FOR NONINVASIVE PD DETECTION AND LOCALIZATION

6.1 INTRODUCTION

In this chapter, a chipless radio-frequency identification (RFID)-based passive sensor is proposed for partial discharge (PD) detection and faulty source identification. The sensor consists of an ultra-wide band (UWB) antenna to capture PD signals and a number of cascaded resonators for signifying identification data. A unique combination of the resonators in each sensor embeds a distinct frequency signature within the frequency spectrum of the captured PD signal. By analyzing the modified PD signal, both the PD level and the source identification can be retrieved. A prototype sensor is designed, fabricated, and tested for performance verification. The organization of this chapter is as follows:

- Firstly, the introduction section discusses the limitations of current technology for PD detection and localization. It summarizes the main features of frequency modulation-based PD sensors and the suitability of the proposed sensor for HV condition monitoring.
- Next, the theory section details the operating principle of the proposed RF sensor and illustrates the configuration of the sensor system. In addition, the theory regarding simultaneous PD detection for multiple faulty HV equipment using time–frequency analysis is presented in this section.
- Then, a frequency modulation-based chipless RFID PD sensor system is developed and experimentation is performed to validate data encoding with UWB PD signals.

Chipless RFID Sensors, First Edition. Nemai Chandra Karmakar, Emran Md Amin and Jhantu Kumar Saha.

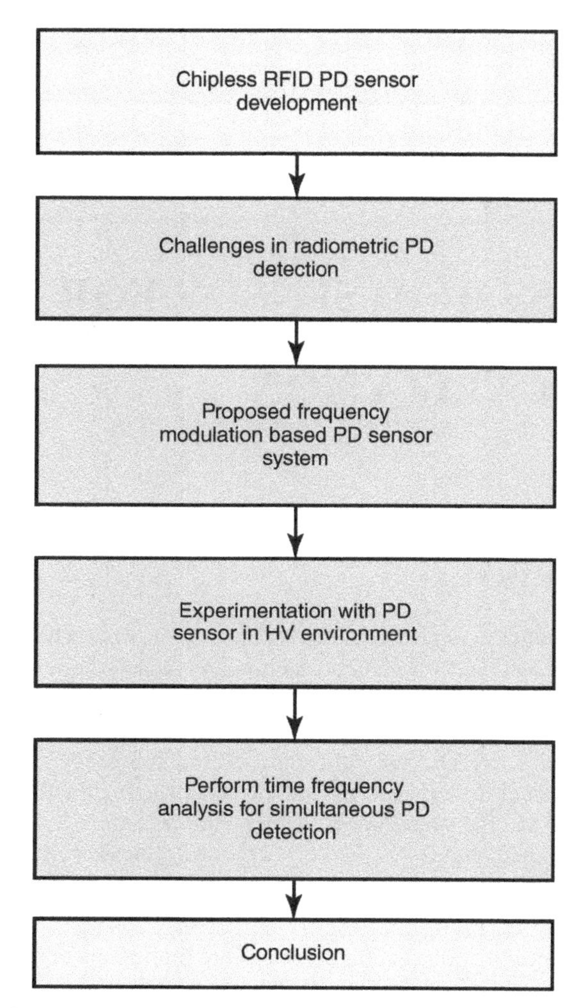

Figure 6.1 Organization of the chapter on noninvasive PD detection and localization

- Subsequently, time–frequency analysis of captured PD signals is carried out to investigate simultaneous PD detection in 2-D time–frequency plots.
- Finally, the conclusion highlights the outcomes and future directions of this research.

Figure 6.1 illustrates the organization of this chapter in a flow chart.

This research is a part of Australian Research Council's (ARC) Linkage Project Grant (LP0989355: Smart Information Management of Partial Discharge in Switchyards Using Smart Antennas). The goal of this project is to develop a fully passive, nonintrusive sensor for HV condition monitoring. To fulfill the project aim, a comprehensive review is performed on conventional radiometric PD detection systems. The next section details limitations of current technologies and proposed sensor system.

6.1.1 Radiometric PD Detection

The condition monitoring of HV equipment in a switchyard or substation is a routine check throughout the whole lifespan of the equipment. The insulation in HV equipment may be damaged due to overvoltages, manufacturing defects, or aging. This can eventually result in electrical breakdown, causing power failure, equipment casualty, and loss of life and property. PD is the energy dissipation caused when the electric field across a dielectric exceeds a threshold breakdown value [1]. PD detection is used in power systems to monitor the state of the dielectric insulation of HV equipment. As PD is an early warning of insulation deterioration, failure to detect PD efficiently and on time can lead to the catastrophic disruption of the equipment. Reliable online PD detection is of significant interest for power distributors to ensure personnel safety and reduce potential loss of service.

In radiometric detection, the RF emission due to PD is captured and analyzed for PD level and location finding [2]. The frequency span of the emitted RF waves is found between 300 MHz and 3 GHz on the UHF band showing highest power spectrum between 750 and 850 MHz [3]. The main challenges associated with radiometric PD detection are as follows: (i) PD signals are spontaneous in nature and (ii) PD events can take place at any time. Moreover, PD localization techniques are limited by external interferences, the presence of metallic substances, and multipath effects in a switchyard. This makes exact identification of a PD source very difficult [4].

In Ref. [5], an ultrasonic receiving planar array transducer is proposed for the detection of PD location in power transformers. The transducer consists of 16×16 elements to form digital beam-scanning using planar array theory. It is reported to locate multiple PDs within a transformer. A novel VHF–UHF radio interferometer system (VURIS) for PD localization is proposed in Ref. [6]. The system captures the emitted EM wave using two identical antennas and calculates the angle of arrival from the phase difference. This requires fewer antennas for location finding; however, the feasibility has been tested only in the laboratory. In Refs. [7–9], the finite difference time domain (FDTD) method is used to simulate the propagation of EM waves from PD sources, and PD location is achieved through the calculation of time of flight. The FDTD method improves the accuracy of location finding as the optimum 3-D positioning of the UHF sensors is analyzed. In Ref. [10], PD source localization using maximum likelihood estimation is presented.

In Ref. [11], a four-element wide-band antenna array was used to receive PD signals from a distance and a time difference of arrival (TDOA) algorithm was developed to localize the faulty sources [12]. For TDOA analysis, the receiving antennas have to capture the initial part of the radiated wideband PD signals. The direct PD signal is distorted due to multipath reflection while propagating through the environment; thus, this method has limitations in the location of the exact source of a fault. The efficacy of the PD localization technique using time-of-arrival method is significantly affected by the switchyard environment.

Radiometric detection has advantages when the equipment is electrically isolated as the field dispersion is reduced. A UHF sensor-based PD detection technique is presented in Ref. [13] for monitoring a number of items of shielded HV equipment. Each sensing unit was connected to an oscilloscope via a cable loop, and the time information of a PD occurrence was used to identify faulty sources. However, this method failed to differentiate between PDs happening at the same instance.

The above study reveals that radiometric detection has emerged as an important technique for online nonintrusive PD diagnosis of HV power apparatuses. However, there remain challenges in localizing the faulty source by using the TDOA method. Other challenges in PD localization are due to the external neighboring interferences, the presence of metallic substances, multipath, and reflections, and above all the fact that PD pulses have very low intensity and short duration. We, therefore, propose a chipless RFID-based PD sensor for exact localization and condition monitoring of HV apparatus in a distribution switchyard.

The proposed sensor introduces a frequency modulation-based PD localization technique, which is a novel concept in the field of online radiometric PD detection. This localization technique is more immune to external noises than TDOA methods are. Moreover, PD localization techniques using time information have fundamental limitations in distinguishing simultaneous PD occurrences. As in a switchyard, the possibility of simultaneous PD occurrences is greater than that of random occurrences; the most significant feature of the proposed sensor system is its ability to detect simultaneous PDs in a substation. To realize simultaneous PD detection, time–frequency analysis is performed on the captured PD signals. The results verify that the proposed sensor can identify multiple faulty sources emitting PDs at the same instance.

6.2 THEORY

6.2.1 Proposed PD Sensor

The proposed PD sensor comprises of cascaded stepped impedance resonators (SIR) to denote distinct frequency signatures within the power spectrum. The general working principle of the sensor is presented in Chapter 3 (Section 3.3.2.1) [14]. We use the UWB nature of short-span PD pulses to encode data in frequency spectra. The resonators attenuate certain frequencies within the frequency band of the RF signal. A particular resonator's configuration embeds a distinct frequency signature to encode data as identification bits. The presence of a magnitude null resembles logic "0," whereas the absence of a magnitude null at a distinct frequency resembles logic "1." In addition, the combination of data bits when no filter is present is omitted in our proposed sensor system. Therefore, N number of cascaded resonators can produce $K = 2^N - 1$ different frequency signatures to detect K number of objects. The general structure of the proposed PD sensor is depicted in Figure 6.2. Each sensor

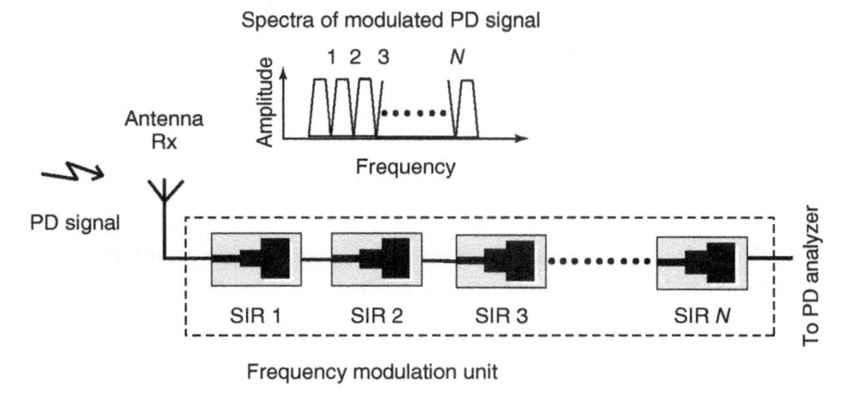

Figure 6.2 Proposed cascaded multiresonator-based chipless RFID PD sensor

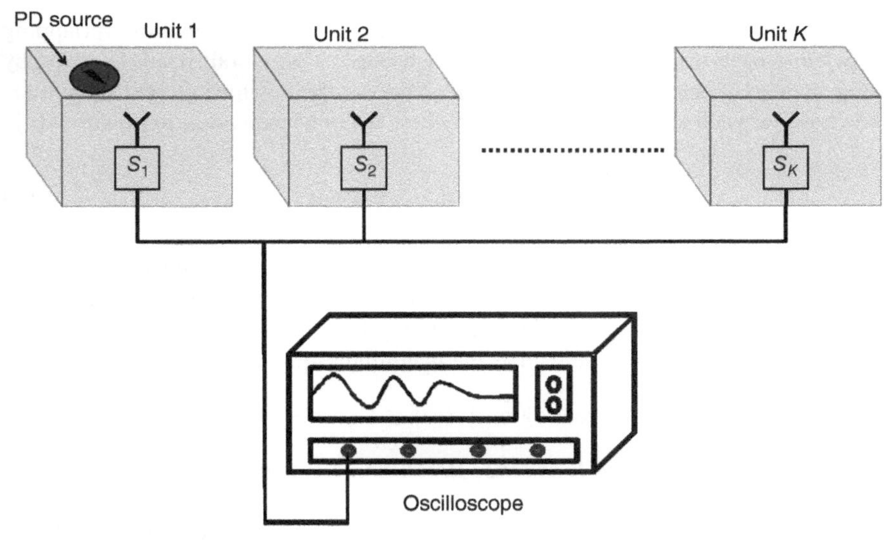

Figure 6.3 Overview of proposed PD sensor system

block is installed in an individual unit and the unique frequency pattern is used to identify that particular HV plant.

6.2.2 PD Sensor System Overview

Figure 6.3 illustrates the overall PD sensor system. A number of metal-enclosed HV units are monitored in a substation. The units are mounted with K number of sensors $S_1, S_2, \dots, S_K$. Cables coming from each sensor are connected to a single channel of an oscilloscope. Here the oscilloscope is operating as a PD analyzer. In the event of PD in an HV unit, the captured signal transmits through the corresponding sensor toward the oscilloscope. The signal is analyzed to extract information about the PD level and detect the faulty unit, and the responses of individual sensors are fully independent and uncorrelated. Therefore, multiple units can be monitored by the proposed sensor system to provide a comparative analysis of the equipment producing more severe PD activity.

6.2.3 Simultaneous PD Detection

6.2.3.1 Time–Frequency Analysis Time–frequency analysis is introduced to perform a detailed investigation of the frequency signature produced by the cascaded resonators. A basic tool for time–frequency analysis is short time frequency transform (STFT). STFT is a Fourier representation of a signal that determines the frequency and phase content of local sections as it varies over time [15–18]. One of the shortcomings of the Fourier transform is that it does not give any information on the time at which a frequency component occurs. On the other hand, STFT gives the time resolution of a frequency spectrum.

If a discrete time signal $x(n)$ is to be transformed, it is broken up into overlapping frames and each time frame is Fourier transformed. The overall result is added by using an overlap–add (OLA) algorithm [19,20] to a matrix, which records magnitude and phase for each point in time and frequency. It can be expressed in Equation 6.1.

$$X_m(\omega) = \sum_{n=-\infty}^{n=\infty} x(n)w(n - mR)e^{-j\omega n} \tag{6.1}$$

where $x(n)$ is the input signal at time n; $w(n)$, window function having length M; $X_m(\omega)$, discreet time Fourier transform (DTFT) of windowed data centered about time mR; R, hop size or step size in time samples between two adjacent DTFTs.

Also, the spectrogram is defined as the magnitude square of the STFT [21],

$$\text{Spectrogram}\{x(n)\} \equiv |X_m(\omega)|^2$$

6.2.3.2 Effect of Time and Frequency Resolution The time resolution Δt and frequency resolution Δf are two essential parameters in performing STFT analysis. The time and frequency resolution depend on the window type, window length M, and hop size R. For instance, in the case of a fixed window type, the window length M determines the frequency resolution and the value of R corresponds to the number of oversamples in the time dimension. If the hop size is equal to the window length, then there is no overlapping between two windows, which results in poor time resolution [22]. On the other hand, increasing the number of overlapping windows entails a greater number of frames in a given time compared with the nonoverlapping window setting. This increasing spectral frame for a given time enhances the time resolution. The time resolution Δt is related to the hop size R given by

$$\Delta t = \frac{R}{f_s}$$

where f_s is the sampling frequency.

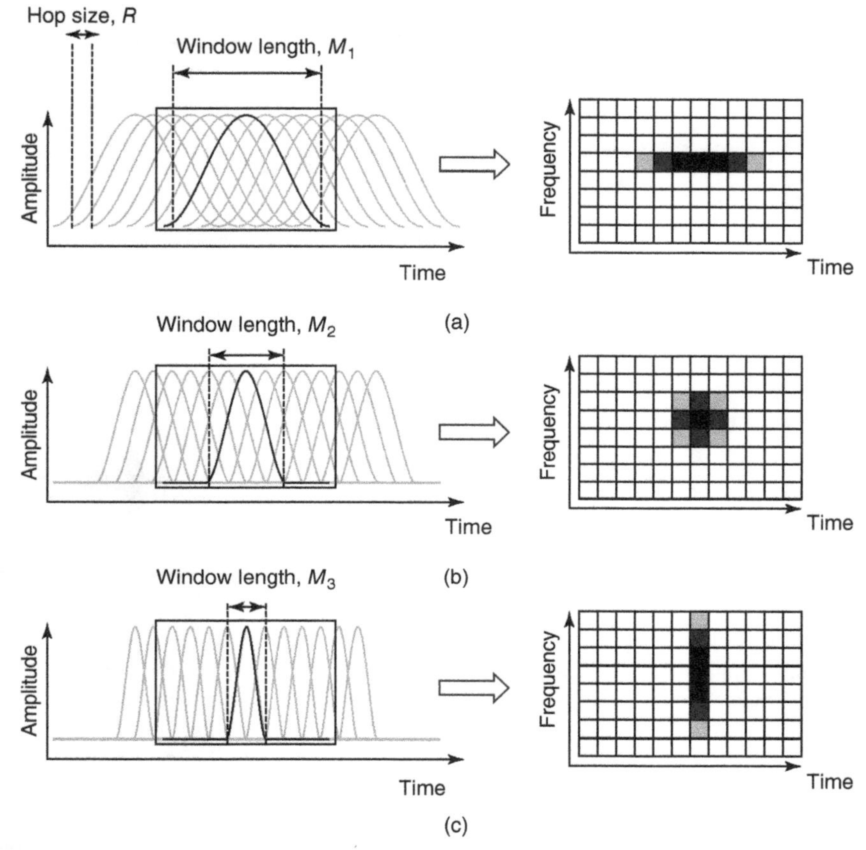

Figure 6.4 Illustration of multiresolution effect in STFT analysis. The window length for the three cases is (a) M_1, (b) M_2, and (c) M_3. Here, $M_1 > M_2 > M_3$

The multiresolution effect is due to the uncertainty principle, which implies that a signal cannot be localized exactly in both time and frequency domains simultaneously [23,24].

The effect of windowing parameters on time and frequency resolution is illustrated in Figure 6.4. In this figure, three distinct cases are taken into account. In each case, the window type is constant. However, the window size M is decreased from $M_1 > M_2 > M_3$. The effect of shrinking the window size is prominent in the STFT results. A wide time window, M_1, localizes the signal in the frequency domain sacrificing time resolution (Figure 6.4(a)). As the window size is decreased from M_1 to M_2, both time and frequency resolutions are optimum (Figure 6.4(b)). Furthermore, a narrow window (M_3) for STFT requires the signal to be localized in the time domain, as shown in Figure 6.4(c). In this case, the time resolution is maximum. The localization of a signal in the time domain using STFT has a remarkable effect on the detection of multiple concurrent PD events, as described in the next section.

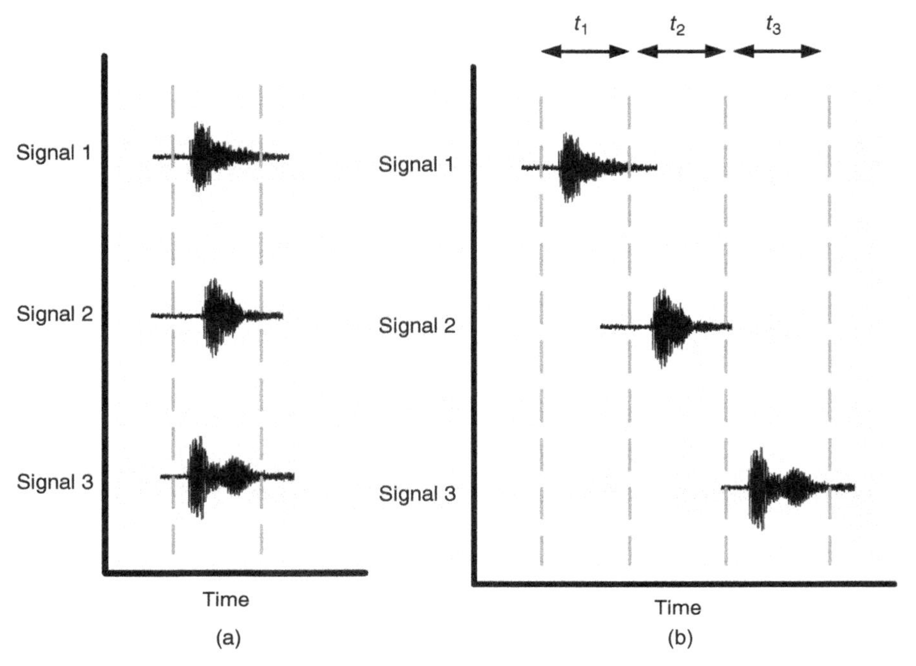

Figure 6.5 Time-domain concurrent signals (a) without time delay and (b) with time delay. Here the signals from sensor S_1, S_2, and S_3 are separated by t_1, t_2, and t_3 seconds

6.2.3.3 Simultaneous PD Detection Incorporating Time Delay

In our proposed PD sensor system, a faulty unit is identified by analyzing the frequency signature embedded in the RF signal. Each sensor is connected to a single channel of an oscilloscope through an RF power combiner, as shown in Figure 6.3. In the case of multiple PD occurrences at the same instance, signals from all sensor blocks combine at the oscilloscope. The signals cannot be identified using frequency analysis as they are added in the time domain. An illustration of this scenario is presented in Figures 6.5 and 6.6. In Figure 6.5(a), three PD signals originating from different HV units are shown in the time domain. The signals are concurrent and their time—frequency analysis represents the frequency signature of the combined signals (Figure 6.6(a)). As the signals are not separated in the time domain, the combined frequency signature does not provide correct identification data. In contrast, in Figure 6.5(b), the signals from each sensor are separated in the time domain, and the signals from S_1, S_2, and S_3 are delayed by t_1, t_2, and t_3 seconds using fixed time delay circuits. The effect of introducing time delay circuits is evident in the time-frequency representation of the signals (Figure 6.6(b)), where the frequency information is separated in time and individual sensor data can be retrieved. Thus, simultaneous PD detection from multiple sources is possible by incorporating time delay circuits within the sensors and localizing the captured signals in the time domain.

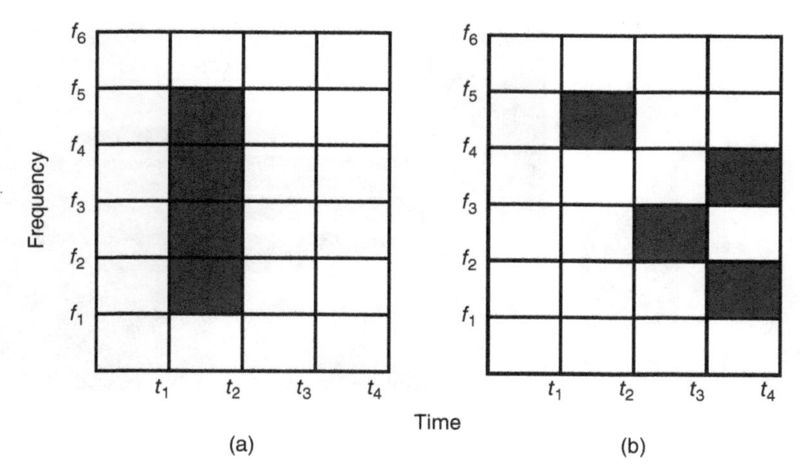

Figure 6.6 Time–frequency representation of the signals shown in (a) panel (a) of Figure 6.5 and (b) panel (b) of Figure 6.5

6.3 PD LOCALIZATION USING CASCADED MULTIRESONATOR-BASED SENSOR

6.3.1 PD Sensor

For the prototype of our PD sensor, a cascaded multiresonator-based chipless RFID sensor is used. Detailed design and development of a PD sensor is presented in Chapter 3 (Section 3.4.3). In the following section, we have carried out experimentation to verify frequency signature-based PD detection and faulty source identification.

6.3.2 Experimentation with PD Signal

Experimentation has been undertaken to validate the response of the fabricated PD sensor. The whole experiment is performed in an anechoic chamber, where external EM waves do not interfere. For the conduct of experiments with PD signals, a short-duration PD pulse generator is used as a source. The source is a PD calibrator CAL2B from Power Diagnostix [25]. The PD source is compliant with the IEC 60270 standard and emits PD current pulses with rise time $\leq$ 200 ps and bandwidth $\geq$ 1.5 GHz. The calibrator voltage level is set to 50 V throughout the experiment (Figure 6.7(a)). A high-speed oscilloscope displayed the captured signal is used as PD analyzer. The oscilloscope is a Tektronix DSA72004 digital serial analyzer, which features high bandwidth up to 20 GHz matched across four channels. The sampling frequency is set to 25 GS/s and the record length is 200 ns (refer to Figure 6.7(b)).

Figure 6.8(a) shows the time-domain PD pulse captured with the antenna without being transmitted through an SIR filter. Figure 6.8(b) shows the power spectrum of

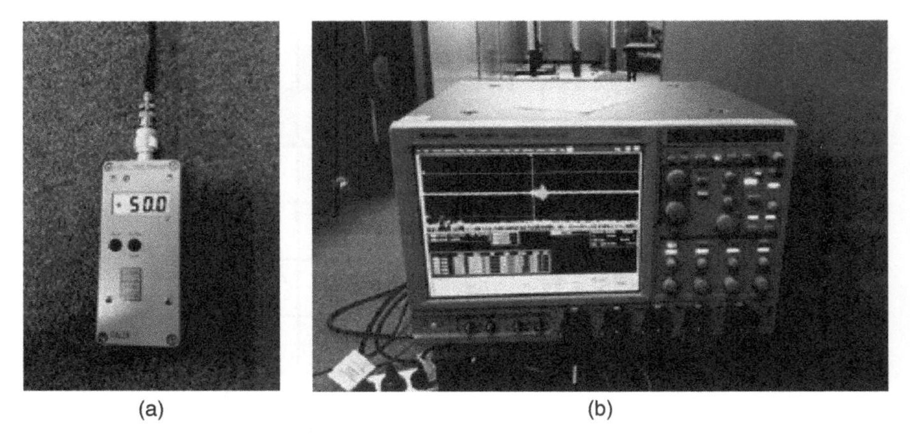

Figure 6.7 (a) PD calibrator CAL2B and (b) oscilloscope DSA 72004

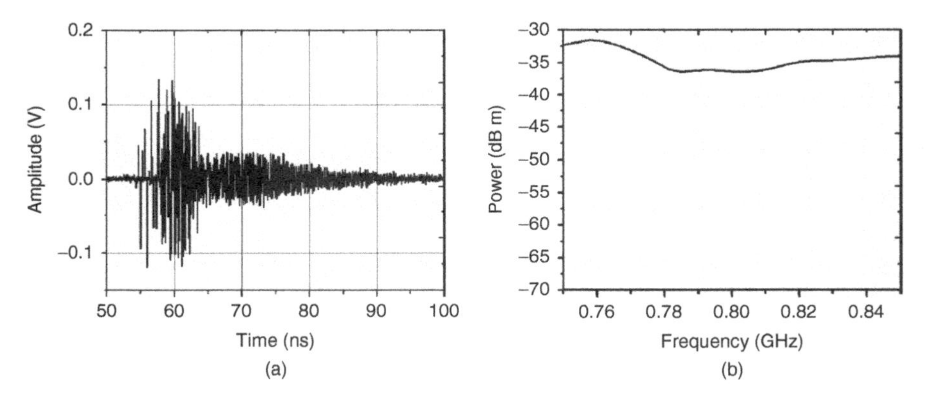

Figure 6.8 (a) Time-domain PD signal and (b) power spectrum of time-domain PD signal in (a)

the PD pulse over the frequency band of interest. As expected, the spectrum is almost constant over the frequency band.

6.3.3 Data Encoding in PD Signal

Next, we performed an experiment to investigate chipless RFID-based PD sensor operation. To validate the sensor performance, the experimental setup illustrated in Figure 6.9 is used. Three HV units were considered and the sensors installed in these units had the SIR filter combination S_1, S_2, and S_3 (see Table 3.2).

For experimental purposes, each HV unit is modeled as an EM-shielded enclosure with a PD source and a sensor mounted on it. Figure 6.9 shows a block diagram of

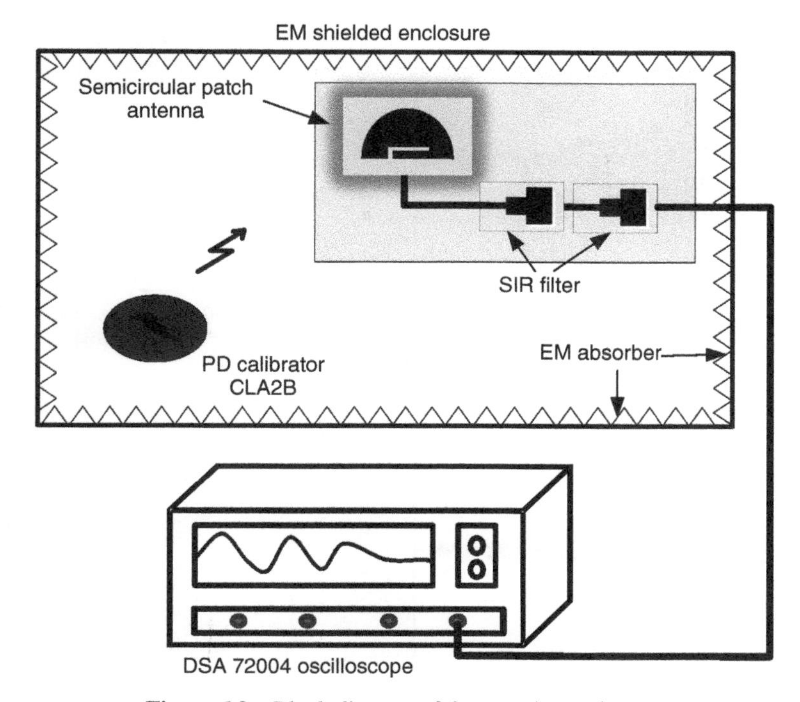

Figure 6.9 Block diagram of the experimental setup

a single unit. The sensor comprises of an antenna and a unique combination of SIR filters. The output of the sensor is wired to an oscilloscope channel. The distance between the PD source and the RF sensor is kept at 1 m in all three units, and the PD source intensity is set to 50 picocoulomb (pC) throughout the experiment.

Figure 6.10 shows the captured PD signal at the oscilloscope when the PD source in each unit is energized separately. Consequently, Figure 6.10(a–c) corresponds to time-domain signals through sensors S_1, S_2, and S_3, respectively. The PD signal in the time domain does not provide any information regarding the HV unit from which it originated.

In Figure 6.10(d–f), the power spectra of captured PD signals of Figure 6.10(a–c) are plotted. In Figure 6.10(d), the power spectrum has a sharp attenuation at 775 MHz, which suggests that the signal originated from the unit with sensor S_1. Similarly, the signal in Figure 6.10(e) has a dip in the power spectrum at 825 MHz, and the signal in Figure 6.10(f) has two distinct dips at 775 and 825 MHz. These facts suggest that the PD signals in Figure 6.10(e) and (f) originated from HV units mounted with S_2 and S_3, respectively. Hence, the sensors can identify PD sources separately.

Moreover, the sensor provides information about the PD signal level by calculating the cumulative energy [26]. Here, we have measured a correlation between the PD

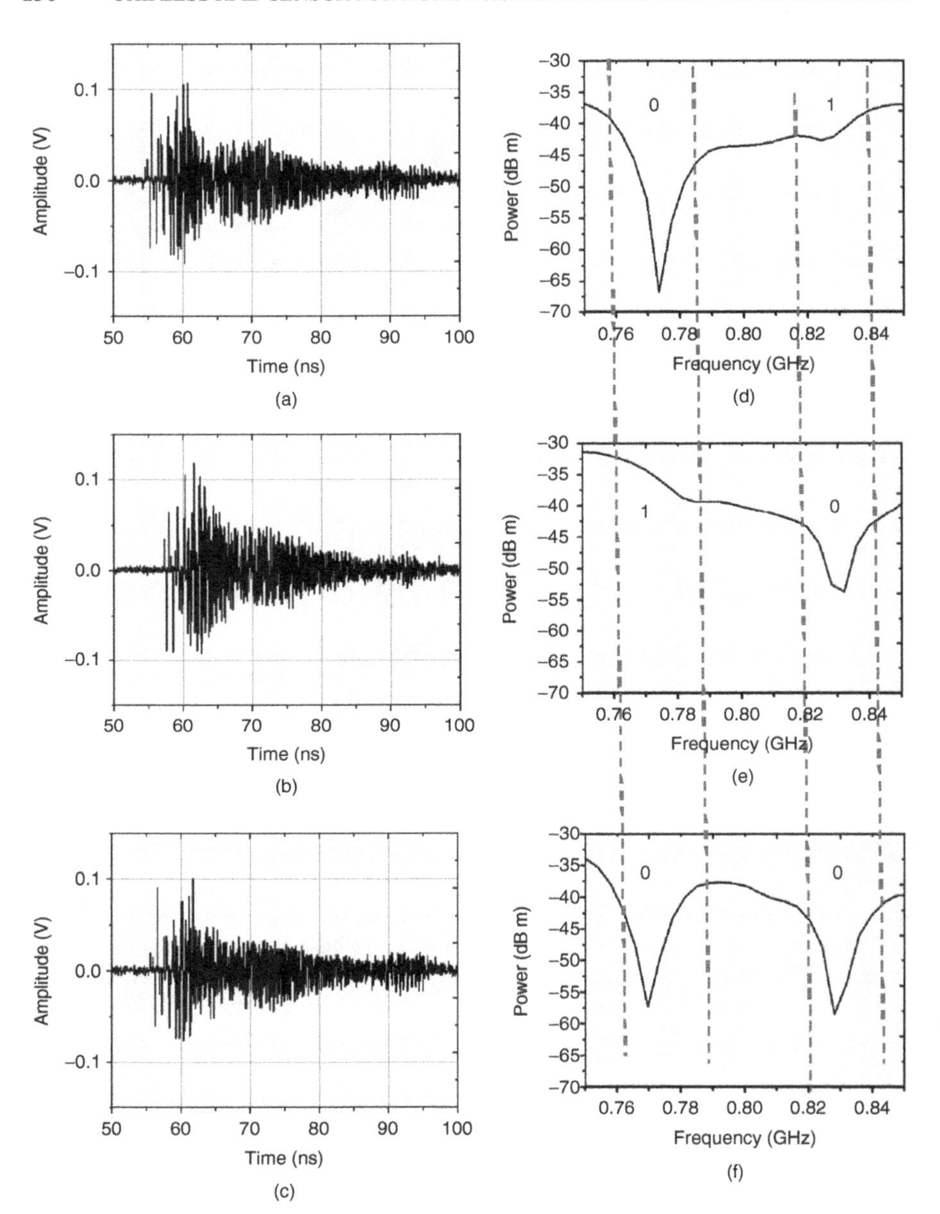

Figure 6.10 Captured PD signal in time domain transmitting through sensor (a) S_1, (b) S_2, and (c) S_3. Power spectrum of captured PD signal transmitting through (d) S_1, (e) S_2, and (f) S_3

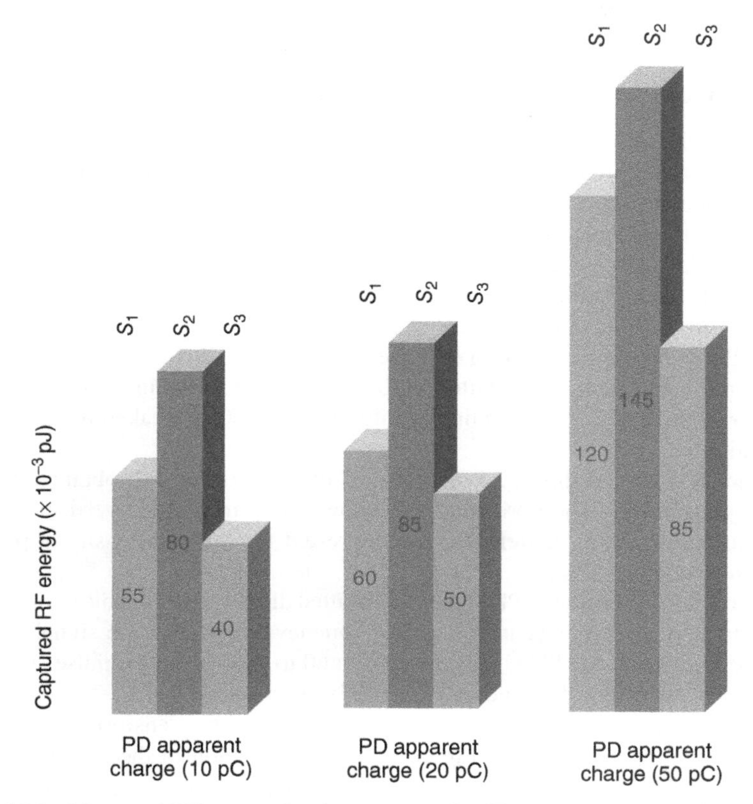

Figure 6.11 Measured RF energy for three sensors for PD apparent charge of 10, 20, and 50 pC

apparent charge emitted from the source and captured energy. PD apparent charge is an International Electrotechnical Commission (IEC) Standard PD measurement technique. IEC 60270 standard can give apparent charge quantity of PD, which is widely used in preventive test of many HV power equipment [27,28]. The widely accepted unit for PD apparent charge is in picocoulomb.

To correlate the PD apparent charge to the captured RF energy, we plotted the cumulative energy for different PD levels emitted from the source (refer to Figure 6.11). All the sensors show increased captured energy as the PD apparent charge increases. However, at PD apparent charges 10, 20, and 50 pC, S_1 and S_2 have higher RF energies than S_3. This is due to the presence of both the filters in series resulting in higher attenuation. However, in frequency spectral analysis the encoded data bits can be clearly identified, irrespective of the emitted PD energy level. Moreover, in real occurrences of PD events in HV equipment, the PD signal strength is many thousand times greater than that of the PD calibrator. Therefore, the developed chipless RFID sensor is a robust technology for PD detection in faulty power apparatuses in switchyards.

6.4 SIMULTANEOUS PD DETECTION

6.4.1 Time–Frequency Analysis

As discussed above, time–frequency analysis of a PD signal gives the time information as well as the frequency signature embedded in it. Moreover, a microwave filter has a particular group delay, which causes a physical time delay of the amplitude envelope at the resonant frequency. The group delay of the SIR filter is in nanoseconds, and this time lag causes a certain frequency band to concentrate at a lower intensity in the spectrogram plot. This effect is manifested in the STFT analysis described in the following section.

The PD signal captured during the experiment is analyzed in the time domain using Equation 6.1. For each signal with $N = 5000$ samples, a Hamming window [29] with a window size of $M = 1024$ and time resolution $\Delta t = 960\,\mu s$ is taken for spectrogram calculation.

Figure 6.12(a)–(d) shows 2-D projections of the surface plots obtained from the spectrogram results. The horizontal axis shows the time and the vertical axis plots frequency information. Different colors represent the power intensity at particular times and frequency spans.

Figure 6.12(a) shows the PD signal transmitted directly to the oscilloscope without filter attenuation. It represents the time–frequency analysis of the signal shown in Figure 6.8(a). Here, the PD signal spans from 60 to 100 ns and the pulse contains all the frequencies within the band 750–850 MHz, as expected.

Figure 6.12(b)–(d) shows the STFT results obtained by transforming the signals plotted in Figure 6.10(a–c). In Figure 6.12(b), a variation in the spectral intensity at around 775 MHz frequency band is evident, and the color map of this 2-D plot depicts approximately 20 dB attenuation at this frequency within the time span of the captured PD signal. This frequency attenuation identifies that this PD signal is transmitted through S_1, or as explained previously, this spectral modification is denoted by "01." Similarly, the results plotted in Figure 6.12(c) and (d) correspond to frequency signatures denoting "10" and "00." Therefore, the STFT plots provide additional time information of the captured PD signals compared with the FFT results shown in Figure 6.10(d–f).

6.4.2 Effect of Time and Frequency Resolution

In analyzing the PD signal using spectrograms, a detailed investigation of the frequency resolution, Δf and time resolution, Δt is performed. Figure 6.4 indicates that in performing STFT, increasing time localization negates frequency resolution. Similarly, to achieve a higher frequency resolution, time resolution has to be sacrificed.

This phenomenon is verified from the following observation. Here, the captured signal from S_2 shown in Figure 6.10(b) is analyzed using spectrogram having different parametric values. Here, three observations are presented and the values of window length M and time resolution Δt for each observation are given in Table 6.1. The results of 2-D spectrogram are shown in Figure 6.13(a–c). In case 1, the frequency

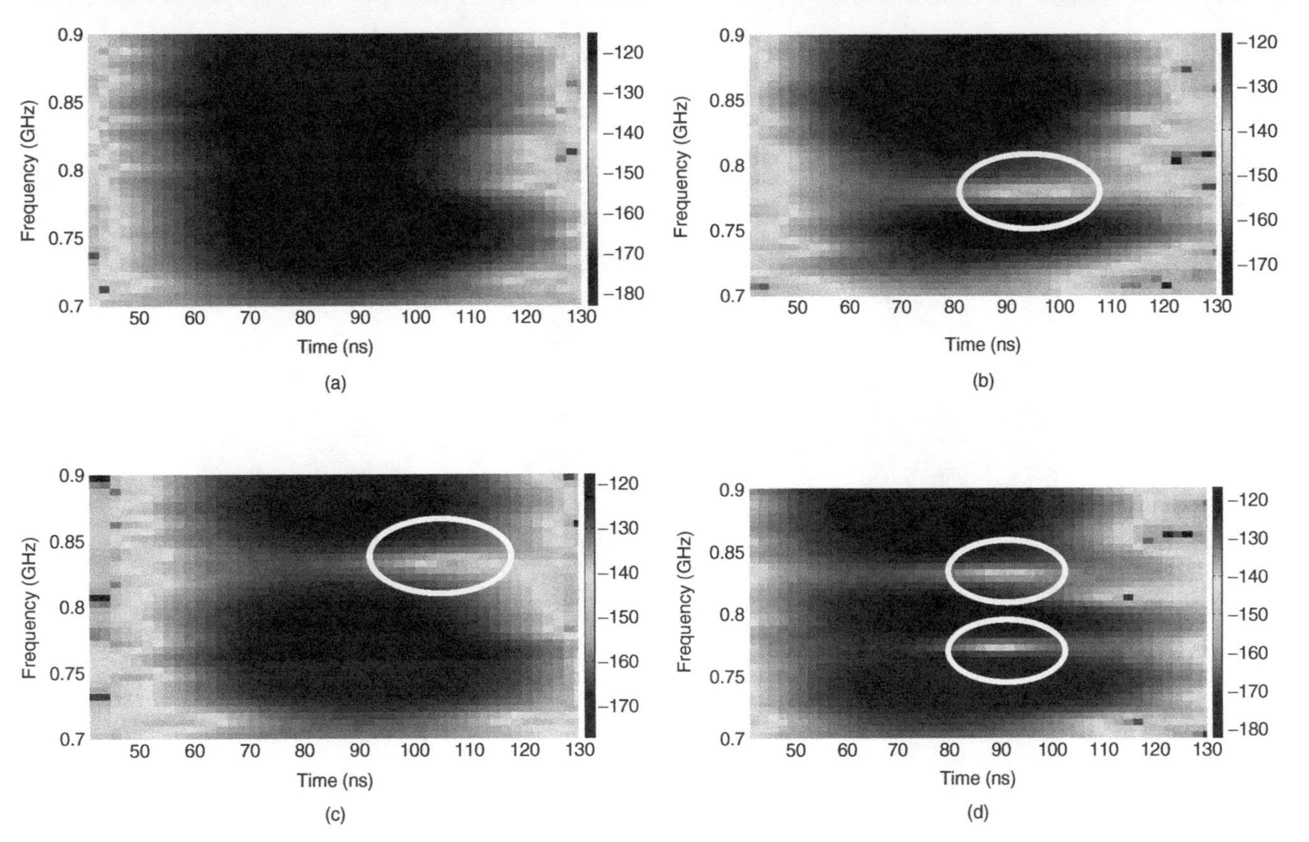

Figure 6.12 2-D projection of surface plot obtained from spectrogram analysis on the PD signals captured (a) directly shown in Figure 6.8(a) and through the sensors (b) S_1, (c) S_2, and (d) S_3 (the location of signal attenuation is highlighted)

TABLE 6.1 Values of Window Length and Time Resolution for Different Observations in Figure 6.13

Observation	Window Length, M	Time Resolution, Δt (ns)
Case 1	2048	1.9
Case 2	512	0.48
Case 3	128	0.32

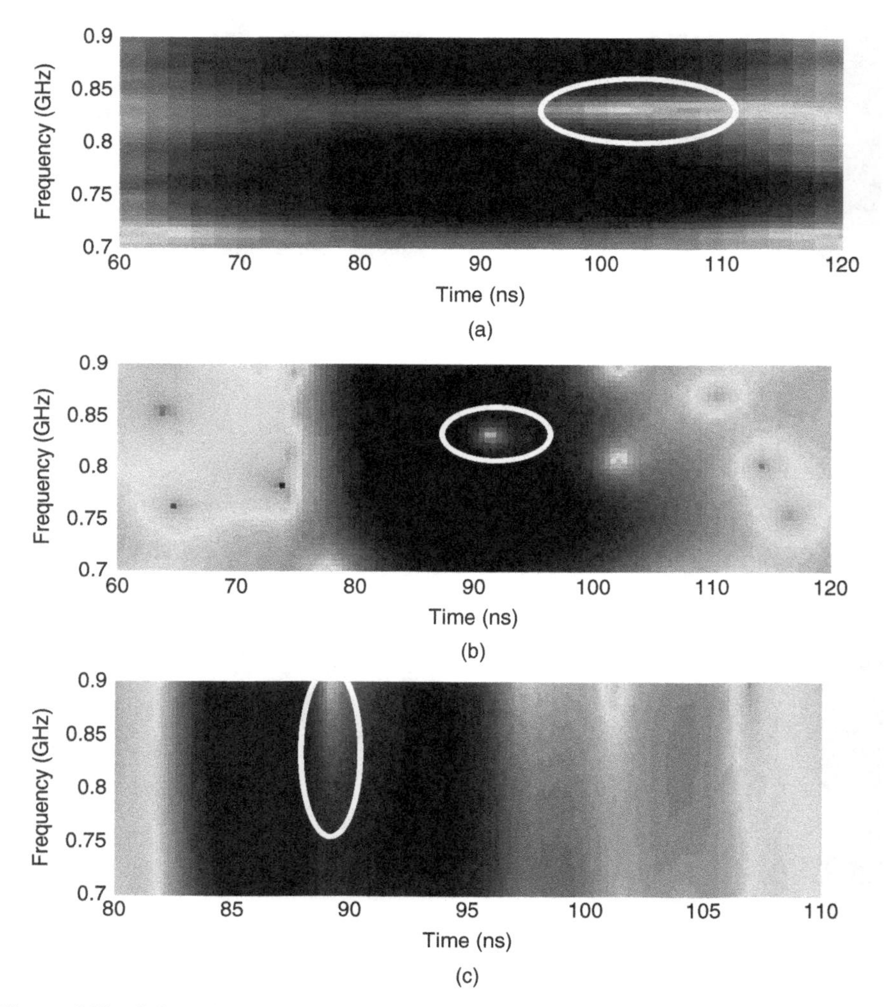

Figure 6.13 2-D projection of spectrogram for signal transmitted through sensor S_1 (a)–(c) refer to cases 1–3 in Table 6.1 (here the location of signal attenuation is highlighted)

resolution is high but the signal spans through a large time width. In this case, the time of frequency attenuation is not reflected from the STFT result (Figure 6.13(a)). This is due to the group delay of SIR filter, which is in nanoseconds range.

In Figure 6.13(b), both the time and frequency information is optimized as the frequency attenuation is bounded in both time and frequency domains. Here, the time resolution is enhanced compared to case 1. This causes the attenuated frequencies to localize in the time domain. On the other hand, further reduction in M, as in case 3, gives more accurate time information (Figure 6.13(c)). Using this time information, exact time of occurrence can be calculated. However, in this case, the attenuated frequency identification is not possible as the signal becomes unbounded in the frequency domain.

The above results imply that using multiresolution effect, it is possible to extract both the time information and the frequency signature within the PD signal. The localization of PD signal in time domain has a remarkable aspect in detecting multiple PD events occurring at the same instance. By increasing the time resolution, we can separate two concurrent PD signals that are differed by defined time delays. Therefore, the signals can be identified using the frequency signature embedded within the spectrum. The experiment carried out in the next section validates this feature of time–frequency analysis for detecting simultaneous PD events.

6.4.3 Simultaneous PD Detection Incorporating Time Delay

In Figure 6.3, the configuration of proposed multiple PD source is illustrated. Here, multiple PD signals originating at the same instance will arrive as a combined signal at the oscilloscope channel. As described in the theory (Section 6.2.3), the resultant frequency signature of a number of simultaneous PD signals does not provide individual source ID. In contrast, by incorporating time delay within the sensor, we can separate the frequency signature in time domain. This enables simultaneous PD detection.

The experimental setup shown in Figure 6.14 is used to detect two simultaneous PD events. The sensor installed in Unit 1 and Unit 2 has filters operating at 775 and 825 MHz. Moreover, the cable length from Unit 1 and Unit 2 is kept different to give defined time delays T_1 and T_2. The calculated time delay difference $|T_2 - T_1|$ is given in Table 6.2.

When the PD source of both units is switched on, the oscilloscope channel captures the combined signal originated from each unit. The time domain data plotted in Figure 6.15 has a higher amplitude compared to the signals captured in Figure 6.10(a–c).

This time domain signal is analyzed in time and frequency domains to identify separate faulty sources (Figure 6.16). Here, three observations are considered for STFT analysis as given in Table 6.2. In case 1 (Figure 6.16(a)), there are two distinct frequency dips within the frequency band 750–850 MHz, which confirms the presence of PD. Here, the frequency signature can be interpreted as "00" as previously found in Figure 6.12(d). However, this is incorrect due to poor time resolution in STFT

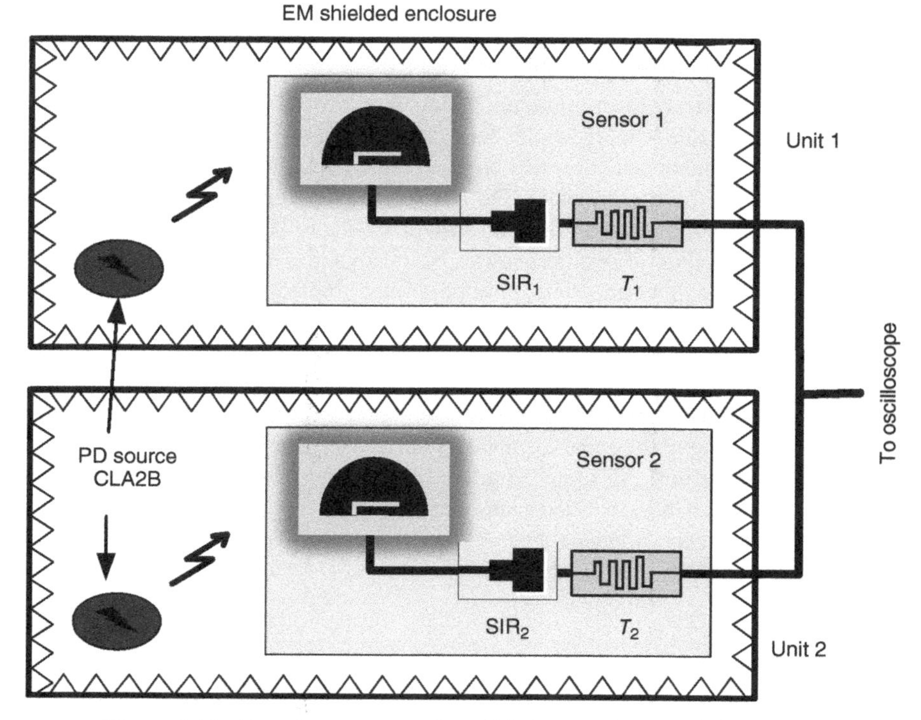

Figure 6.14 Experimental setup for validating simultaneous PD detection

TABLE 6.2 Calculated and Measured Time Delay for Different Observations for The Experimental Setup in Figure 6.16

Observation	Window Length, M	Time Resolution, Δt (ns)	Calculated Time Delay Difference $\lvert T_2 - T_1 \rvert$	Measured Time Delay (ns)
Case 1	2048	1.9		–
Case 2	512	0.48	15	3
Case 3	128	0.32		8

analysis. Thus, STFT analysis in Figure 6.16(a) could not separate the PD signals in the time domain for appropriate source identification.

In Figure 6.16(b), there is an obvious time difference between the two attenuated frequencies that indicate the PD originating from two different sources. However, in case 3, where the time resolution is maximum, the arrival time delay between two PD signals can be measured (Table 6.2). The 2-D plot of STFT analysis shown

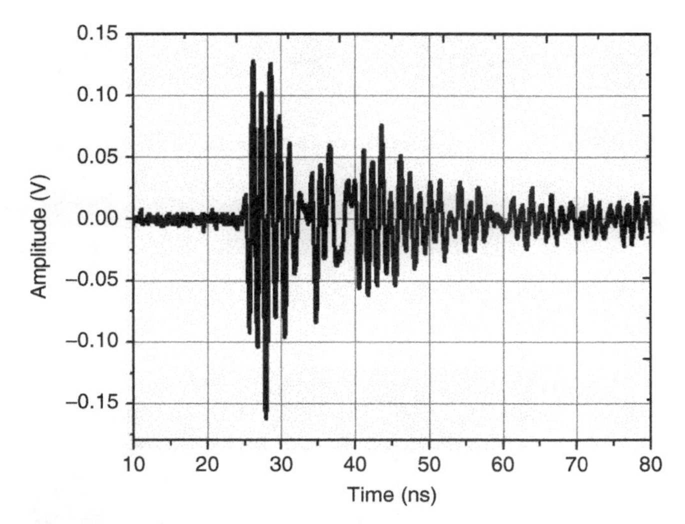

Figure 6.15 Captured combined time-domain signal from HV Unit 1 and Unit 2 shown in Figure 6.14

in Figure 6.16(c) accurately detects the two PD signals originated from Unit 1 and Unit 2. However, there is difference between the calculated time delay and the measured delay found in case 3. This can be compensated by further increase of the time resolution, which in turn deteriorates the frequency information.

The results obtained in this section clearly demonstrate the applicability of time–frequency analysis for the detection of concurrent PD events. Furthermore, simultaneous PD detection can be extended for K number of separate units connected to a single channel, as shown in Figure 6.3. Here, each unit has a designated frequency signature and a unique time delay, so that it can be identified in both frequency and time domains. In future research, time–frequency-based PD localization will be extended to realize more data bits to investigate the potential for monitoring PDs in large substations. Furthermore, the time delay introduced by different cable lengths increases system complexity. To overcome this, a passive meander delay line [30,31] will be introduced within the RF sensor to incorporate the defined time delay from each unit. Finally, the prospects of our proposed sensor for monitoring HV units in distribution substations will be evaluated using practical experiments.

6.5 CONCLUSION

A passive RF sensor for radiometric PD detection has been developed, which provides low-cost, automated condition monitoring of HV equipment in substations. A prototype sensor system has been constructed and verified by identifying the faulty

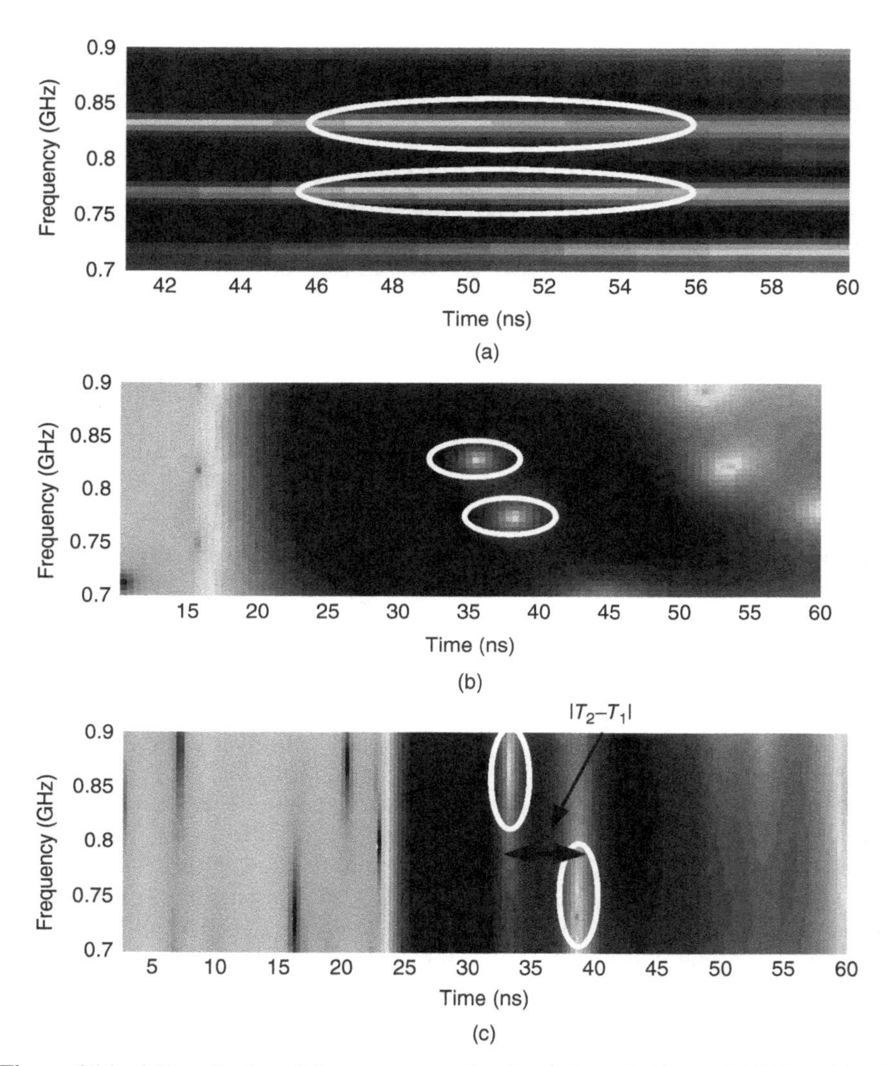

Figure 6.16 2-D projection of the spectrogram for signal shown in Figure 6.15. Plots (a)–(c) refer to cases 1–3 in Table 6.2 (here the location of signal attenuation is highlighted)

unit in the event of PD. The novel contributions of the proposed PD sensor in the field of radiometric PD detection can be summarized as follows:

- A fully passive, maintenance-free chipless RFID sensor has not been investigated earlier for HV condition monitoring. This sensor gives signal strength in picocoulomb as well as faulty source ID within the frequency spectrum of captured PD signal. Results show that it can detect frequency signature in nanosecond-duration PD signal.

- Simultaneous PD detection for multiple faulty HV apparatus is another novel attribute of our proposed sensor system. In this chapter, we have presented experimental validation for simultaneous PD detection using STFT analysis. To the best of our knowledge, simultaneous PD detection and localization has not been investigated yet in radiometric PD detection techniques.

The future development of PD sensor will be focusing on a fully planar UWB antenna design and development, introducing passive meander delay line in sensor circuit and investigating more number of bits to provide a sensor system that can be scaled up for large substations for mass deployment.

The next chapter explores an RCS scatterer-based chipless RFID sensor for real-time environment monitoring, presents a detailed comparative study of smart materials for humidity sensing, and outlines the chipless RFID humidity sensor development and the associated experimentation. Finally, critical parameters are identified for a highly sensitive chipless RFID tag sensor.

REFERENCES

1. F. H. Kreuger, *Partial Discharge Detection in High-Voltage Equipment*, 1st ed.: Butterworths, 1989.
2. S. Xiao, P. J. Moore, M. D. Judd, and I. E. Portugues, "An investigation into electromagnetic radiation due to partial discharges in high voltage equipment," in *Power Engineering Society General Meeting, 2007. IEEE*, 2007, pp. 1–7.
3. W. R. A. Bojovschi and A. K. L. Wong, "Electromagnetic Field Intensity Generated by Partial Discharge in High Voltage Insulating Materials," *Progress In Electromagnetics Research*, vol. 104, pp. 167–182, 2010.
4. A. J. Reid, M. D. Judd, B. G. A. Stewart, and R. A. Fouracre, "Partial Discharge Current Pulses in SF 6 and the Effect of Superposition of Their Radiometric Measurement," *Journal of Physics D: Applied Physics*, vol. 39, p. 4167, 2006.
5. Y. Luo, S. Ji, and Y. Li, "Phased-Ultrasonic Receiving-Planar Array Transducer for Partial Discharge Location in Transformer," *IEEE Transactions on Ultrasonics, Ferroelectrics and Frequency Control*, vol. 53, pp. 614–622, 2006.
6. M. Kawada, "Fundamental Study on Location of a Partial Discharge Source with a VHF–UHF Radio Interferometer System," *Electrical Engineering in Japan*, vol. 144, pp. 32–41, 2003.
7. X. Hu, M. D. Judd, and W. H. Siew, "A study of PD location issues in GIS using FDTD simulation," in *Universities Power Engineering Conference (UPEC), 2010 45th International*, 2010, pp. 1–5.
8. Y. Tian, M. Kawada, and K. Isaka, "Locating Partial Discharge Source Occurring on Distribution Line by Using FDTD and TDOA Methods," *IEEJ Transactions on Fundamentals and Materials*, vol. 129, pp. 89–96, 2009.
9. Y. Tian, M. Kawada, and K. Isaka, "Visualization of electromagnetic waves emitted from multiple PD sources on distribution line by using FDTD method," in *Electrical Insulating Materials, 2008. (ISEIM 2008). International Symposium on*, 2008, pp. 95–98.

10. H. Ishimaru and M. Kawada, "Localization of a Partial Discharge Source Using Maximum Likelihood Estimation," *IEEJ Transactions on Electrical and Electronic Engineering*, vol. 5, pp. 516–522, 2010.
11. P. J. Moore, I. E. Portugues, and I. A. Glover, "Radiometric Location of Partial Discharge Sources on Energized High-Voltage Plant," *IEEE Transactions on Power Delivery*, vol. 20, pp. 2264–2272, 2005.
12. B. G. Stewart, A. Nesbitt, and L. Hall, "Triangulation and 3D location estimation of RFI and partial discharge sources within a 400 kV substation," in *Electrical Insulation Conference, 2009. EIC 2009. IEEE*, 2009, pp. 164–168.
13. A. Reid, I. Dick, and M. Judd, "UHF monitoring of partial discharge in substation equipment using a novel multi-sensor cable loop," in *20th International Conference and Exhibition on Electricity Distribution – Part 1, 2009. CIRED 2009.* 2009, pp. 1–4.
14. S. Preradovic, I. Balbin, N. C. Karmakar, and G. F. Swiegers, "Multiresonator-Based Chipless RFID System for Low-Cost Item Tracking," *IEEE Transactions on Microwave Theory and Techniques,* vol. 57, pp. 1411–1419, 2009.
15. J. B. Allen and L. R. Rabiner, "A Unified Approach to Short-Time Fourier Analysis and Synthesis," *Proceedings of the IEEE*, vol. 65, pp. 1558–1564, 1977.
16. J. Allen, "Short term spectral analysis, synthesis, and modification by discrete Fourier transform," *IEEE Transactions on Acoustics, Speech and Signal Processing*, vol. 25, pp. 235–238, 1977.
17. I. Daubechies, "The wavelet transform, time–frequency localization and signal analysis," *IEEE Transactions on Information Theory*, vol. 36, pp. 961–1005, 1990.
18. L. B. Almeida, "The fractional Fourier transform and time–frequency representations," *IEEE Transactions on Signal Processing*, vol. 42, pp. 3084–3091, 1994.
19. B. G. Lawrence and R. Rabiner, *Theory and Application of Digital Signal Processing*: Prentice-Hall, 1975.
20. J. O. Smith, Spectral Audio Signal Processing. Available: https://ccrma.stanford.edu/~jos/sasp/sasp.html.
21. L. Stankovic, "A Method for time–frequency Analysis," *IEEE Transactions on Signal Processing*, vol. 42, pp. 225–229, 1994.
22. T. H. Park, *Introduction to Digital Signal Processing: Computer Musically Speaking*: World Scientific, 2010.
23. H. G. Feichtinger, *Gabor Analysis and Algorithms: Theory and Applications*, 2003.
24. P. M. Bentley and J. T. E. McDonnell, "Wavelet Transforms: an Introduction," *Electronics and Communication Engineering Journal*, vol. 6, pp. 175–186, 1994.
25. Power Diagnostix Systems GmbH. Available: http://www.pd-systems.com/accessories.html#mail.
26. A. J. Reid, M. D. Judd, R. A. Fouracre, B. G. Stewart, and D. M. Hepburn, "Simultaneous Measurement of Partial Discharges Using IEC60270 and Radio-Frequency Techniques," *IEEE Transactions on Dielectrics and Electrical Insulation*, vol. 18, pp. 444–455, 2011.
27. A. Cavallini, G. C. Montanari, and M. Tozzi, "PD apparent charge estimation and calibration: a critical review," *IEEE Transactions on Dielectrics and Electrical Insulation*, vol. 17, pp. 198–205, 2010.
28. L. Hongjing, J. Wenjie, L. Jianyin, Z. Zhong, C. Yangchun, Y. Hong, *et al.*, "The effectiveness assessment of calibration and measurement of PD apparent charge quantity," in *Electrical Insulation Conference (EIC), 2013 IEEE*, 2013, pp. 129–132.

29. S. Kadambe, "On the window selection and the cross terms that exist in the magnitude squared distribution of the short time Fourier transform," in *Statistical Signal and Array Processing, 1992. Conference Proceedings, IEEE Sixth SP Workshop on*, 1992, pp. 22–25.
30. B. J. Rubin and B. Singh, "Study of Meander Line Delay in Circuit Boards," *IEEE Transactions on Microwave Theory and Techniques,* vol. 48, pp. 1452–1460, 2000.
31. W. Ruey-Beei and C. Fang-Lin, "Laddering Wave in Serpentine Delay Line," *IEEE Transactions on Components, Packaging, and Manufacturing Technology, Part B: Advanced Packaging,* vol. 18, pp. 644–650, 1995.

7

CHIPLESS RFID SENSOR FOR REAL-TIME ENVIRONMENT MONITORING

7.1 INTRODUCTION

In this chapter, a chipless radio-frequency identification (RFID) sensor is proposed for environment monitoring. The radar cross-section (RCS) scatterer-based resonator developed in Chapter 3 is used in the sensor development. The main focus of this chapter is to integrate smart humidity sensing materials with RCS scatterers to incorporate environment sensing in frequency spectral analysis. The structure of this chapter is illustrated in Figure 7.1.

As shown in Figure 7.1, chipless RFID sensor development is carried out into two phases. The aim of the first phase is to perform RF characterization of two humidity sensing polymers Kapton and polyvinyl alcohol (PVA) to compare their humidity sensitivity. In this phase, we also identify a number of RF sensing parameters for calibrating humidity in real environments.

In the second phase, a chipless RFID humidity sensor is developed based on RCS scatterer. Detailed experimentation is carried out to verify humidity sensor performance. The measured sensor response is analyzed to exploit calibration curve for unknown humidity measurement. Finally, the effect of hysteresis, repeatability, and time response are discussed.

7.2 PHASE 1. HUMIDITY SENSING POLYMER CHARACTERIZATION AND SENSITIVITY ANALYSIS

7.2.1 Theory of Dielectric Sensor

A coplanar waveguide (CPW) line consists of a dielectric substrate with an infinite length center strip line on the top surface. The strip line is separated by a narrow

Chipless RFID Sensors, First Edition. Nemai Chandra Karmakar, Emran Md Amin and Jhantu Kumar Saha.

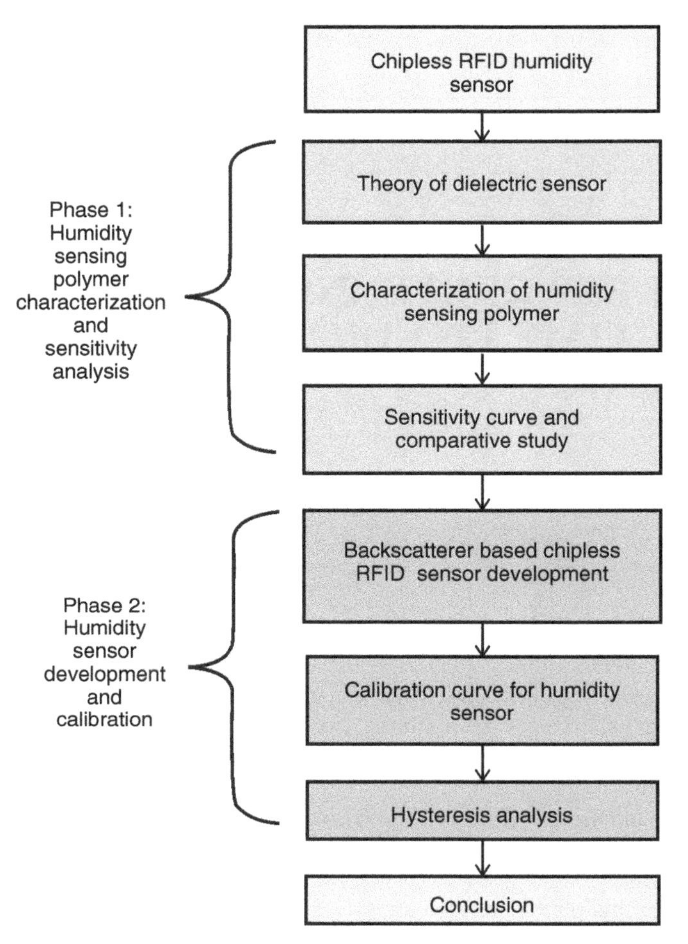

Figure 7.1 Organization of the chapter on real-time environment monitoring

gap from two ground planes on either side. In this arrangement, the top plane of the CPW line is filled with air with permittivity ε_0. The size of (i) the center strip (S), (ii) the gap (W), (iii) the height (h_1), and (iv) the relative permittivity (ε_{r1}) of the substrate determine the effective dielectric constant and characteristic impedance of the line (Figure 7.2). However, the transmission properties of a CPW line sandwiched between two dielectric substrates depend also on the height (h_2) and relative permittivity (ε_{r2}) of the superstrate material [1]. The characteristic impedance (Z_0) of a CPW line between two dielectrics is given by

$$Z_0 = \frac{30\pi}{\sqrt{\varepsilon_{\text{eff}}}} \frac{K(k_0^{'})}{K(k_0)} \tag{7.1}$$

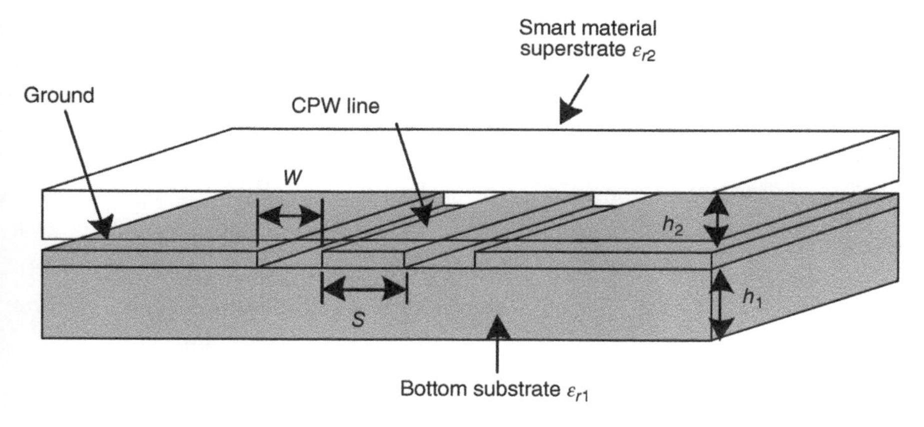

Figure 7.2 CPW line with smart material as superstrate

Here, ε_{eff} is the effective dielectric constant and $K(k_0)$ is the modulus of complete elliptic integrals

$$k_0 = \frac{S}{S + 2W} \tag{7.2}$$

$$k_0' = \sqrt{1 - k_0^2} \tag{7.3}$$

Also, the effective dielectric constant is given by

$$\varepsilon_{\text{eff}} = 1 + q_1(\varepsilon_{r1} - 1) + q_2(\varepsilon_{r2} - 1) \tag{7.4}$$

Here, q_1 and q_2 are the partial filling factor depending on the structural parameters of a CPW line [2].

From Equations 6.2 to 6.4, an important conclusion can be drawn about the sensing mechanism. For the CPW line shown in Figure 7.2, the characteristic impedance is related to the superstrate properties. Hence, a superstrate with hydrophilic/hydrophobic nature can be used to incorporate relative humidity (RH) change in the CPW line parameters. This principle is used in our proposed microwave resonator to incorporate humidity sensing. As the top dielectric changes its relative permittivity (ε_{r2}) with humidity, the resonant condition varies accordingly. The variation of resonant frequency and Q factor variation at resonance can be calibrated against the humidity for monitoring real-time environmental conditions.

7.2.2 Characterization of Humidity Sensing Polymers

In this research, two moisture-absorbing polymers have been investigated for passive humidity sensing: Kapton and PVA. In Chapter 4, a summary of humidity sensitivity for these two materials was presented. In this section, we perform RF characterization of these materials using two-port insertion loss (S_{21}) measurement.

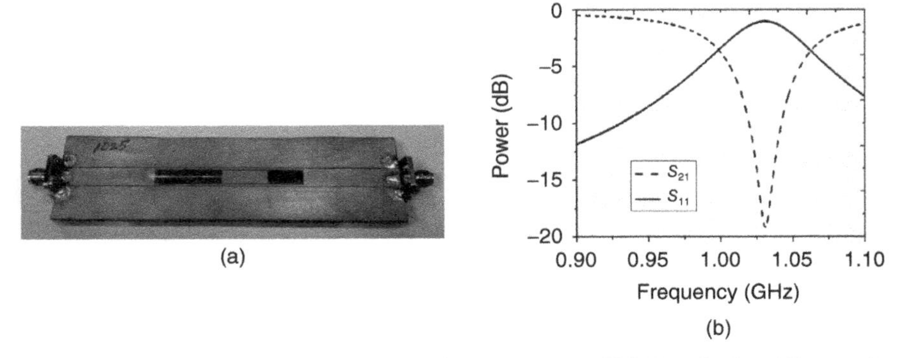

Figure 7.3 (a) Photograph of fabricated SIR resonator at 1025 MHz for humidity sensing polymer characterization (b) measured insertion loss (S_{21}) and reflection loss (S_{11}) of the SIR resonator

Here, a CPW SIR filter operating at 1025 MHz (refer to Figure 7.3(a)) is used for dielectric loading. Sensing polymers are used as the superstrate (Figure 7.2) that changes the CPW characteristic impedance Z_0. This will shift the SIR resonant frequency and phase response. By calibrating the resonant shift, we can determine the humidity sensitivity of each polymer. The measured insertion loss (S_{21}) and reflection loss (S_{11}) are shown in Figure 7.3(b).

7.2.2.1 Experimental Setup In this research, Miller Nelson humidity and temperature controller HCS-501 [3] was used to ensure a stable and regulated environment. The controller provides conditioned and controlled air and water flow for laboratory testing, where it is necessary to direct regulated air with known temperature and humidity toward a test object. It has a mass flow controller to regulate air flow through a water reservoir and heater. The outlet of moist air is controlled by a humidity and temperature sensor, which forms a feedback loop to the controller circuit. Thus, it actuates the heater in the water reservoir and the humidified air stream to attain and maintain the desired temperature and humidity conditions.

A block diagram of the experimental setup is shown in Figure 7.4(a). The humidity controller is connected to an Esky chamber through a water flow sensor. The Esky chamber has an airtight lid to ensure stable temperature and humidity. Our sensor and a DIGITECH QP-6013 data logger are placed inside the chamber. The data logger reads and stores the temperature and humidity at regular time intervals. A vector network analyzer (VNA) is used for frequency response measurement. By changing the set temperature and humidity of the Miller Nelson controller, we measured the responses of the resonators for different environment conditions. Figure 7.4(b) shows a photograph of the overall setup. In Figure 7.5, temperature and humidity records captured from the data logger during the experiment are plotted. The figure shows an almost constant temperature of about 22.5 °C throughout the total time span. However, the relative humidity (RH) changed from 50% to 90% inside the chamber.

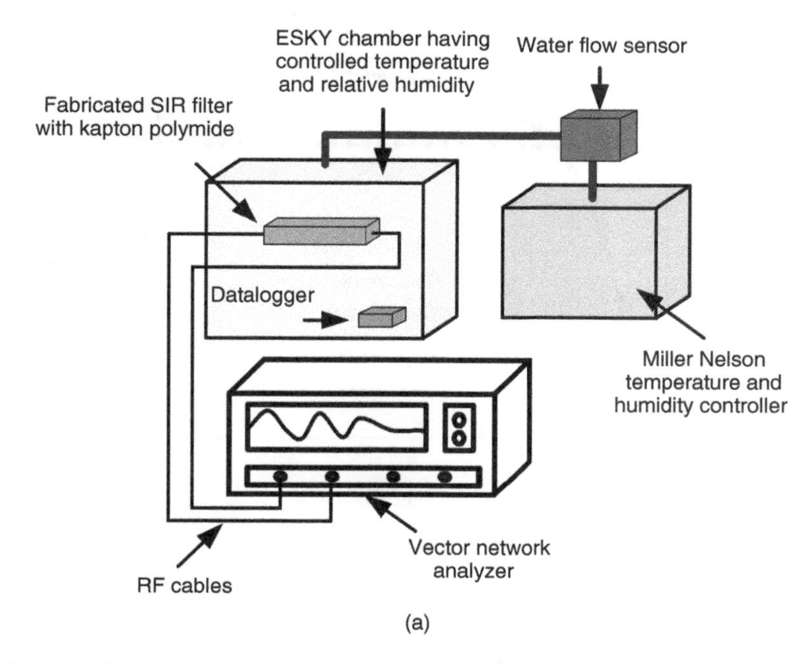

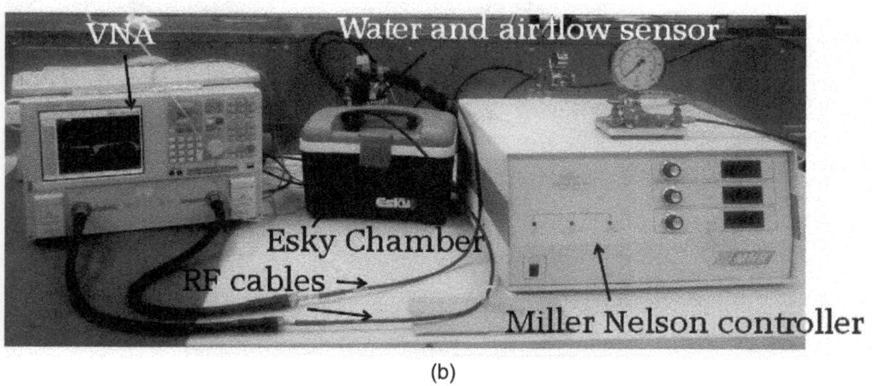

Figure 7.4 (a) Block diagram of overall experimental setup, (b) photograph of humidity sensing experiment carried out at Monash Microwave Antennas and RFID Sensors laboratory

7.2.2.2 Humidity Effect on SIR Having Kapton Superstrate To incorporate humidity sensing using Kapton, a Kapton HN [4] adhesive tape of 0.1 mm thickness was attached to the top surface of CPW SIR resonator. The resonator was then connected to the two ports of the VNA inside the Esky chamber and its S_{21} was measured for various humidity conditions (Figure 7.6). The results show a steady shift of the resonant frequency with RH. The total frequency shift is 25 MHz for an RH change of 50–90%.

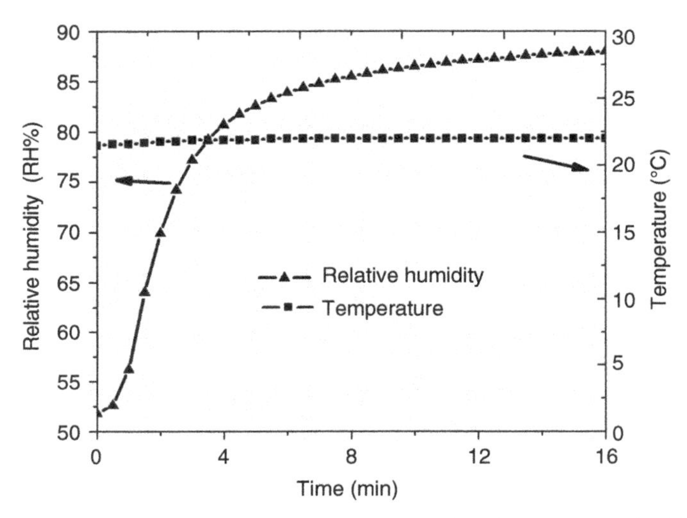

Figure 7.5 Plot of relative humidity and temperature against time measured inside the chamber using data logger

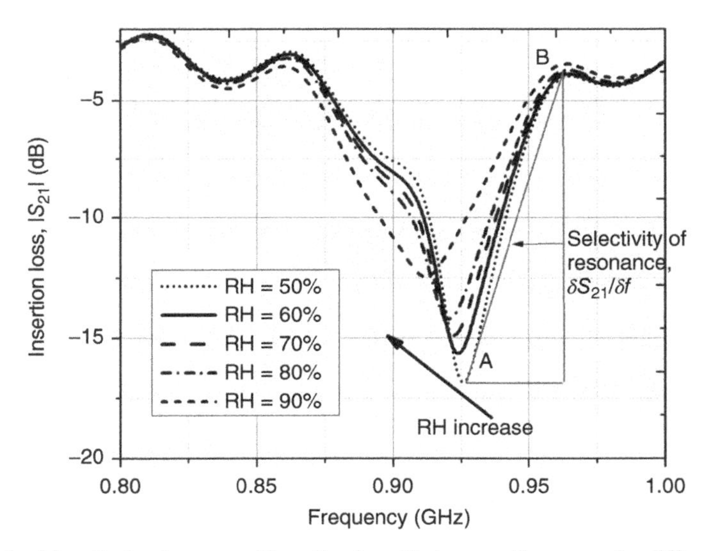

Figure 7.6 Magnitude of measured insertion loss (S_{21}) versus frequency for different humidity conditions with Kapton as superstrate

The increased moisture in air increases the permittivity of Kapton film. This enhances the total capacitance of the SIR filter on the CPW line, which is observed as resonance shift. Also, the Q factor of the resonator is changed at high humidity conditions. This is due to the imaginary permittivity $\varepsilon_r^{'}$ variation with RH. To quantify the effect of Q factor variation, a parameter "selectivity of resonance" is defined as $\delta S_{21}/\delta f$. Here, δS_{21} is the power difference at resonant frequency (A) and at the immediate maximum point (B) (refer to Figure 7.6). Also, δf is the frequency deviation between points A and B. Hence, this slope $\delta S_{21}/\delta f$ gives an indication of Q factor variation, which decreases with RH increase.

Figure 7.7 shows the measured phase (φ) of insertion loss S_{21}. The phase response corresponds to resonance shift as the slope $\delta\varphi/\delta f$ changes for different humidity conditions. Now, from the theory of filter group delay we know, group delay of a filter is defined as [5]

$$\tau_g = -\frac{\partial\varphi}{\partial\omega} = -\frac{\partial\varphi}{2\pi\partial f}$$

Here, ω is the angular frequency and $\omega = 2\pi f$. Hence, the change of phase slope $(\delta\varphi/\delta f)$ at resonance frequency is a measurement of group delay. For ease of analysis and comparative study, we have denoted group delay, $\tau_g = \delta\varphi/\delta f$ in this study. In Figure 7.8, we plot group delay against frequency. As expected, the peak value of differential phase change with respect to frequency changes from a positive maximum value to a negative minimum as humidity increases. Hence, the maximum group delay at each humidity condition corresponds to a certain humidity level.

Detailed analysis of resonance selectivity and group delay for humidity sensitivity measurement is presented in Section 6.2.3.

7.2.2.3 Humidity Effect on SIR Having PVA Superstrate To incorporate humidity sensing using PVA polymer, a thin layer of PVA 31-50000 was coated on top of

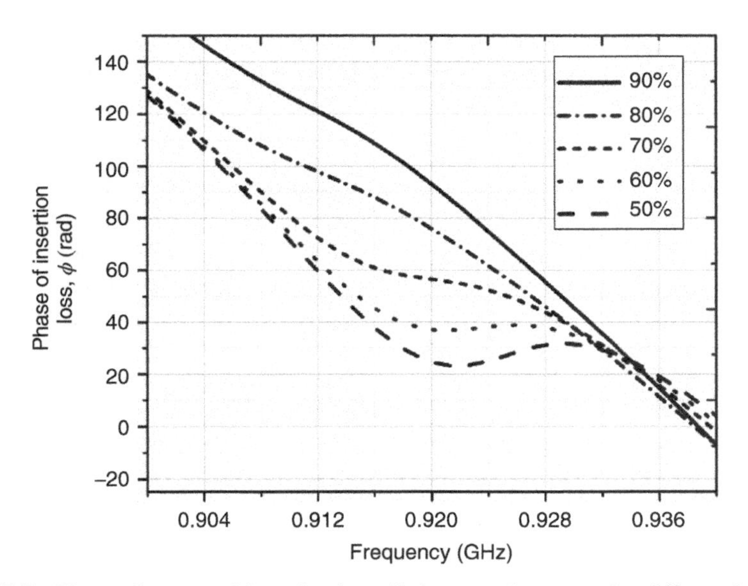

Figure 7.7 Phase of measured insertion loss (S_{21}) versus frequency for different humidity conditions with Kapton as superstrate

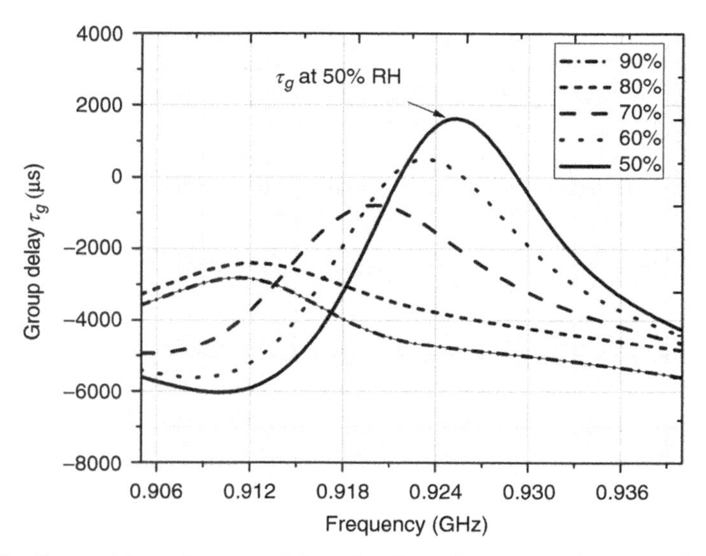

Figure 7.8 Group delay of measured insertion loss (S_{21}) versus frequency for different humidity conditions with Kapton as superstrate

the SIR resonator. The PVA polymer was acquired from Sigma-Aldrich and it was dissolved in a solution of H_2O/ethanol 3/1 for about 3 h by magnetic stirring until it became completely soluble and transparent. It was then carefully poured on top of the SIR structure using fine droplets. It was then dried to create a 0.1 mm coating of PVA on top of the SIR structure. Figure 7.9 shows the measured S_{21} in dB versus frequency for the SIR resonator. Compared to Kapton, PVA shows strong sensitivity to environmental moisture and the total frequency shift measured is 95 MHz for RH change of 50–90%. In addition, the resonance selectivity ($\delta S_{21}/\delta f$) variation is more prominent in the case of PVA compared to Kapton for equivalent change in RH.

Figure 7.10 shows the measured phase of insertion loss (S_{21}) for humidity variation. The slope of S_{21} phase corresponds to the humidity increase similar to Kapton. At high humidity (90%), the S_{21} phase has almost no phase reversal at resonant frequency. Hence, it shows a straight line in the group delay curve with a large negative value (Figure 7.11). This corresponds to the S_{21} magnitude plot, where at 90% humidity the resonance dip is largely detuned. This is to be expected, as under high humidity conditions, hygroscopic PVA absorbs water, resulting in large dielectric loss in EM propagation.

Detailed analysis of resonance selectivity and group delay for humidity sensitivity measurement is presented in Section 6.2.3.

7.2.3 Sensitivity Curve and Comparative Study

In this study, the sensitivity curve for humidity sensing polymers was analyzed taking three parameters into account: (i) resonant frequency (f_r), (ii) selectivity of resonance

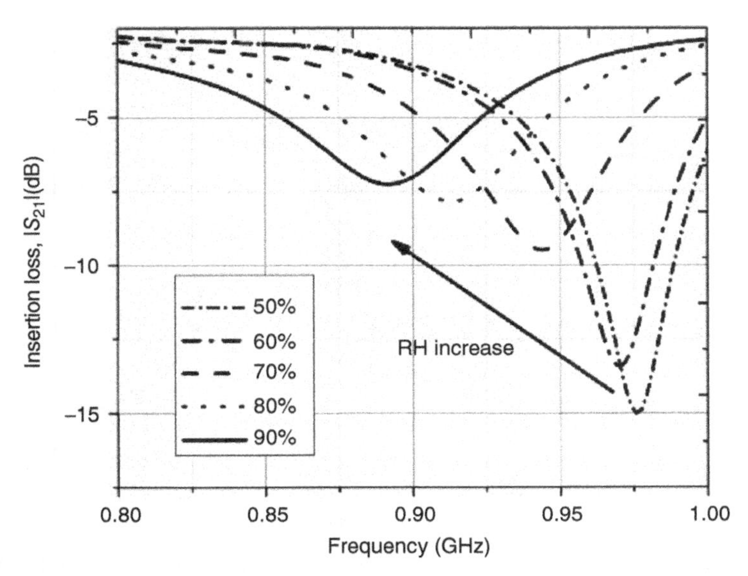

Figure 7.9 Magnitude of measured insertion loss (S_{21}) versus frequency for different humidity conditions with PVA as superstrate

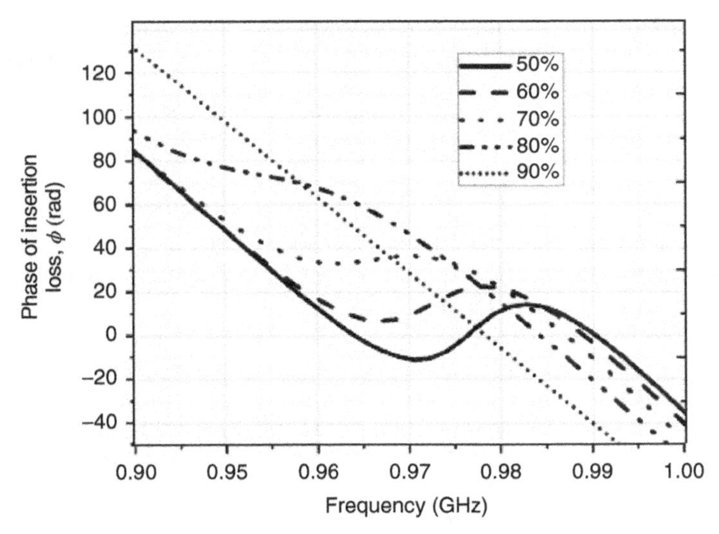

Figure 7.10 Phase of measured insertion loss (S_{21}) versus frequency for different humidity conditions with PVA as superstrate

($\delta S_{21}/\delta f$), and (iii) maximum group delay τ_{gmax}. The sensitivity curve for f_r plots the measured resonant frequency at different RHs, whereas the sensitivity curve for ($\delta S_{21}/\delta f$) and τ_{gmax} plots the normalized parametric values at different RHs. Here, the slope at minimum RH is taken as the reference for normalization.

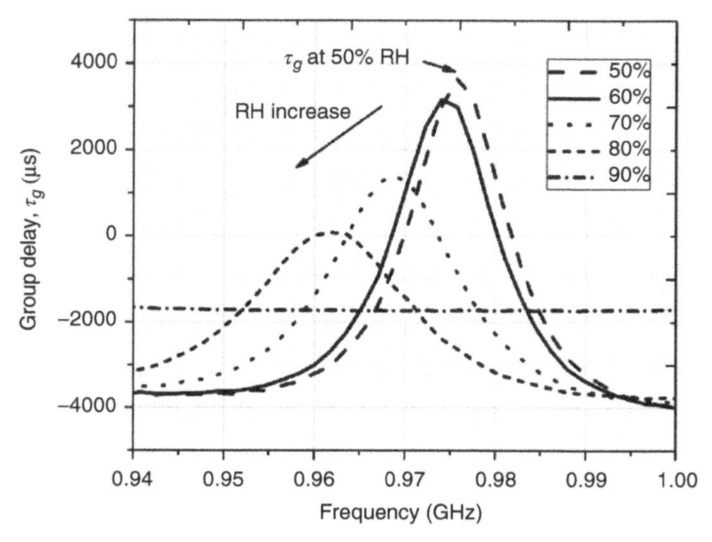

Figure 7.11 Group delay of measured insertion loss (S_{21}) versus frequency for different humidity conditions with PVA as superstrate

The sensitivity curves of these two parameters for Kapton and PVA are shown in Figures 7.12–7.14. To calculate sensitivity for 1% RH change, the following formula is used:

$$S_\gamma = \frac{|\Delta\gamma|}{|\Delta \mathrm{RH}|} = \frac{|\gamma(\mathrm{RH_{high}} - \mathrm{RH_{low}})|}{|\mathrm{RH_{high}} - \mathrm{RH_{low}}|}$$

where γ denotes the three sensing parameters mentioned above.

Table 7.1 summarizes the sensitivities S_{fr}, $S_{\delta S21/\delta f}$, *and* $S_{\tau g\max}$ for the two polymers with different resonators. The sensitivity values in Table 7.1 are calculated from the slope of the curves shown in Figures 7.12–7.14.

7.2.3.1 Kapton Polyamide Kapton polyamide is a hydrophobic organic material and operates as a capacitive humidity sensor. It has linear dielectric response when it absorbs water as the weight is proportional to RH. In Ref. [4], the linear dielectric behavior of Kapton film with RH is reported. For 100% RH variation, its dielectric constant changes by about 25%. The enhanced dielectric constant induces capacitance, which is reflected as a frequency shift in the S_{21} response. The resonant frequency variation is 0.63 MHz and the normalized maximum group delay $\tau_{g\max}$ varies by 2.25 μs for 1% RH change. Moreover, Kapton exhibits a linear increase of dissipation factor with humidity.

This implies that at higher humidity Kapton has more dielectric loss, which degrades the Q factor of the resonators. This effect was calibrated by determining the normalized resonance selectivity ($\delta S_{21}/\delta f$) at various RHs, as shown in Figure 7.13. The resonance parameter value decreases by about 28% at maximum humidity

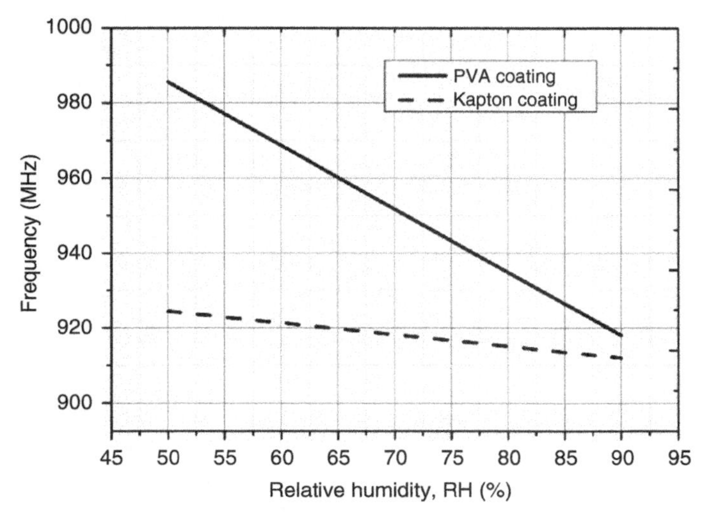

Figure 7.12 Measured sensitivity curve of the SIR resonator for resonant frequency (f_r) versus relative humidity (RH)

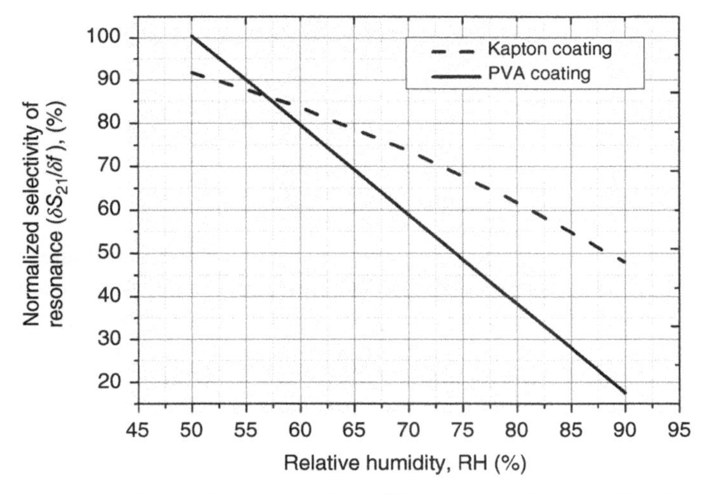

Figure 7.13 Measured sensitivity curve of the SIR resonator for normalized selectivity of resonance versus relative humidity (RH)

compared to its initial value, and $\delta S_{21}/\delta f$ for 1% RH change is 1.02%. This suggests that, at high humidity, the Q factor variation is prominent for Kapton.

7.2.3.2 ***PVA Polymer*** PVA is hydrophilic in nature and can, therefore, be used as a polyelectrolyte-based resistive sensor. In Refs. [6, 7], the microwave frequency

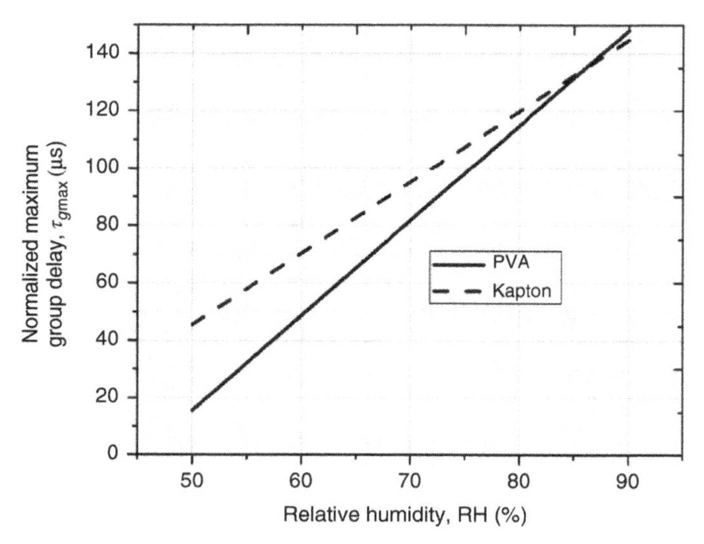

Figure 7.14 Measured sensitivity curve of the SIR resonator for normalized maximum group delay versus relative humidity (RH)

TABLE 7.1 Sensitivity Parameters for Kapton and PVA Dielectrics

Humidity sensitivity	S_{fr} (MHz/RH)	$S_{\delta S21/\delta f}$ (%/RH)	$S_{\tau gmax}$ (μs/RH)
Kapton	0.63	1.02	2.25
PVA	2.38	2.1	3.25

characteristics of PVA in aqueous solution were reported. In these studies, the dielectric behavior of water is investigated as the temperature and PVA concentration are changed. The results show that, as the PVA concentration in water increases, the real part of permittivity ε_r' decreases at any frequency (0.2–20 GHz). The sensitivity curve for resonant frequency corresponds to the increased permittivity of PVA sensor with RH. Compared to Kapton, PVA sensor exhibits higher resonant frequency shift for 1% RH change. Calculated sensitivity of frequency (S_{fr}) and normalized maximum group delay τ_{gmax} are 2.38 MHz/RH and 3.25 μs/RH. These values are higher than corresponding values measured for Kapton.

In addition, the normalized selectivity of resonance ($\delta S_{21}/\delta f$) of the PVA sensor varies significantly (refer to Figure 7.13). The results show that the slope parameter value decreases by about 50% of its initial value at maximum humidity, and the selectivity of resonance for PVA is about two times that of Kapton. This verifies that, for equivalent RH change, PVA exhibits greater dielectric loss and essentially higher sensing resolution. The change of Q factor is due to the imaginary relative permittivity (ε_r'') variation with water absorption. In Ref. [6], a study of dielectric dispersion of PVA in aqua solution was presented. It was reported that, in PVA aqua solution, the ε_r

changes due to the breaking and reforming of the hydrogen bonds of water molecules with the "–OH" groups present in the PVA chain. This causes high dielectric loss (ε_r'') as PVA absorbs water.

The variation of resonance selectivity parameter gives an indication of the Q factor change, which can be calibrated as a sensing parameter. Along with resonance frequency and group delay, this parameter can carry additional information about RH and increase sensing accuracy and reliability. From Table 7.1, it is evident that PVA shows better humidity sensitivity than Kapton polyamide shows. The measured humidity sensitivity of PVA at microwave frequency indicates its potential for integration in passive RF sensors to monitor environmental humidity. The results presented here also illustrate the superior sensitivity of PVA polymer compared to previous reported work on humidity sensors [8, 9].

In the next phase, we develop a backscatterer-based chipless RFID humidity sensor taking PVA as sensing polymer. Also, the three sensing parameters resonant frequency, selectivity of resonance, and group delay is taken into account for calibration data measurement.

7.3 PHASE 2. CHIPLESS RFID HUMIDITY SENSOR

7.3.1 Backscatterer-Based Chipless RFID Humidity Sensor

An RCS scatterer-based chipless RFID humidity sensor integrating a U-shaped slot loaded rectangular tag and the ELC resonator is used for humidity sensing. Figure 7.15 shows the general structure of chipless RFID humidity sensor. Here, a multislot patch resonator is designed to carry tag ID. Moreover, an ELC resonator has a superstrate material that changes dielectric property with humidity. This change

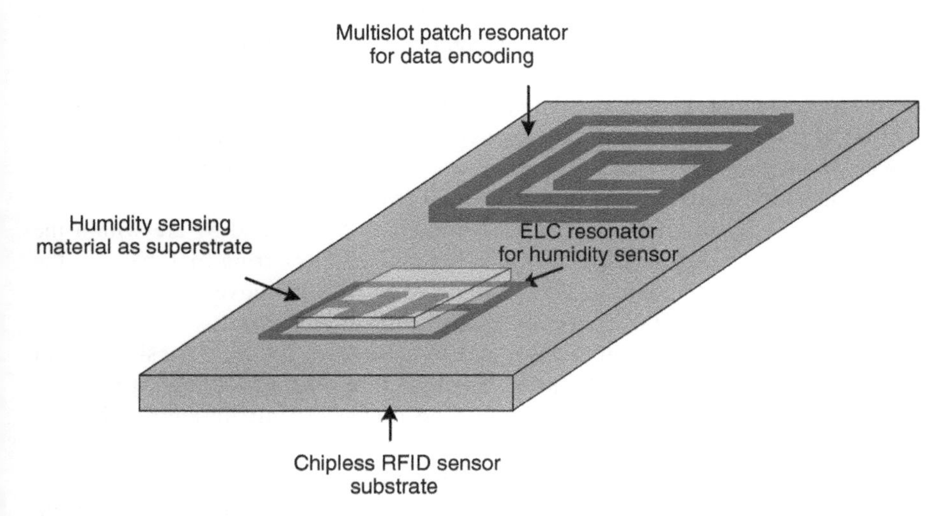

Figure 7.15 General structure of backscatterer-based chipless RFID humidity sensor

manifests into measurable resonant shift and can be calibrated for RH measurement in an unknown environment.

In this chapter, we develop a chipless RFID humidity sensor based on the design presented in Section 3.4.6. Three slots, S_1, S_2, and S_3 are designed to encode a specific data bit within 7–10 GHz depending on its frequency shifting parameters. We selected the frequency shifting parameters such that the 6-bit data ID is "111001." Moreover, the sensing ELC resonator is designed to operate between 6 and 7 GHz. It has a large functional band as the resonant frequency will shift with environmental humidity. The overall size of our 6-bit tag sensor is very compact and requires a footprint area of 15 mm × 6.8 mm.

7.3.2 Experimentation and Results

The tag was fabricated by etching copper from the Taconic substrate, and a thin layer of PVA 31-50000 was placed on top of the ELC resonator. The layer of PVA had a thickness of 0.1 mm. We used a Miller Nelson humidity and temperature controller setup as shown in Figure 7.4(b) for a controlled ambient humidity change. Figure 7.16 shows the overall measurement setup inside the box. The tag sensor was placed inside the Esky chamber for wireless measurements. Two horn antennas were connected to a VNA for frequency response measurement. The tag sensor was placed between the two antennas for transmission coefficient measurement. As the tag is one sided with no metal ground plane, it can be read using two antennas on either side.

By changing the set temperature and humidity of the Miller Nelson controller, we measured the transmission coefficient (S_{21}) of our sensor tag for different environmental conditions. During the experiment, the temperature was almost constant at around 22.5 °C. However, the relative humidity changed from 35% to 85% inside the chamber.

Figure 7.17 shows the calibrated transmission coefficient (S_{21}) versus frequency for different humidity conditions. The calibration was done using the "No tag" condition as reference. In the VNA, the "No tag" response (where the response is from the environment only) is measured and saved as "Memory data." Later, the normalized RCS response of the tag for various humidity conditions was measured and saved, which gave "Tag data"/"Memory data." Finally, the calibrated tag response was recalculated using the normalized data and "Memory data."

In Figure 7.17, the resonant frequencies of slots S_1, S_2, and S_3 are within the bands 7.9–8.1, 8.5–8.7, and 9.1–9.3 GHz, respectively. Referring to Table 3.3, the data encoded for the three slots are "11", "10", and "01," respectively. Hence, the tag sensor ID can be successfully retrieved from the measured transmission coefficients.

It is observed that although the positions of resonances for the three slots do not vary considerably, the resonant frequency of the ELC resonator is significantly shifted to a lower frequency with RH. By calibrating this frequency shift, the ambient humidity can be determined.

The changes in resonance of slots with RH increase are due to the dielectric property variation of the Taconic substrate. However, this shift is insignificant compared to the resonance shift of the ELC resonator. It was reported in Ref. [10] that Taconic

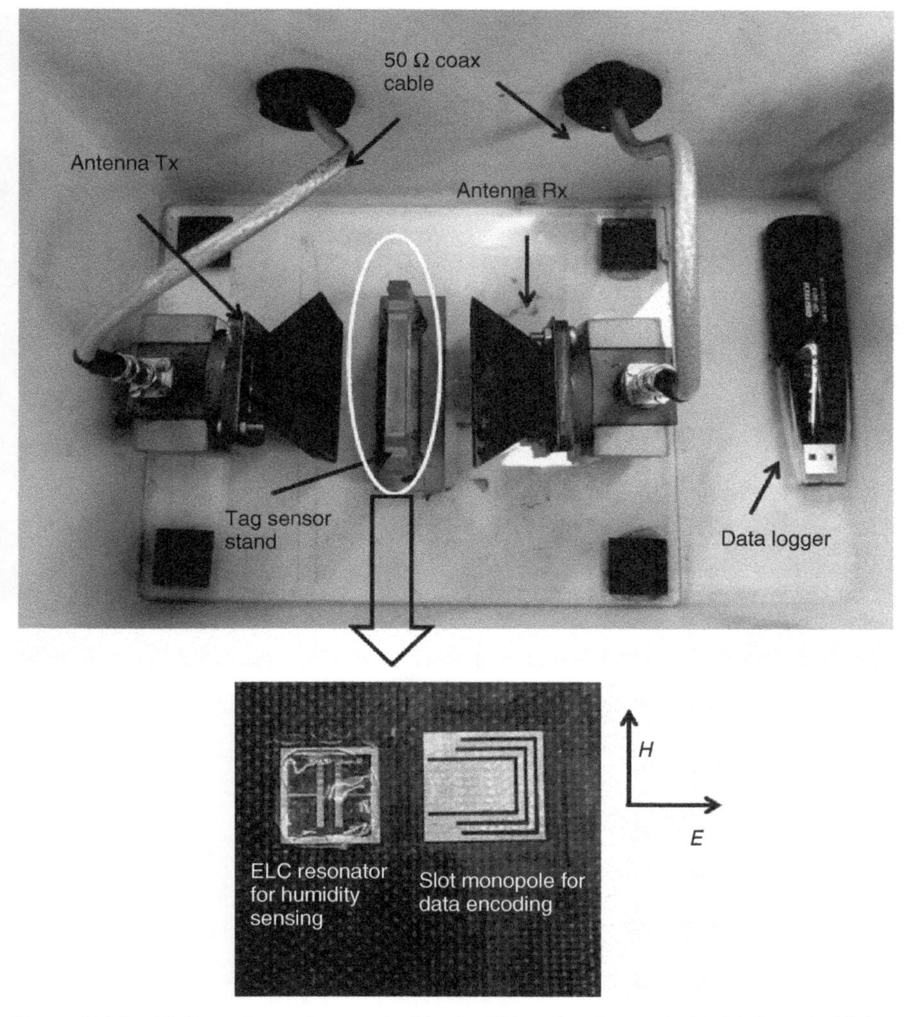

Figure 7.16 (a) Experimental setup inside the Esky chamber and (b) fabricated chipless RFID tag sensor on TLX_0 substrate

substrates have a typical relative permittivity change of 5–10% for relative humidity change from 0% to 100%. In contrast, PVA exhibits about 70% increase of its dielectric constant value for 25% increase of water content in a water–PVA hydrogel [7].

7.3.3 Calibration Curve for Humidity Sensor

To determine a calibration curve for the humidity sensor, a detailed analysis was performed with the ELC resonator with PVA coating.

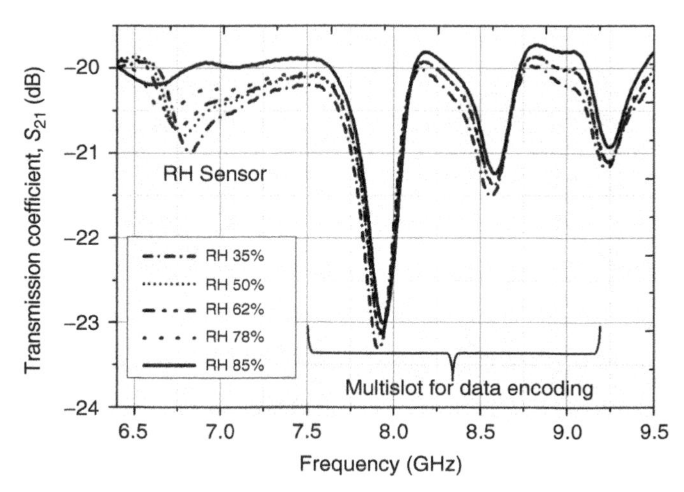

Figure 7.17 Measured transmission coefficient (calibrated) (S_{21}) versus frequency for the chipless RFID humidity sensor with PVA coating

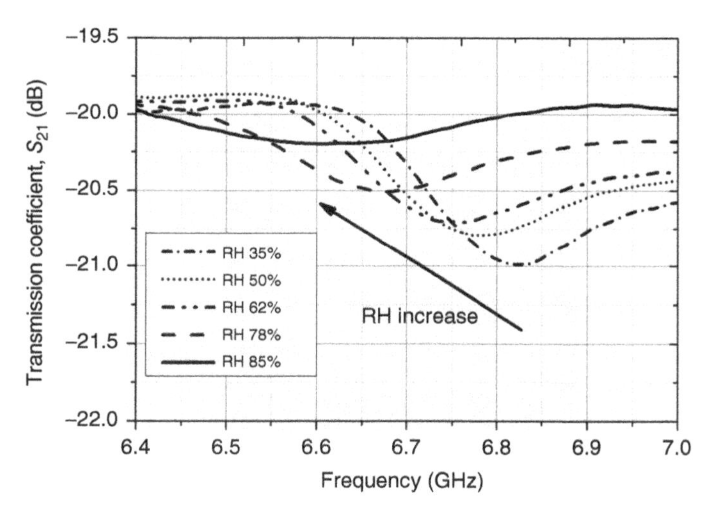

Figure 7.18 Magnitude of measured insertion loss (S_{21}) versus frequency of ELC resonator with PVA superstrate for various RHs

Figures 7.18 and 7.19 show the calibrated transmission coefficient (S_{21}) versus frequency during this experiment for the ELC resonator only. The obtained S_{21} magnitudes show a consistent shift of resonance frequency from right to left. The total frequency shift for 35–85% RH change is 320 MHz, and the S_{21} phase reversal at each resonant frequency diminishes as humidity increases. By plotting the rate of phase change with frequency in Figure 7.20, we find the maximum group delay τ_{gmax} changes from 40 μs at 35% RH to 10 μs at 85% RH.

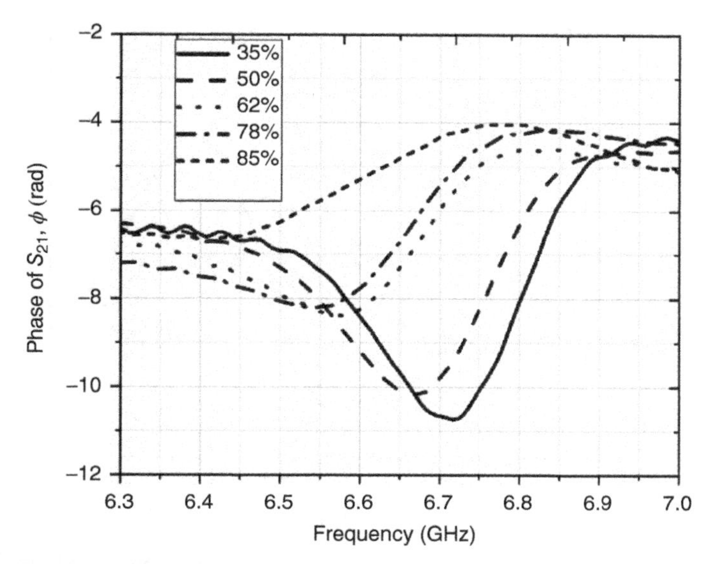

Figure 7.19 Phase of measured insertion loss (S_{21}) versus frequency of ELC resonator with PVA superstrate for various RHs

Three separate sensitivity curves for resonant frequency shift, selectivity of resonance ($\delta S_{21}/\delta f$), and maximum group delay τ_{gmax} are shown in Figures 7.21–7.23. The resonant frequency at 35% is 6.87 GHz and shifts to 6.55 GHz at 85%. This gives a 6.4 MHz frequency shift for 1% RH change. In addition, τ_{gmax} changes by 0.6 µs for 1% RH increase (Figure 7.23). Finally, the S_{21} plot in Figure 7.18 shows a steady decrease in minimum S_{21} at resonance. The minimum S_{21} change for 35% to 85% RH variation is about 1 dB. This change is due to the imaginary ε_r'' variation with water absorption, as discussed in the previous section. The three sensitivity curves can be used as calibration data for measuring RH of an unknown environment.

The three sensitivity curves give superior RH measurement compared to single sensing parameter-based chipless RFID sensors presented in Refs. [11, 12]. In these studies, only the phase information of retransmitted EM signal is calibrated for sensing environment parameters. Therefore, the accuracy of these sensors is prone to ambient conditions. Moreover, the dynamic range of sensor data is reduced. In this regard, our chipless RFID sensor with three sensing parameters gives enhanced degrees of freedom and is less affected by the environmental noise.

7.3.4 Hysteresis Analysis

One of the challenges of passive smart material-based environment sensors is their slow response time, which manifests in delay in electrical/chemical property changes as physical parameters vary. In particular, after reaching maximum humidity most hydrophilic polymers require a large time to dehydrate. For many materials, an

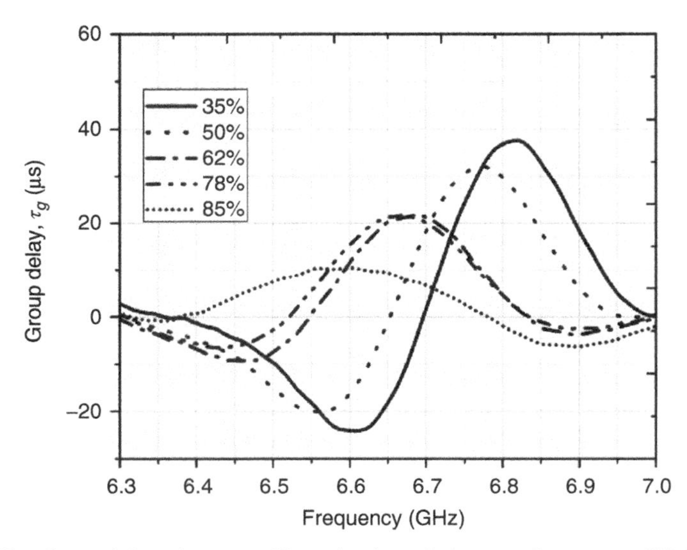

Figure 7.20 Group delay of measured insertion loss (S_{21}) versus frequency of ELC resonator with PVA superstrate for various RHs

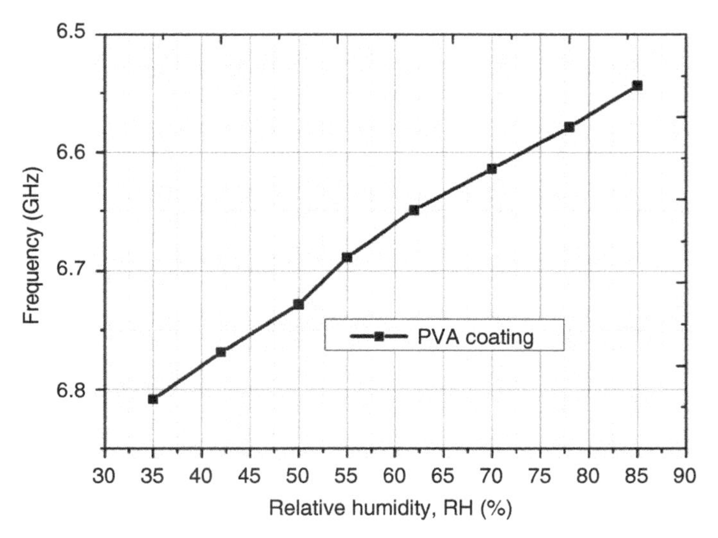

Figure 7.21 Measured sensitivity curve of the ELC resonator for resonant frequency (f_r) versus relative humidity (RH)

external dehydration process (i.e., heating, air flow) is carried out to separate water molecules from the polymer chain. In addition, traditional hygroscopic materials are limited by the hysteresis effect. These materials do not show similar dielectric property changes during water absorption and desorption.

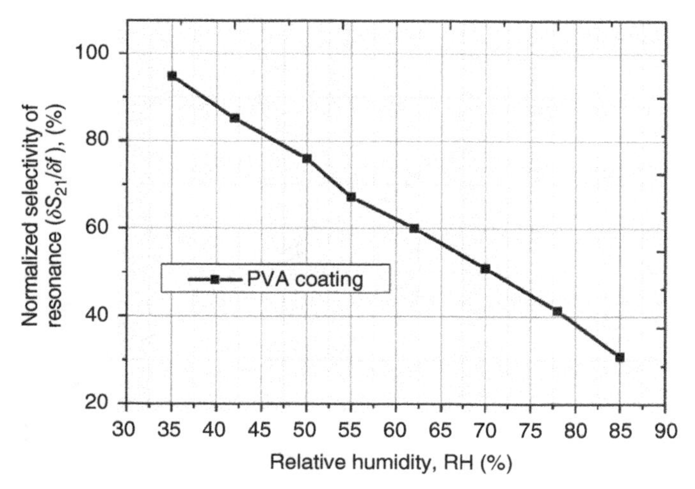

Figure 7.22 Measured sensitivity curve of the ELC resonator for normalized selectivity of resonance versus relative humidity (RH)

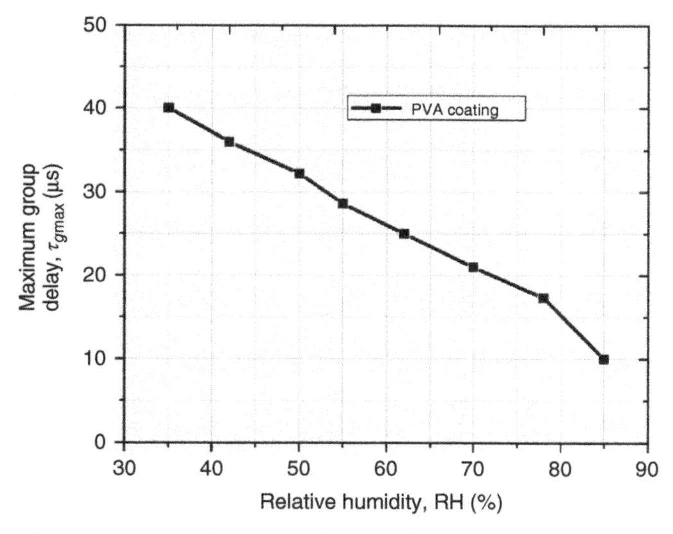

Figure 7.23 Measured sensitivity curve of the ELC resonator for maximum group delay versus relative humidity (RH)

One of the benefits of PVA is that it has low hysteresis and a fast response time [13]. To investigate the hysteresis effect and response time, we performed an experiment using the Miller Nelson humidifier. In this experiment, the humidity inside the Esky chamber was increased from 35%to 85% using the regular procedure discussed earlier. When the inside RH reached 85%, controlled dry air is blown in

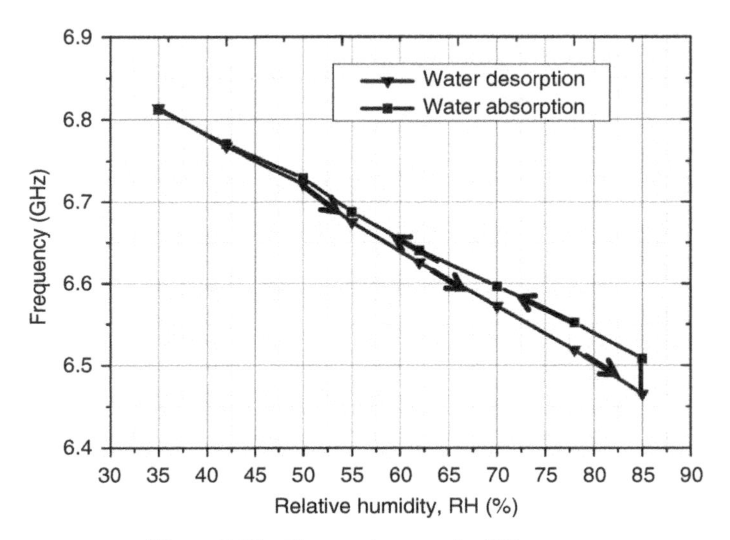

Figure 7.24 Hysteresis curve for RH sensor

the Esky chamber to reduce the humidity at the initial state. Figure 7.24 shows the measured resonant frequencies of the ELC resonator for water absorption and desorption. The total process took only 3 h and the resonant frequency showed minimum hysteresis at high humidity. However, at low humidity, the frequency response was identical.

7.4 CONCLUSION

This chapter presents a novel chipless RFID humidity sensor for real-time environment monitoring. In the first phase of this chapter, a comparative study of dielectric sensitivity between Kapton and PVA was reported. The results show that PVA has pronounced humidity sensing capability in the UHF range, and our sensor can detect change in humidity both in the magnitude and phase response of backscattered RCS. We also explored three potential humidity sensing parameters to calibrate RCS magnitude and phase data into measurable RH. In the second phase, an RCS scatterer-based chipless sensor was developed and rigorous humidity sensing experiments were carried out. Results show three prominent calibration curves can carry humidity information for environment monitoring.

Next chapter presents another novel application of chipless RFID sensor considered in this book. We aim to develop a chipless RFID memory sensor for event detection. The final goal is to develop a chipless RFID tag having both real-time humidity and temperature threshold sensor.

REFERENCES

1. R. N. Simons, *Coplanar Waveguide Circuits, Components, and Systems*: Wiley, 2001.
2. S. Gevorgian, L. J. P. Linner, and E. L. Kollberg, "CAD Models for Shielded Multilayered CPW," *IEEE Transactions on Microwave Theory and Techniques*, vol. 43, pp. 772–779, 1995.
3. Miller-Nelson Test Atmosphere Generators. Available: http://www.assaytech.us/mr_instr.htm
4. Kapton HN polyimide film datasheet. Available: http://www2.dupont.com/Kapton/en_US/.
5. J.-S. G. Hong and M. J. Lancaster, *Microstrip Filters for RF/Microwave Applications*, 2nd ed.: Wiley-Interscience, 2011.
6. R. J. Sengwa and K. Kaur, "Dielectric Dispersion Studies of Poly(vinyl alcohol) in Aqueous Solutions," *Polymer International*, vol. 49, pp. 1314–1320, 2000.
7. Y. T. Chen and H. L. Kao, "Humidity Sensors Made on Polyvinyl-Alcohol Film Coated Saw Devices," *Electronics Letters*, vol. 42, pp. 948–949, 2006.
8. M. Penza and G. Cassano, "Relative humidity sensing by PVA-coated dual resonator SAW oscillator," *Sensors and Actuators B: Chemical*, vol. 68, pp. 300–306, 2000.
9. E. McGibney, J. Barton, L. Floyd, P. Tassie, and J. Barrett, "The High Frequency Electrical Properties of Interconnects on a Flexible Polyimide Substrate Including the Effects of Humidity," *IEEE Transactions on Components, Packaging and Manufacturing Technology*, vol. 1, pp. 4–15, 2011.
10. Z. Abbas, Y. K. Yeow, K. Khalid, and M. Z. A. Rahman, "Improved Dielectric Model for Polyvinyl Alcohol-Water Hydrogel at Microwave Frequencies," *American Journal of Applied Sciences*, vol. 7, pp. 270–276, 2010.
11. S. Shrestha, M. Balachandran, M. Agarwal, V. V. Phoha, and K. Varahramyan, "A Chipless RFID Sensor System for Cyber Centric Monitoring Applications," *IEEE Transactions on Microwave Theory and Techniques*, vol. 57, pp. 1303–1309, 2009.
12. C. Mandel, H. Maune, M. Maasch, M. Sazegar, M. Schüssler, and R. Jakoby, "Passive wireless temperature sensing with BST-based chipless transponder," in *German Microwave Conference (GeMIC)*, 14–16 March 2011, pp. 1–4.
13. M. Penza and V. I. Anisimkin, "Surface Acoustic Wave Humidity Sensor Using Polyvinyl-Alcohol Film," *Sensors and Actuators A: Physical*, vol. 76, pp. 162–166, 1999.

8

CHIPLESS RFID TEMPERATURE MEMORY AND MULTIPARAMETER SENSOR

8.1 INTRODUCTION

This chapter presents a chipless radio-frequency identification (RFID) memory sensor for temperature threshold detection. In supply chain management, it is frequently desirable to detect a certain event rather than continuous monitoring. For example, the violation or increase of a certain temperature threshold value can be of great significance during the transport and storage of products, chemicals, pharmaceuticals, and explosive materials. With these applications in mind, a novel chipless RFID temperature threshold sensor tag is presented for event detection. The sensor is a passive RCS resonator-based tag that embeds the frequency signature within the backscattered signal. However, a particular resonator in the tag is modified by using a smart memory material to trigger at the violation of a temperature threshold. Furthermore, the memory sensor is integrated with a real-time humidity sensor to develop a multiparameter sensor. The chipless sensor can independently carry humidity data and information on temperature threshold violation together with tag ID. This low-cost tag sensor can be used to monitor individual items for critical temperature violation.

A number of research approaches can be found on chipless RFID tag integrating temperature sensors [1, 2]. In Ref. [3], a temperature threshold sensor using a UHF RFID tag is presented based on shape memory polymer for actuation. However, the overall tag size is bulky and it is unsuitable for robust item-level tagging. A magnetic material-based temperature threshold sensor principle is reported in Ref. [4]. However, a prototype temperature sensor based on this method has not been reported in the chipless RFID domain. Hence, a fully printable, one-sided chipless

Chipless RFID Sensors, First Edition. Nemai Chandra Karmakar, Emran Md Amin and Jhantu Kumar Saha.

RFID tag sensor for detecting critical temperature violation has not been reported to date. Also, multiparameter sensing is a novel feature of the chipless RFID sensor platform proposed in this book. This is due to the independent operation of backscatterers in the proposed sensor. Reported works on chipless RFID sensor [2, 5] lack the ability to sense more than one physical parameter. Multiple parameter sensing within these chipless sensors will require changing the principle of operation or overall tag configuration.

This chapter presents chipless RFID memory sensor and multiparameter sensor development in two phases. Figure 8.1 shows the overall structure of this chapter.

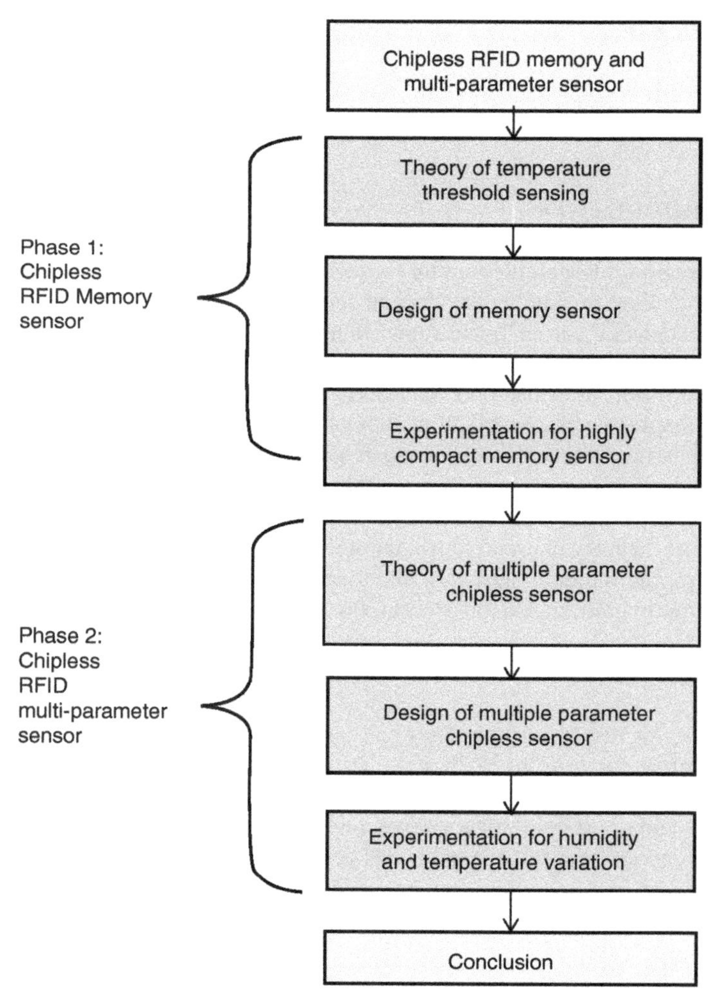

Figure 8.1 Organization of the chapter on chipless RFID memory and multiparameter sensor

The organization of this chapter is as follows:

- In Phase 1, a chipless RFID memory sensor is presented. Firstly, the theory of temperature threshold detection and memory material characteristics are discussed. Next design of a single ELC resonator loaded with memory material is presented. Finally, detailed experimentation is carried out for a highly compact chipless RFID memory sensor.
- In Phase 2, a chipless RFID multiparameter sensor is presented. Here, detailed theory, design, and experimentation are undertaken for multiparameter sensor development. Finally, the practical challenges of chipless tag with multiple parameter sensing are discussed.

8.2 PHASE 1: CHIPLESS RFID MEMORY SENSOR

8.2.1 Theory

The principle of our temperature sensor with memory effect is explained in Figure 8.2. Here, we utilize a dielectric material with irreversible dielectric change once a critical temperature is reached. Traditional temperature-sensitive polymers have a reversible dielectric property change with temperature (Figure 8.2(a)). Hence, a microwave resonator of this material would have resonant frequency shift during both endothermic and exothermic processes. However, there are certain dielectric materials that permanently change dielectric properties when a critical temperature is attained. Once the critical temperature is reached, these materials show a constant dielectric property, even if the temperature goes below the critical temperature.

These smart materials are suitable for temperature threshold sensors to give a memory effect (Figure 8.2(b)). A microwave resonator or RCS scatterer with such a memory material would exhibit a permanent resonant frequency shift upon temperature threshold violation. Thus, a permanent memory sensor is designed.

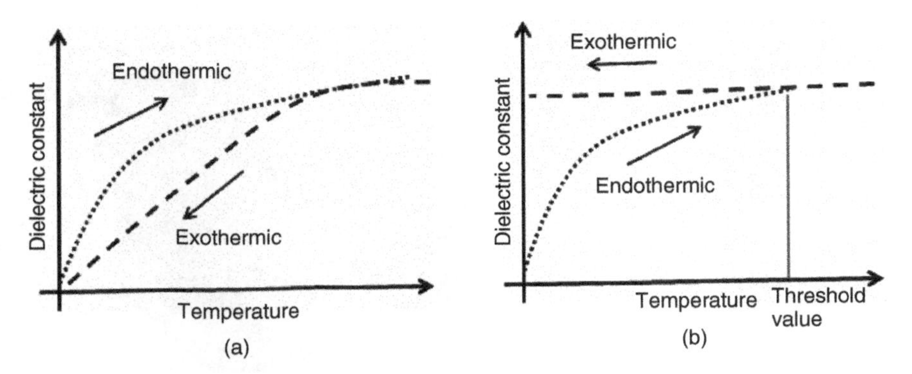

Figure 8.2 Dielectric behavior of (a) reversible temperature-sensing material and (b) irreversible temperature-sensing material

Phenanthrene is a sublimate material from the polycyclic hydrocarbon group with a transition temperature $T_c = 72\,°C$. Details of the dielectric behavior of phenanthrene are presented in Chapter 4. Here, it is used as a dielectric loading to our chipless RFID tag to realize a temperature threshold sensor. Section 8.2.2 presents the design of memory sensor using ELC resonator.

8.2.2 Design of Memory Sensor with ELC Resonator

8.2.2.1 Phenanthrene Material Preparation A disadvantage of phenanthrene is its insolubility in water. This creates difficulty in preparing a solution of phenanthrene that can be used as a superstrate layer for sensing. In this research, we used tetrahydrofuran (THF) to make the phenanthrene solution, as phenanthrene is easily soluble in THF. Since THF being a noxious and hazardous material, we undertook the experiment at Monash University Material Engineering Laboratory. The whole experiment was done inside a fume hood to ensure proper ventilation.

To prepare 1 mole of phenanthrene and THF solution, a chemical container was filled with 1.78 gm (exactly measured on a micro balance) of phenanthrene powder and 200 ml of THF. Then, it was heated at around 60 °C and magnetically stirred for about 10–15 min. When phenanthrene was completely dissolved in THF, the solution was used for the experiments. Figure 8.3 shows a photograph of the experimental setup for the preparation of phenanthrene: THF solution.

8.2.2.2 ELC Resonator Loaded with Phenanthrene The solution was poured on top of an ELC resonator by using a fine droplet and masking technique. Figure 8.4(a)

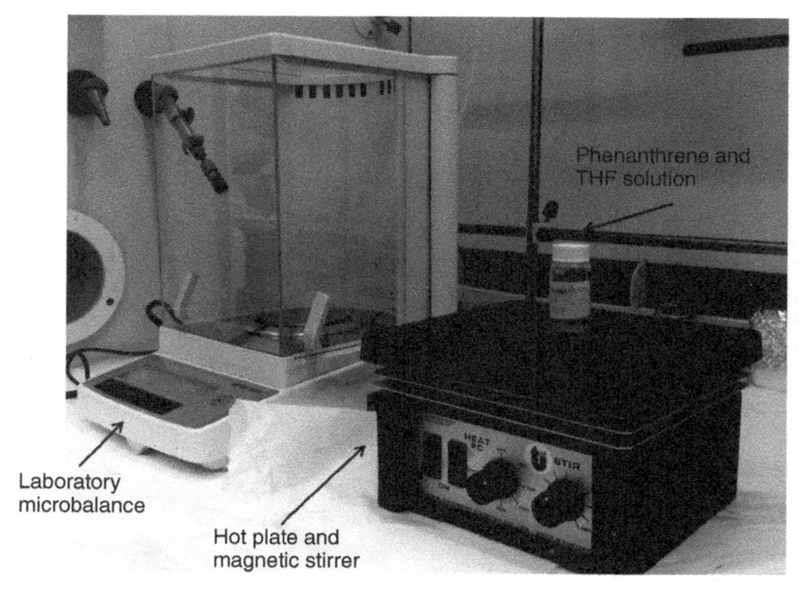

Figure 8.3 Experimental setup in Monash Material Engineering Department Laboratory for preparation of 1 mole phenanthrene: THF solution

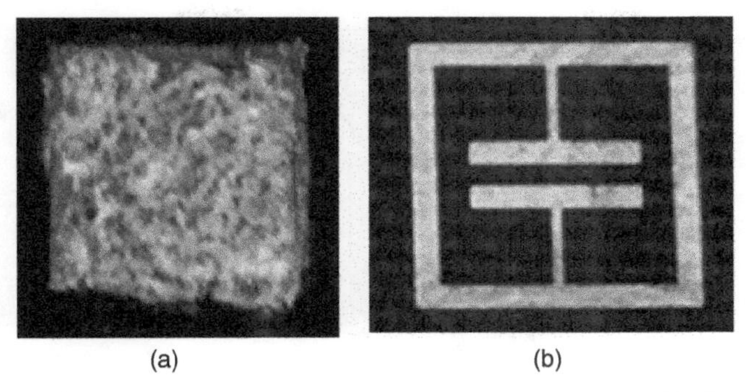

Figure 8.4 (a) Photograph of phenanthrene-loaded ELC resonator. (b) ELC resonator after sublimation

shows a photograph of ELC resonator loaded with phenanthrene. The resonator was then heated at a low temperature (around 40 °C) to evaporate THF and a crystal of phenanthrene formed on the ELC resonator. The thickness of the phenanthrene film was 0.2 mm. Figure 8.4(b) shows a photograph of the ELC resonator without phenanthrene loading.

Next, a highly compact ELC coupled chipless RFID tag sensor presented in Figure 3.43 is used for chipless RFID memory sensor. Detailed experimentation and results of temperature threshold sensing is presented in the following section.

8.2.3 Experimentation for Chipless RFID Memory Sensor

8.2.3.1 Experimentation with Phenanthrene-Loaded ELC Resonator The experiments were performed in an enclosure where the temperature could be controlled. The transmission coefficient (S_{21}) of the ELC resonator with phenanthrene loading was measured at different temperature conditions (65, 75, 85, and 95 °C), as shown in Figure 8.5. At each temperature, the ELC resonator was kept for 90 min, and the resonant frequency of the ELC resonator was plotted against measurement time. At room temperature, the resonant frequency was constant for all temperatures (6.45 GHz). However, for 65 and 75 °C, the resonant frequency shift was negligible for the whole period of time. In contrast, at 85 and 95 °C, there was a drastic shift of about 320 MHz. This corresponds to a temperature threshold violation at around 80 °C. As phenanthrene is sublimated at this temperature, the resonant shift remains permanent once triggered.

Figure 8.6 shows the transmission coefficient (S_{21}) magnitude of the ELC resonator measured at 30-min intervals, while the temperature is at a constant 85 °C. The figure shows, at the initial condition (0 min), the resonant frequency is 6.45 GHz, which shifts gradually to 6.77 GHz after 90 min. Both the S_{21} magnitude and phase correspond to the frequency shift (Figure 8.7). This permanent frequency shift is a mark of temperature threshold violation.

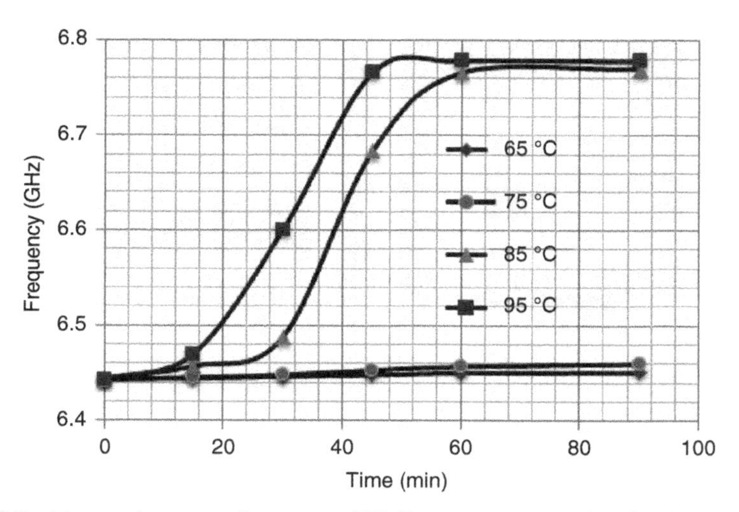

Figure 8.5 Measured resonant frequency of ELC resonator versus time for the set temperatures 65, 75, 85, and 95 °C

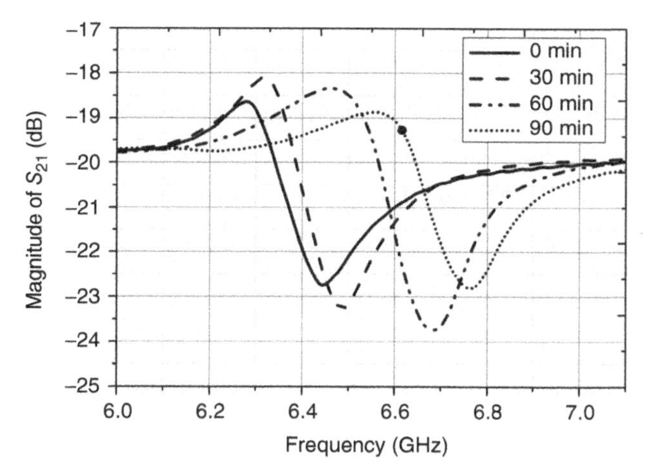

Figure 8.6 Measured magnitude of transmission coefficient (S_{21}) of ELC resonator for different times at 85 °C

8.2.3.2 Experimentation with ELC-Coupled Chipless RFID Sensor A highly compact ELC coupled multislot resonator-based chipless RFID tag sensor is shown in Figure 3.43. Here, the tag sensor was used to show temperature threshold sensing. Figure 8.8 (inset) shows a photograph of the sensor. Here, the ELC is loaded with phenanthrene material to perform temperature threshold sensing.

The temperature of the tag sensor was increased from room temperature to the critical temperature T_c and kept for a certain time span. Afterward, the tag sensor was kept at the ambient condition. Figure 8.8 shows the measured transmission

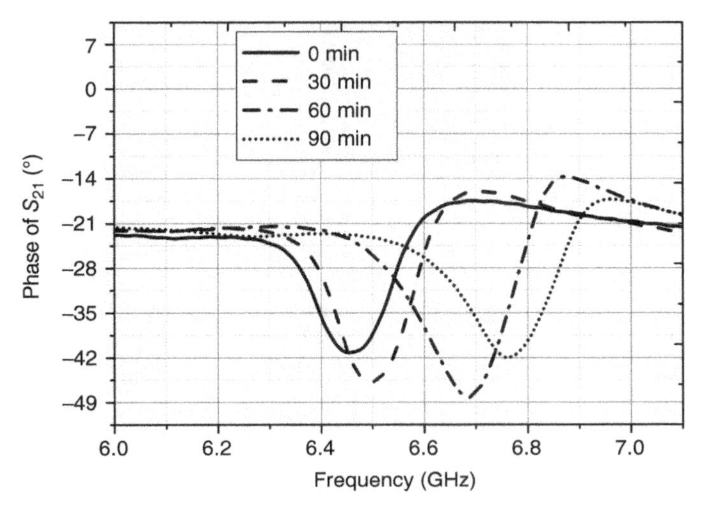

Figure 8.7 Measured phase of transmission coefficient (S_{21}) of ELC resonator for different times at 85 °C

coefficient of the tag sensor before and after the critical temperature was achieved. As expected, the resonances of the slot resonators are not greatly affected by the temperature increase. However, the ELC resonator has about a 270 MHz shift of frequency for the temperature threshold violation.

The minor shifts of the slot resonators are due to the effect of temperature on the sheet resistance of copper and the dielectric constant change of the Taconic substrate. With temperature change, copper has increased the sheet resistance. However, the temperature coefficient of copper is only 0.003862 K^{-1} and the change is not pronounced. On the other hand, Taconic substrate exhibits a 2.5% increase of ε_r over the temperature range of −60 to 200 °C [6]. Hence, in the S_{21} spectrum, we see a slight left shift of the resonant frequency of slot resonators as the temperature is above the critical temperature. However, the frequency shift of slot resonators is insignificant compared to the permanent frequency shift of the ELC resonator.

8.2.3.3 Discussion From the above experiments, we observed that the sensor triggers only in the event that the temperature exceeds a critical temperature (T_c) for a definite time referred as "Sublimation time." Transition from one phase to another is not instantaneous and requires a certain time span to reform atomic structures. This time is reflected in the sensor response as the minimum time to trigger from one state to another. The sublimation time depends on the dielectric loading and ambient temperature. Essentially, a particular mole of phenanthrene requires a certain enthalpy change ΔH to reach triple point for phase transition given by $\Delta H = mC_p\Delta T$. Here, m and C_p are the mass and specific heat capacity of a substance, respectively. For various amounts of substance, the required enthalpy change to sublimate is not constant. Therefore, samples with different phenanthrene layers show varying sublimation times. In addition, the measured critical temperature (T_c) is higher than the

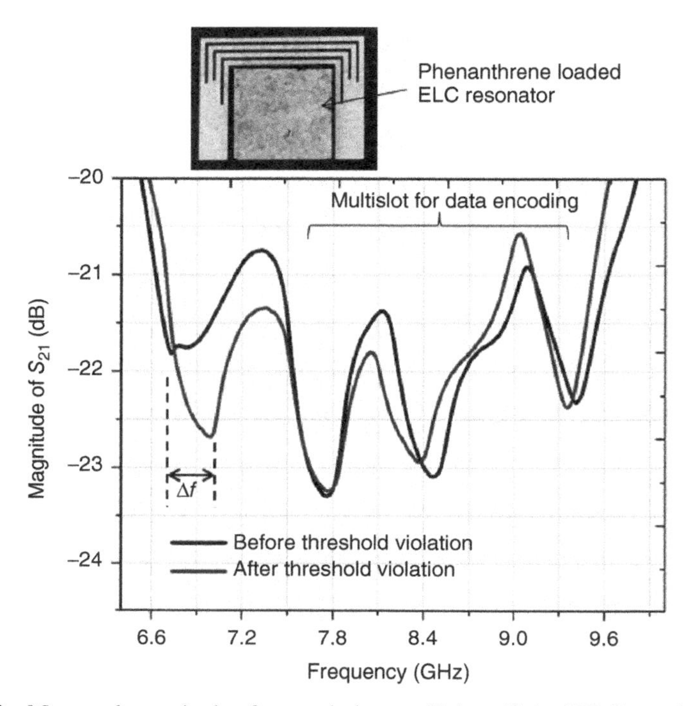

Figure 8.8 Measured magnitude of transmission coefficient (S_{21}) of ELC-coupled chipless RFID tag (inset photo) for temperature threshold sensing

theoretical transition temperature of phenanthrene due to the excess heat required for decrystallization.

The above result demonstrates that an ELC resonator loaded with phenanthrene operates as a memory sensor for a particular temperature violation. Phenanthrene is chosen as an example of the polycyclic hydrocarbon-based sublimate materials group. Similar chemical composite materials with different transition temperatures T_c can be used to realize threshold sensing. For instance, Naphthalene has a transition temperature of 25.15 °C. Therefore, it can be used for many real-world applications, where the critical temperature is near room temperature [7].

8.3 PHASE 2: CHIPLESS RFID MULTIPARAMETER SENSOR

8.3.1 Theory

In this section, the multiple parameter sensing feature of our chipless RFID sensor is presented. This is a novel attribute of frequency modulation-based chipless sensors and has been reported in an invention disclosure [8]. Multiple parameter sensing enables a single chipless tag with a distinct ID to sense a number of physical parameters (i.e., temperature, humidity, crack, strain, and pH) independently. Figure 8.9

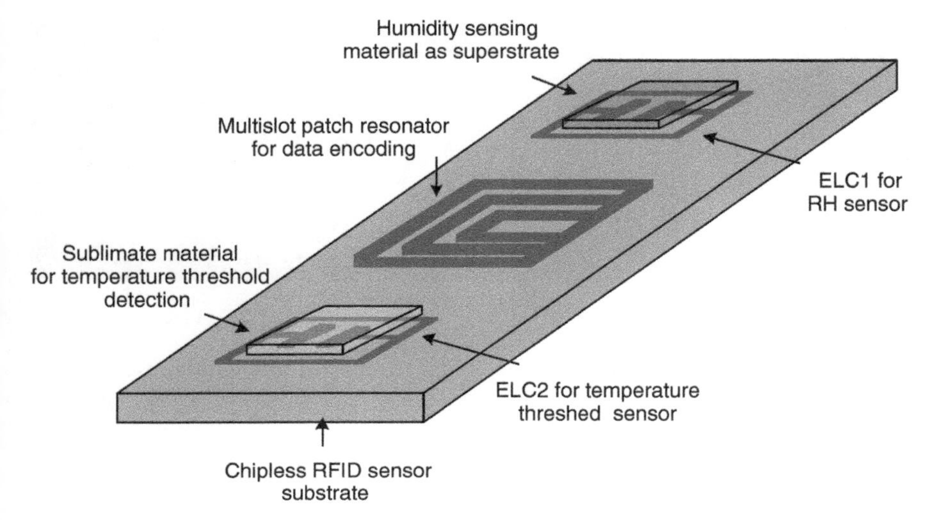

Figure 8.9 Illustration of a backscatterer-based chipless RFID multiparameter sensor

shows a backscatterer-based chipless RFID multiparameter sensor. Here, two ELC resonators ELC1 and ELC2 are dedicated for RH and temperature sensing, respectively. Also, multislot patch resonator carries tag ID.

A number of real-world applications require monitoring two or more physical parameters to accurately determine the status of an object. Moreover, to analyze the cross-relation trend between multiple parameters, it is advantageous to have multiple sensing features in a single tag. Finally, multiple sensing chipless technology assurances a minimal cost per sensor as it does not require multiple IC or chip sets for data encoding and sensing. Therefore, this feature provides potential commercial viability in low-cost item tagging and sensing.

8.3.2 Design

Here, a chipless RFID tag with both humidity sensing and temperature threshold detection is presented. The tag sensor has three independent RCS scatterers to carry (i) tag ID, (ii) real-time RH data, and (iii) information on temperature threshold violation. However, this tag sensor can be extended for N number of physical parameter sensing applications.

In Chapter 3, a chipless tag for multiple parameter sensing was presented. As shown in Figure 8.10, this tag has a multislot resonator for tag ID and two spatially separated ELC resonators for sensing. Figure 8.11 shows the measured transmission coefficient (S_{21}) of the sensor, where ELC resonators are operating from 6–7 GHz (ELC1) and 7–8 GHz (ELC2). These are dedicated for RH sensing (ELC1) and temperature sensing (ELC2), respectively. Also, three slot resonators are operating between 8 and 11 GHz for data encoding. As expected, the resonance for data ID and sensing can be clearly identified from the S_{21} plot.

Figure 8.10 Photograph of humidity and temperature sensing tag sensor

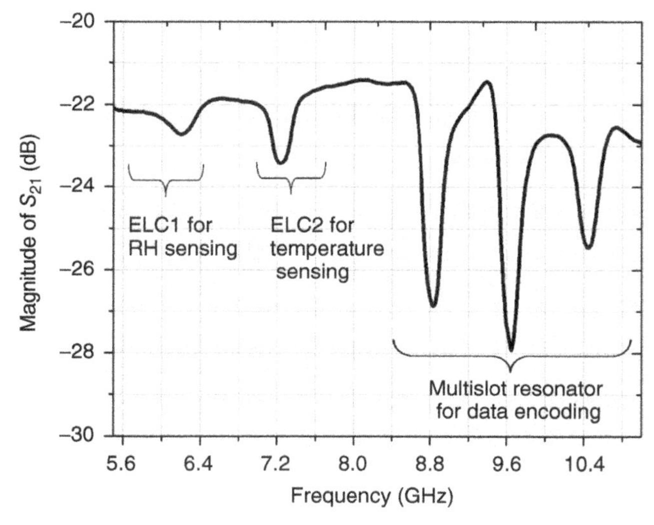

Figure 8.11 Measured magnitude of transmission coefficient (S_{21}) of multiple parameter chipless RFID tag

To incorporate the sensing mechanism, the tag sensor shown in Figure 8.10 was loaded with appropriate smart materials. ELC1 was coated with PVA, as described in Chapter 6. Similarly, ELC2 was coated with phenanthrene (0.2 mm thick) for temperature threshold detection.

8.3.3 Experimentation for Multiple Parameter Sensing

Experimentation on humidity sensing and temperature threshold violation was performed in two steps. In the first step, humidity variation before temperature threshold violation was investigated. Only ELC1 was expected to change according to humidity, whereas ELC2 would retain its resonance. In the second step, we examined humidity sensing after the temperature exceeded the threshold. Therefore, both ELC1 and ELC2 were expected to shift according to the assigned physical parameter.

Step 1: Humidity Sensing before Temperature Threshold Violation In this experiment, the tag sensor was placed inside the Esky chamber at room temperature and the ambient humidity of the chamber was varied from 35% to 70%. The transmission coefficient (S_{21}) of the tag sensor was recorded for humidity variation. As shown in Figure 8.12, the resonance of the multislot resonators was not affected by humidity variation, and they retained the initial position for data encoding. The sensing resonances are shown in Figure 8.13. The figure shows that for a humidity variation of 35% the total frequency shift is 200 MHz, and the minimum S_{21} at resonance ($S_{21\text{min}}$) drops relative to the humidity increase. ELC2 does not exhibit any frequency variation. The resonance of ELC2 confirms that the temperature has not exceeded the threshold temperature of 80 °C.

Step 2: Humidity Sensing after Temperature Threshold Violation In this experiment, the temperature was increased beyond 80 °C to trigger ELC2, and the ambient humidity was changed from 35% to 70% to verify the operation of ELC1. Figure 8.14 shows the measured S_{21} magnitude results for the RH and temperature sensing resonators ELC1 and ELC2, respectively. Firstly, ELC2 confirms the temperature threshold violation by shifting 250 MHz. This indicates sublimation of the phenanthrene layer on top of ELC2. Also, ELC1 has resonant frequency shift due to water absorption of PVA. However, the resonance shift and minimum power variation for ELC1 are lower than the results shown in Figure 8.13. This shows that at high temperature, PVA has minimal dielectric change with humidity.

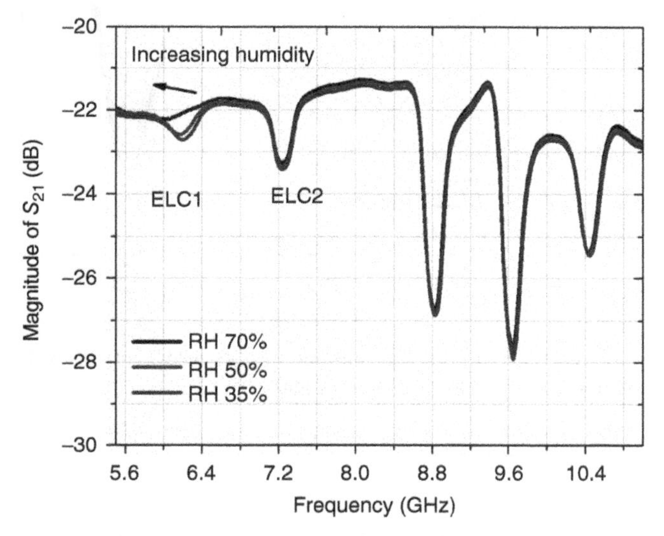

Figure 8.12 Measured magnitude of transmission coefficient (S_{21}) of multiple parameter chipless RFID tag for humidity sensing below threshold temperature (25 °C)

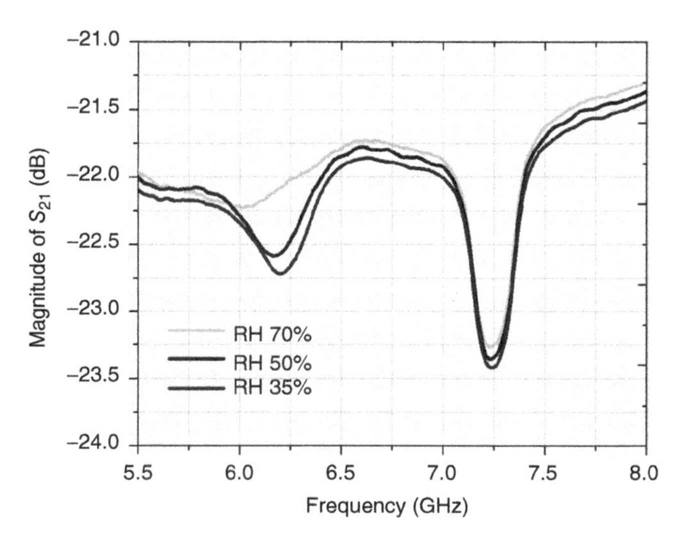

Figure 8.13 Measured magnitude of transmission coefficient (S_{21}) of sensing ELC resonators for humidity monitoring below threshold temperature (25 °C)

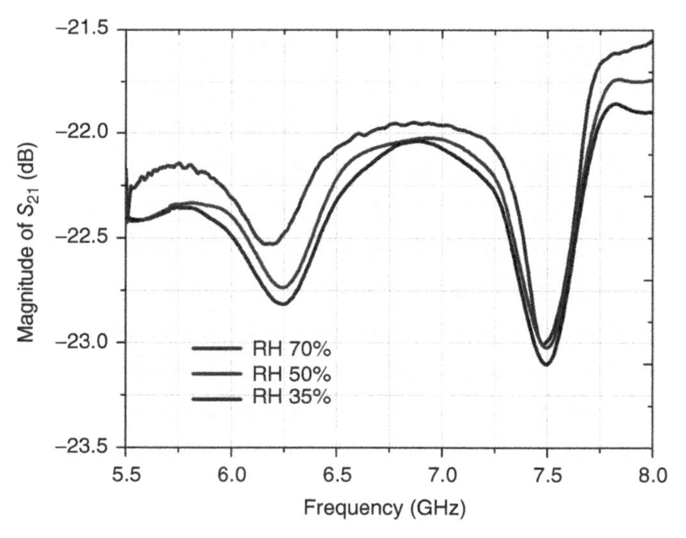

Figure 8.14 Measured magnitude of transmission coefficient (S_{21}) of sensing ELC resonators for humidity monitoring above threshold temperature (85 °C)

The next section discusses the effect of temperature and humidity on each smart material to explain the lower sensitivity of PVA in high temperature conditions.

8.3.4 Practical Challenges of Multiparameter Chipless Sensors

The primary challenge of multiparameter sensing is to take into account the sensitivity of a particular smart material to unwanted physical parameters. For instance, in our temperature and humidity sensor, PVA has a certain dielectric sensitivity to temperature change. Sengwa explains the dielectric dispersion of PVA–water mixtures for varying temperatures in Ref. [9].

The study shows both real and imaginary parts of the relative permittivity of PVA aqueous solution decrease with temperature increase. The dielectric constant (ε_r') decreases by about 8% and the loss factor (ε_r'') decreases by 14% as the temperature increases from 25 to 55 °C. This is due to the breaking and rearranging of the hydrogen bonds of water molecules with –OH groups of the PVA chain. In addition, the temperature increase causes variation in intermolecular association. Hence, an increase in temperature limits the change of relative permittivity due to humidity. At high temperatures, increase in humidity causes less variation in both dielectric constant and loss tangent change. This effect is depicted in Figure 8.14, where ELC1 has reduced sensitivity compared to Figure 8.13. PVA sensitivity to temperature can be reduced by mixing electrolyte materials such as NaCl in the PVA aqueous solution. As shown in Ref. [10], adding particular electrolytes drastically increases the water absorption of PVA solution and the effect is affected minimally by temperature. This will enhance the humidity sensitivity of PVA in high temperature conditions.

On the other hand, phenanthrene has a minimum effect of humidity on its dielectric constant [11]. Therefore, in both cases, ELC2 does not show any frequency variation for humidity change.

Apart from challenges of mutual sensitivity, multiple parameter tag sensor performance can be degraded due to environmental effects. Environmental conditions such as antenna orientation, distance, and presence of metal/liquid and multipath effects can greatly affect sensing data. Careful calibration techniques need to be carried out to overcome these challenges.

8.4 CONCLUSION

This chapter presents a novel application of a chipless RFID sensor for event detection. This feature has advantages over real-time environment monitoring for low-cost perishable item tagging and sensing. For example, in cold chain management of perishable items, suppliers and consumers need to rely on the predetermined "*Expiry date*" for the identification of out-of-date products. In addition, the spoilage of certain items such as milk not only depends on time but also on ambient temperature, environment, and light exposure. Researchers have concluded that milk needs to be stored at a low temperature (below 7 °C) to prevent early spoilage. Therefore, it is desirable to detect whether milk packs have been exposed to temperature beyond

7 °C during transport and storage. In such applications, consumers are interested in knowing the manifestation of this event (a temperature beyond the threshold of 7 °C) rather than the actual temperature at the time of purchase. Furthermore, it is practically impossible to record temperature data for each and every milk pack going through a full cold chain management cycle. In such scenarios, a low-cost, printable, and disposable chipless RFID temperature threshold sensor has a major role. Being ultra-low-cost, chipless sensors can be attached to every milk pack and the sensor can carry ID information and 1-bit sensing information (whether threshold violation has occurred or not). Therefore, a temperature threshold sensor has great potential in creating a paradigm shift in supply chain management.

In this chapter, a fully printable, compact, high data density temperature threshold sensor is presented. The sublimate material phenanthrene is used to exhibit permanent frequency shift beyond its transition temperature (80 °C). However, other materials such as plastic crystals or materials have hydrocarbon groups that show permanent change at lower temperatures.

Finally, another novel feature of the frequency modulation-based chipless tag is multiparameter sensing. We have presented a 3-bit tag and two physical parameter sensing capabilities, namely, real-time humidity and critical temperature. The results presented in this chapter verify the suitability of chipless sensors in multidimensional applications. However, there remain practical limitations in realizing multiple sensing tags with smart materials. Multiple parameter sensing tags demand smart materials with least sensitivity to unwanted physical parameters. Hence, detailed RF characterization and sensitivity measurement of smart materials need to be carried out for a fully functional chipless tag with multiparameter sensing.

In Chapters 6–8, various sensors have been presented. They are in the proof-of-concept prototype stages and need further fabrication processes for industrial applications. Chapter 9 presents nanofabrication techniques of the sensors that facilitates roll-to-roll printing of the chipless RFID sensor, hence the proof-of-concept laboratory prototype to industrial grade products.

REFERENCES

1. W. Buff, S. Klett, M. Rusko, J. Ehrenpfordt, and M. Goroli, "Passive Remote Sensing for Temperature and Pressure Using SAW Resonator Devices," *IEEE Transactions on Ultrasonics, Ferroelectrics and Frequency Control*, vol. 45, pp. 1388–1392, 1998.
2. S. Shrestha, M. Balachandran, M. Agarwal, V. V. Phoha, and K. Varahramyan, "A Chipless RFID Sensor System for Cyber Centric Monitoring Applications," *IEEE Transactions on Microwave Theory and Techniques*, vol. 57, pp. 1303–1309, 2009.
3. R. Bhattacharyya, C. Floerkemeier, S. Sarma, and D. Deavours, "RFID tag antenna based temperature sensing in the frequency domain," in *RFID (RFID), 2011 IEEE International Conference on*, 2011, pp. 70–77.
4. R. R. Fletcher and N. A. Gershenfeld, "Remotely Interrogated Temperature Sensors Based on Magnetic Materials," *IEEE Transactions on Magnetics*, vol. 36, pp. 2794–2795, 2000.

5. C. Mandel, H. Maune, M. Maasch, M. Sazegar, X. Schu, *et al.*, "Passive wireless temperature sensing with BST-based chipless transponder," in *German Microwave Conference (GeMIC)*, 14–16 March 2011, pp. 1–4.
6. Taconic Advanced Dielectric Division. Available: http://www.taconic-add.com/en--technicaltopics--technology-transitions.php (accessed on 15 October 2015).
7. L. A. Torres-Gómez, G. Barreiro-Rodríguez, and A. Galarza-Mondragón "A New Method for the Measurement of Enthalpies of Sublimation Using Differential Scanning Calorimetry," *Thermochimica Acta*, vol. 124, pp. 229–233, 1998.
8. E. M. Amin and N. C. Karmakar, "Low Cost Chipless RFID Sensor for Multiple, Australian Patent Application (no.PCT/AU2013/001276)" Radio Frequency Transponder, 2013.
9. R. J. Sengwa and K. Kaur, "Dielectric Dispersion Studies of Poly(vinyl alcohol) in Aqueous Solutions," *Polymer International*, vol. 49, pp. 1314–1320, 2000.
10. M.-R. Yang and K.-S. Chen, "Humidity Sensors Using Polyvinyl Alcohol Mixed with Electrolytes," *Sensors and Actuators B: Chemical*, vol. 49, pp. 240–247, 1998.
11. J. Kroupa, J. Fousek, N. R. Ivanov, B. Březina, and V. Lhotská, "Dielectric Study of the Phase Transition in Phenanthrene," *Ferroelectrics*, vol. 79, pp. 189–192, 1988.

9

NANOFABRICATION TECHNIQUES FOR CHIPLESS RFID SENSORS

9.1 CHAPTER OVERVIEW

This chapter firstly presents an overview various fabrication techniques that can be used for the development of various chipless radio-frequency identification (RFID) sensors. It includes a review on innovative micro- and nanofabrication technologies suitable for roll-to-roll chipless RFID sensor printing. In addition to the various printing facilities, state-of-the-art micro-/nanofabrication processes such as hand casting, spin coating, electrodeposition, wet chemical, physical and chemical vapor deposition (CVD), laser ablation, direct pattern writing by photolithography/electron beam lithography (EBL)/ion beam lithography, nanoimprint lithography (NIL) and etching, surface and bulk micromachining suitable for chipless RFID sensor fabrication are presented. A general survey and comparison of the different fabrication techniques are also given. The main aim of the chapter is to highlight the limitations of conventional fabrication processes and provide industrial solutions for on-demand, high-speed printing for flexible, robust, mass productivity of chipless RFID sensor. The aim is augmented with printing, imaging, and characterization procedures of micro- and nanostructures, and their integration into RF sensing devices is a new field; hence, printing of microwave passive design on polymers and organic materials has great research potentials. The contents and organization of this chapter are shown in Figure 9.1.

Chipless RFID Sensors, First Edition. Nemai Chandra Karmakar, Emran Md Amin and Jhantu Kumar Saha.

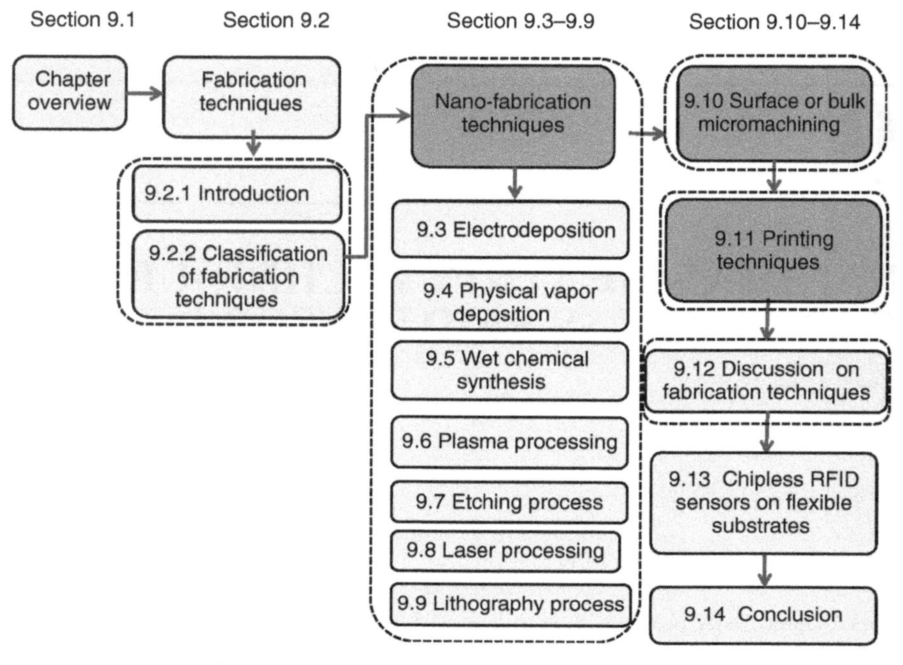

Figure 9.1 Contents flow diagram of this chapter

9.2 FABRICATION TECHNIQUES

9.2.1 Introduction

Fabrication of a multiparameter chipless RFID sensor node is a nontrivial task. This needs significant investigation into syntheses of smart materials for sensing individual physical parameters and fabrication processes. New fabrication and material processing techniques have made it possible to create circuits and components that are of nanometer scales. Combined with fabrication techniques, new materials are emerging that have the potential to challenge well-known fundamental theories in physics.

9.2.2 Classification of Fabrication Techniques

Various fabrication techniques that can be used to develop the chipless RFID sensor are printing, nanofabrication, and surface or bulk micromachining such as micro/nano electro mechanical systems (MEMS/NEMS) as shown in Figure 9.2. Each technique has its own distinctive attributes that can be used for different types of chipless RFID sensor development. Among the three techniques, printing of chipless RFID sensor nodes with roll-to-roll formats will provide the lowest cost tagging identification and sensing solution. Moreover, nanofabrication and micro-electromechanical systems (MEMS) may standout with high data capacity and precision sensing platforms.

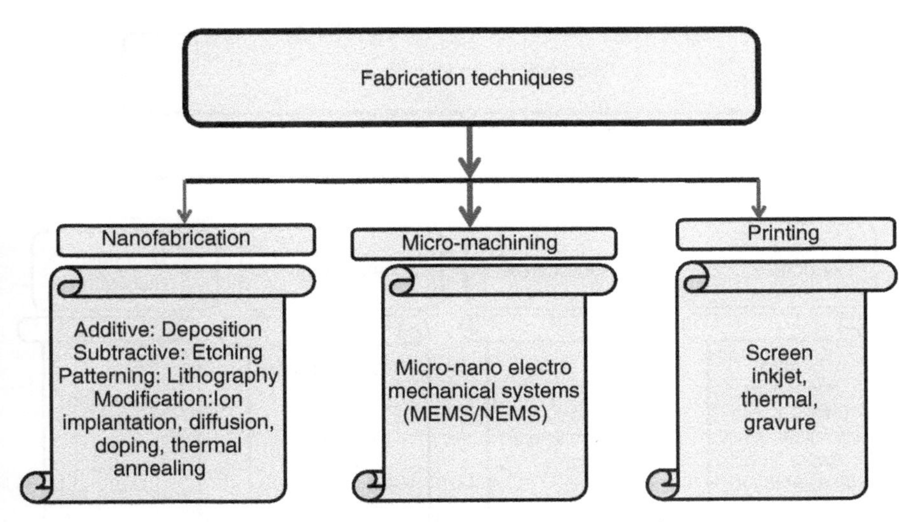

Figure 9.2 Classification of fabrication techniques

Nanofabrication technologies involve advanced physical and chemical processes and tools that enable production of nanostructured materials, devices, and prototypes needed for chipless RFID sensors development. The state-of-the-art micro-/nanofabrication processes that can be used to develop the chipless RFID sensor are electrodeposition, physical and chemical vapor deposition, laser ablation, direct pattern writing by photolithography/EBL/ion beam lithography, NIL and etching, material modification by ion implantation, diffusion, doping, and thermal annealing. Figure 9.3 shows various nanotechniques that can be used for fabrication of chipless RFID sensor.

9.3 ELECTRODEPOSITION

Electroplating is often also called "electrodeposition," a short version of "electrolytic deposition," and the two terms are used interchangeably. As a matter of fact, "electroplating" can be considered to occur by the process of electrodeposition. It is a process using electrical current to reduce cations of a desired material from a solution and coat that material as a thin film onto a conductive substrate surface. Figure 9.4 shows a simple electroplating system for the deposition of copper from copper sulfate solution [1].

9.4 PHYSICAL VAPOR DEPOSITION

Evaporation and sputtering are different kinds of thin films deposition using different techniques. Other PVD techniques are molecular beam epitaxy (MBE) and laser ablation deposition. PVD technique is a line-of-site impingement type deposition.

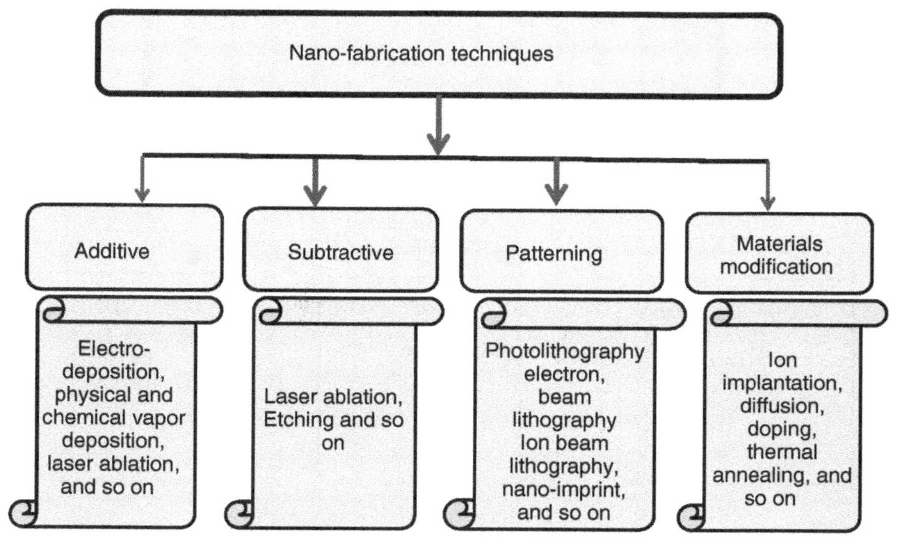

Figure 9.3 Various nanofabrication techniques

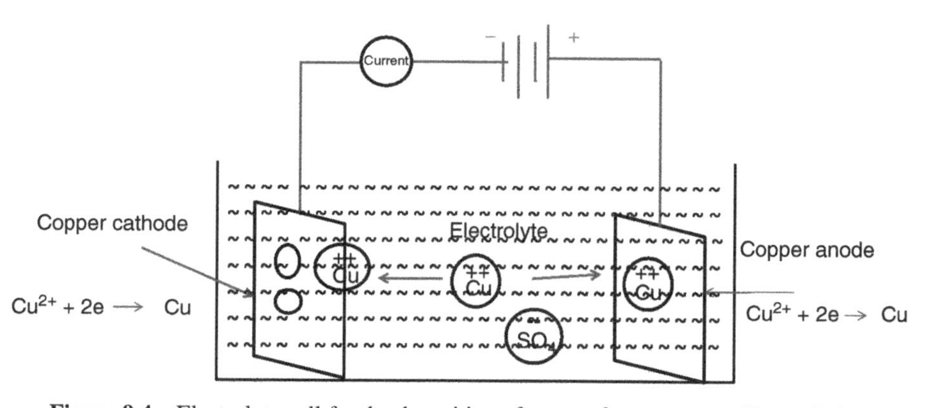

Figure 9.4 Electrolyte cell for the deposition of copper from copper sulfate solution

9.4.1 Thermal Evaporation

Thermal evaporation is based on the boiling off (or sublimating) of a heated material onto a substrate in a vacuum. Figure 9.5 shows the thin-film evaporation setup including traditional as well as electron-beam evaporation setup.

9.4.2 Sputtering

Sputtering is preferred to evaporation: wider choice of materials to work with better step coverage, better adhesion to substrate.

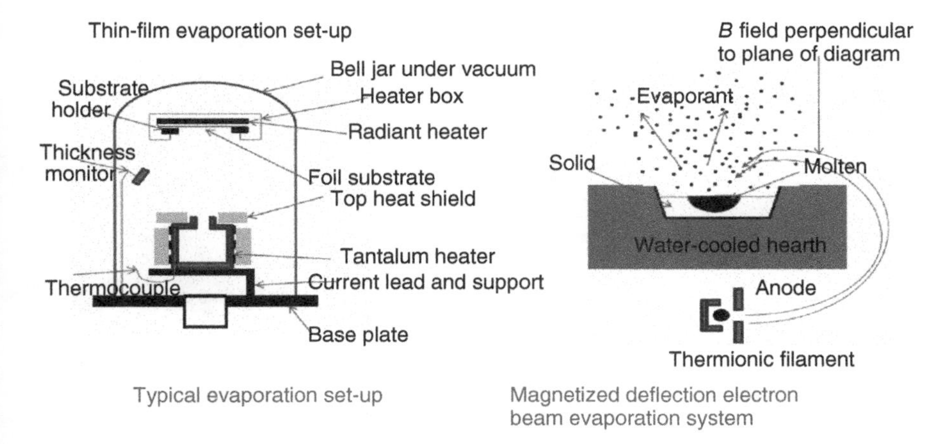

Figure 9.5 Thin-film evaporation setup

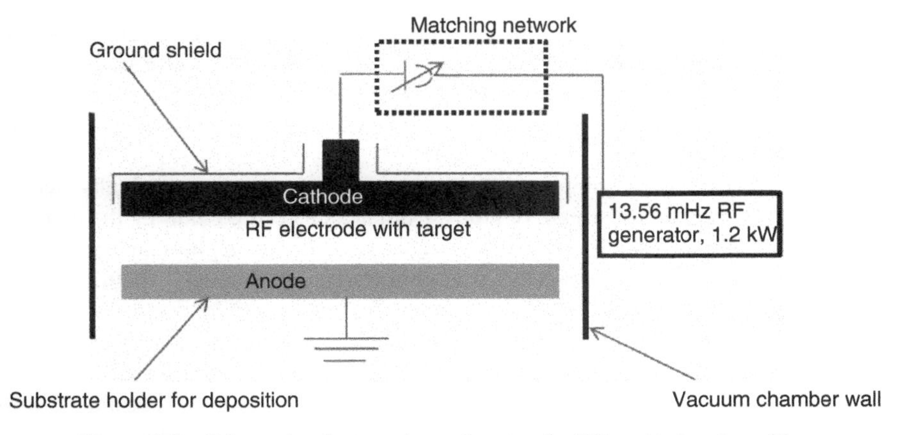

Figure 9.6 Schematic of a two electrode setup for RF sputtering deposition

Sputtering is used to apply films: compact discs, large area active-matrix liquid-crystal displays, and magneto-optic disks. Also, bearing gears and saw blades can be coated with a number of hard, wear-resistant coatings such as TiN, TiC, and TiAlN. Figure 9.6 shows the schematic of a two electrode setup for RF sputtering deposition.

9.4.3 Molecular Beam Epitaxy

Epitaxial techniques arrange atoms in single-crystal manner upon a crystalline substrate acting as a seed crystal so that the lattice of the newly grown films duplicates that of the substrate. Figure 9.7 shows a schematic of MBE growth chamber molecular beam.

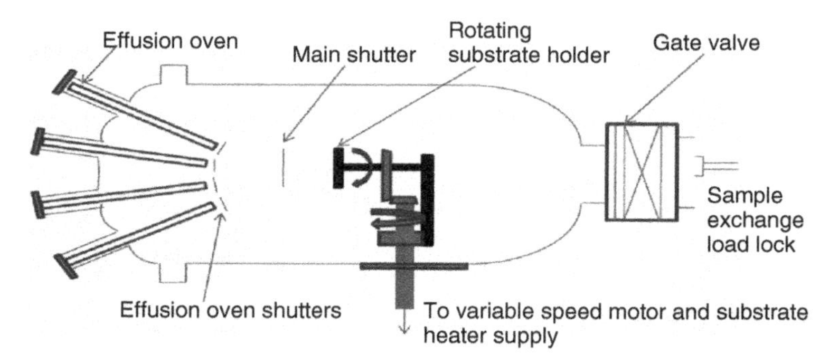

Figure 9.7 Schematics of molecular beam epitaxy growth chamber molecular beam

Deposition film same as substrate is called epitaxy or epi. Epideposition is one of the cornerstone techniques for building micromachines. Si plates with predetermined thickness and doping level can be engineered. Growth rate of an epilayer depends on the substrate crystal orientation. Si (111) planes have the highest density of atoms on the surface and film grows most easily on these planes. If deposit is made on a chemically different substrate, the process is called heteroepitaxy, for example, silicon on insulator (Si on SiO_2), Si on sapphire (Al_2O_3), gallium phosphide, and gallium arsenide.

Heated single-crystal sample (400–800 °C) is placed in an ultrahigh vacuum (10^{-11} Torr) in the path of stream of atoms from heated cells containing the material of interests. These atomic streams impinge, in a line-of-sight manner, on the surface-creating layers with a structure controlled by the crystal structure of the surface, thermodynamics of the constituents, and sample temperature.

MBE is the most sophisticated form of PVD; deposition rate is very slow, 1 μm/h or 1 monolayer/s. Relatively low growth temperatures reduce diffusion and auto doping effects. Precise control of layer thickness and doping profile is possible. Ultrahigh vacuum requirements make MBE operation very expensive—not a production technique.

9.5 WET CHEMICAL SYNTHESIS

Wet chemical methods to prepare metal colloids are usually based on the reduction of a suitable metal salt in the presence of stabilizers. Synthesis of nanoparticles of metals using irradiation is now well established and the mechanism of production is reasonably understood. Because spherical shape being thermodynamically most stable, the formation of spherical nanoparticles is more probable in any wet chemical synthesis method [2]. The synthesis steps for generation of metal nanoparticles by wet chemical method are given in Figure 9.8.

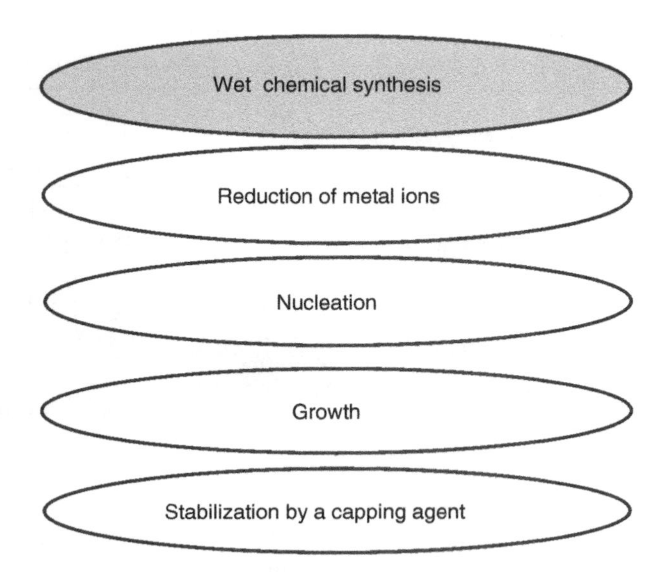

Figure 9.8 Synthesis step of metal nanoparticles in wet chemical method

9.6 PLASMA PROCESSING

CVD is a very versatile and works at low atmospheric pressures and temperatures. It constituents of a vapor phase, often diluted with an inert carrier gas, react at the hot surface in order to deposit a solid film. CVD, a diffusive–convective transport process, involves intermolecular collisions. Amorphous, polycrystalline, epitaxial, and uniaxial oriented polycrystalline layers can be deposited with a high degree of purity, control, and economy. CVD is the most widely used deposition technique in the IC manufacture.

Plasma-enhanced chemical vapor deposition (PECVD) is an RF-induced plasma transferring energy into the reactant gases, allowing the substrate to remain at lower temperatures than Low Pressure Chemical Vapor Deposition (LPCVD) and Atmospheric Pressure Chemical Vapor Deposition (APCVD) do. It uses an RF-induced glow discharge to transfer energy into reactant gases, lower substrate temperature—suitable for substrates with low thermal stability, produces thin films of unique compositions and properties—excellent adhesion, low pin-hole density, good step coverage, compatible with fine line pattern transfer processes.

As it has been reported [3], the excitation frequency of a plasma source has an important effect on the electron acceleration in the plasma, and a high excitation frequency is expected to result in a high electron density and a low electron temperature. Therefore, a new microwave plasma source for thin-film processing application has been developed [4]. The high-density microwave plasma source at low and high pressures using the spoke antenna and wave guide, respectively, is shown schematically in Figure 9.9.

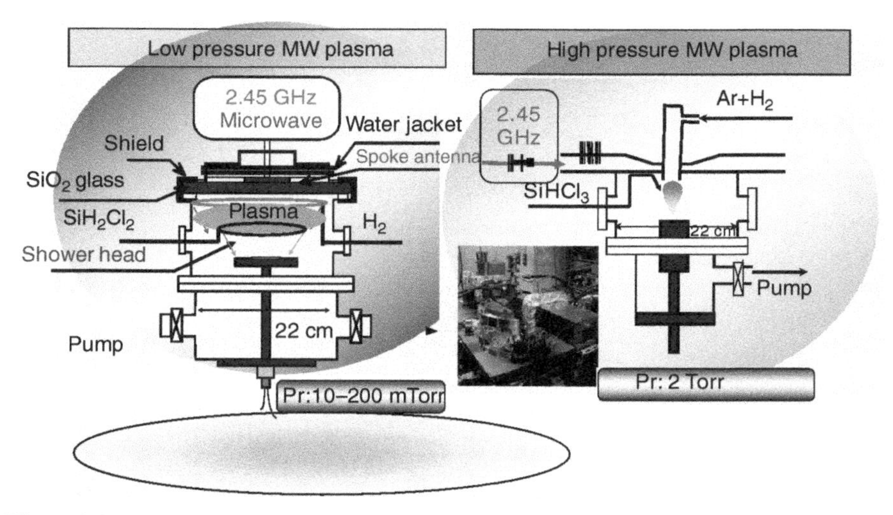

Figure 9.9 The schematic diagram of the low- and high-pressure high-density microwave plasma utilizing the spoke antenna and wave guide, respectively

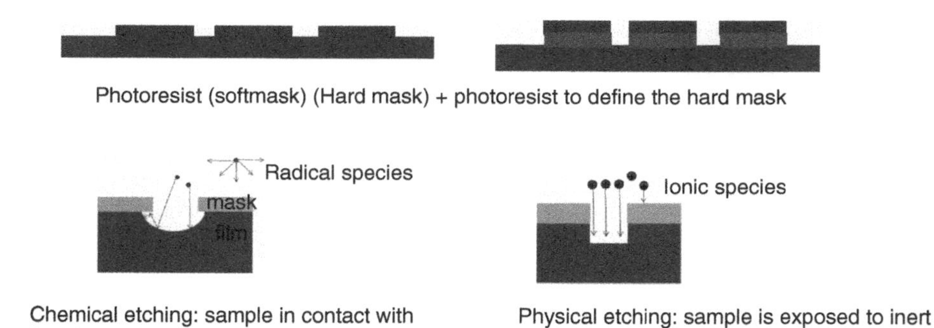

Figure 9.10 Typical etching process

9.7 ETCHING

The etching process removes the unwanted materials from a substrate or laminate, resulting in the originally designed circuit. Etching is the selective removal of deposited films, for example, hydrofluoric (HF) acid to remove native oxides on silicon but not the silicon itself. Etching is often carried out by using a mask to leave the patterned film.

Etching is usually done through a mask and must be done with consideration of prior processes—selectivity of materials must be known as shown in Figure 9.10.

9.8 LASER PROCESSING

Material processing with lasers takes advantage of virtually all of the characteristics of laser light. The high energy density and directionality achieved with lasers permits strongly localized heat or photo treatment of materials with a spatial resolution of better than 10 nm. Figure 9.11 gives an overview of the various applications and parameter regimes employed in laser processing. The intensities and interaction times shown in the figure refer to different types of lasers [5]. Figure 9.11 shows the applications of lasers in material processing. Surface modifications include laser-induced oxidation/nitridation of metals and surface doping. Laser-light intensities exceeding 1016 W/cm^2 generate X-rays that gain increasing importance in nanotechnology.

The PEDOT tag film is patterned according to a microwave design by laser etching on polymer substrates as shown in Figure 9.12(a) design and (b) fabricated tag on polymer substrate by laser etching.

The fabricated sensor is tested for pH sensitivity. We used two horn antennas connected to vector network analyzer (VNA) for wireless measurement. One of the

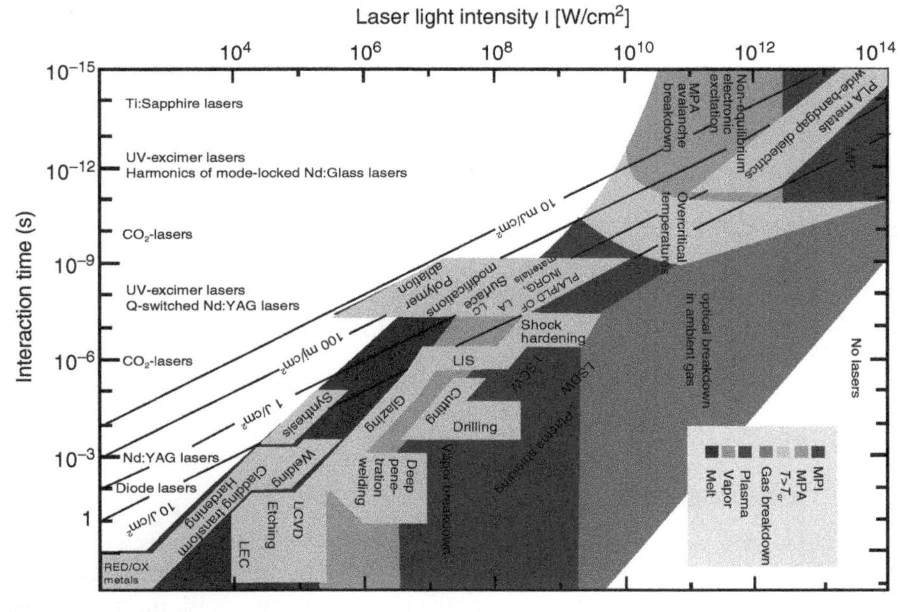

Figure 9.11 Overview of the various applications and parameter regimes employed in laser processing. PLA/PLD, pulsed-laser ablation/deposition; LA, laser annealing; LC, laser cleaning; LIS, laser-induced isotope separation/IR-laser photo chemistry; MPA/MPI, multiphoton absorption/ionization; LSDW/LSCW, laser-supported detonation/combustion waves; LCVD, laser-induced chemical vapor deposition; LEC, laser-induced electrochemical plating/etching; RED/OX, long pulse or CW CO_2-laser-induced reduction/oxidation (*Source:* Adapted from Ref. [6] with permission)

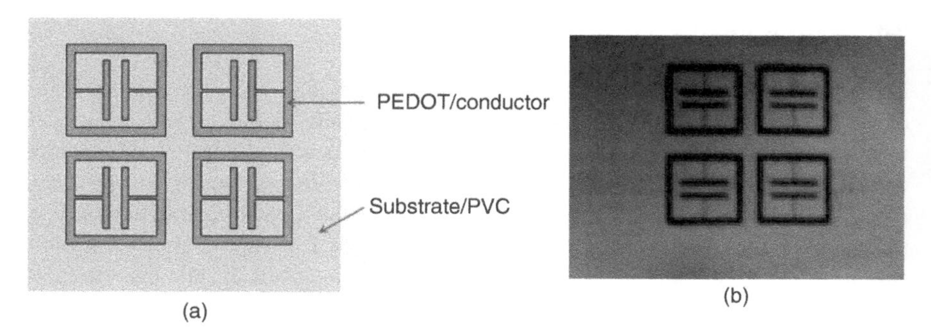

Figure 9.12 (a) Design of tag on polymer substrate by CST software and (b) fabricated tag on polymer substrate by laser etching

antennas transmitted polarized *E* field to the sensor and the second antenna captured the backscattered response. Measurement was taken for two cases. In the first case, the patch array is emerged into acidic solution and then in an alkaline solution. Figure 9.13 shows the measured magnitude and phase of transmission coefficient (S_{21}) for the two cases. As expected, there is a shift in Q factor at resonant frequency as the pH is varied. As PEDOT has increased the sheet resistance at high pH environment), the resonance Q factor increases drastically. Both magnitude and phase response corresponds to pH change. The above results indicate the potential of realizing a chipless RFID pH sensor on flexible substrate using laser patterning. This is a novel technique of passive RF sensor fabrication.

9.9 LITHOGRAPHY

Most widely used form of lithography is photolithography, laser lithography, and electron-beam lithography. In IC industry, most pattern transfer from masks onto thin films is almost carried out by photolithography. Other current lithography techniques are deep-UV; extreme UV and emerging techniques are multibeam, e-beam-lithography, and nanoimprint-lithography.

9.9.1 Photolithography

Microfabrication, microelectronic, and nanoelectronic fabrication starts with lithography [7–9]. This is the technique used to transfer copies of a master pattern onto the surface of a solid material. These techniques are used for mask making, image transfer, and direct writing. Since photolithography has limitations: diffraction limited – prevents the exposures of sub-micron features. Optical Mask is the starting point of any photolithography work as shown in Figure 9.14. It is a "stencil" used to generate patterns on resist-coated wafers repeatedly.

First step in the lithography process—a thin layer of an organic polymer, a photoresist, sensitive to UV radiation is deposited on the substrate surface as shown

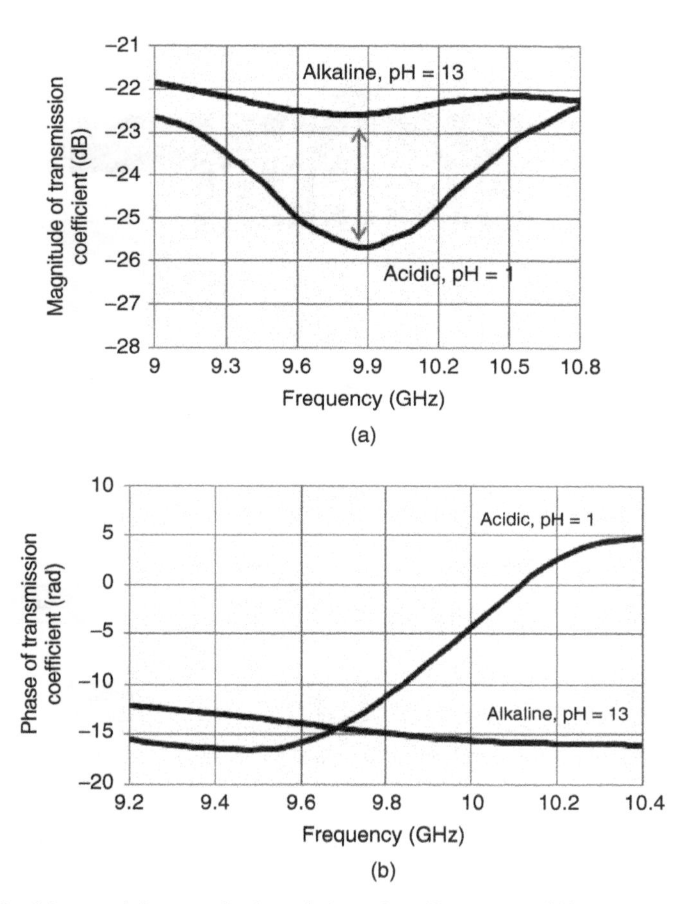

Figure 9.13 Measured S_{21} magnitude and phase for pH sensor at different pH environment

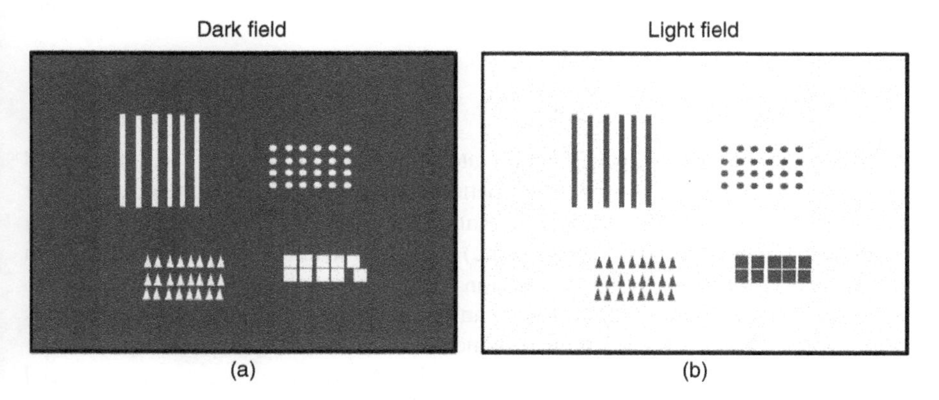

Figure 9.14 Optical mask types

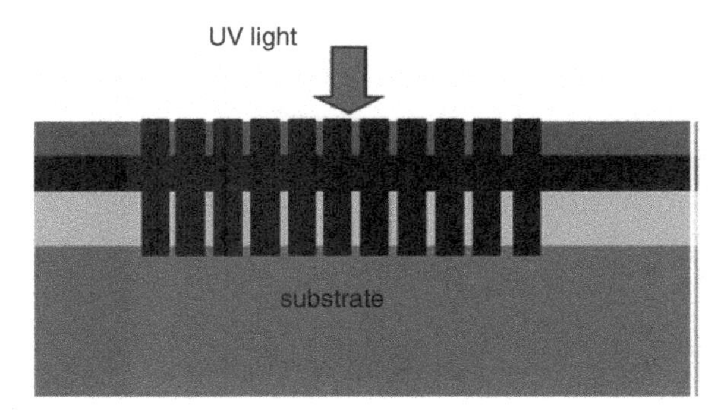

Figure 9.15 UV radiation is deposited on the substrate surface

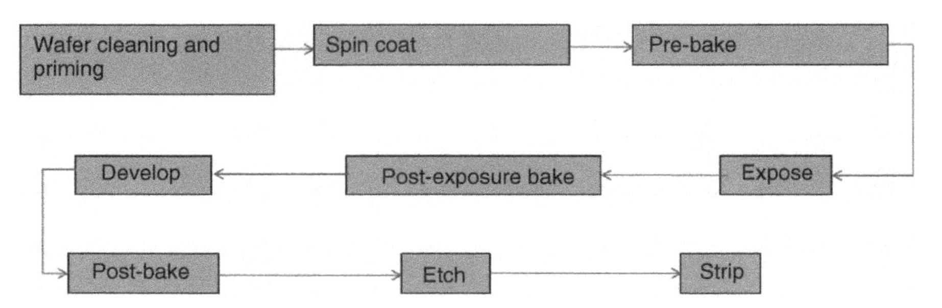

Figure 9.16 Flowchart of a typical photolithography process

in Figure 9.15. The photoresist is dispensed from a viscous solution of the polymer onto the wafer in a resist spinner. Figure 9.16 shows the flowchart of a typical resist-process.

Figure 9.17 shows a lithography technique that can be used for mask making, image transfer, and direct writing directly to any substrate (e.g., polymer and paper substrates).

9.9.2 Electron beam lithography

Electron beam lithography (often abbreviated as e-beam lithography) is a specialized technique for creating extremely fine patterns at the submicron or nanoscale, that is, 100 nm or smaller. The primary advantage of electron-beam lithography is that it can draw custom patterns (direct-write) with sub-10 nm resolution and the form of maskless lithography as high resolution and low throughput, limiting its usage to photomask fabrication. The EBL setup can be found in Ref. [10]. Figure 9.18 shows an example of EBL setup and typical sequential steps of EBL, respectively.

Computer-aided transcription system (CATS) fracturing processes: commercial lithography systems must interpret the CAD pattern as a digital approximation since hardware limits beam placement to discrete points on substrate. This

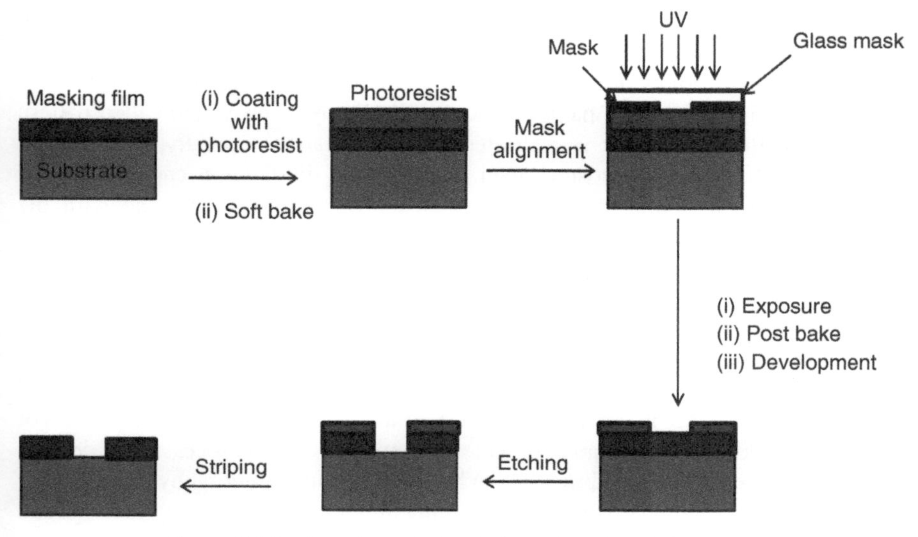

Figure 9.17 Steps in optical printing using photolithography

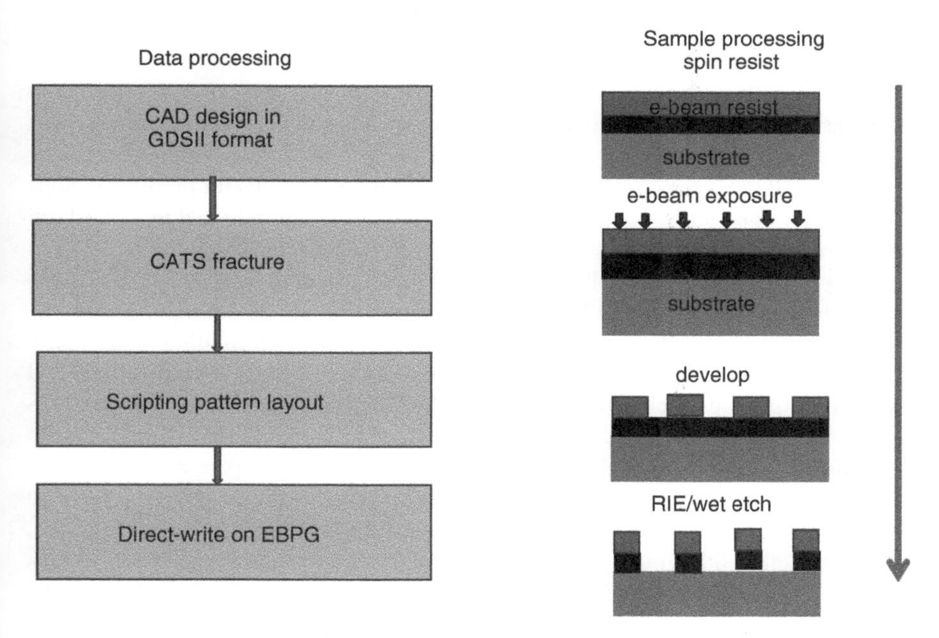

Figure 9.18 Typical sequential steps of EBL

hardware-specific software task, called fracturing, is part of the process of converting the generic CAD data format into the hardware-specific binary exposure format. Discrete exposure points are called picture elements or pixels. Interpretation of a given shape within the CAD pattern is accomplished by a subsystem called pattern generator. Pattern generator is restricted to basic shapes, typically, a trapezoid. Knowledge of intended exposure pixel during CAD digitization can avoid error during fracturing. Smaller resolution and lower roughness are in line with the arbitrary angle.

9.9.3 Ion beam lithography

Ion beam lithography is a process of scanning a focused beam of ions in a patterned fashion across a surface in order to create very small structures [11,12]. The process of lithography involves the formation of patterns for selective area processing of devices at different stages of device fabrication. While conventional lithography is carried out using light for exposing "resists," the continuing miniaturization of integrated circuits has stimulated interest in new exposure techniques. Electrons, X-rays, and ion beams can also deposit energy in a resist to expose it. Ion beams offer ultimate advantages in sensitivity and fineness of feature size because of their penetration properties in materials.

Lithographic patterns can be formed by the use of scanned, finely focused ion beam. Whole lithographic patterns can be transferred by ion optical imaging or by channeled ion lithography. The progress in both types of ion beam lithography and in the development of high-brightness ion sources and ion beam–compatible resists is summarized.

There are three types of ion lithography, that is, three methods for producing alternate irradiated and un-irradiated areas on a surface with minimum features of 0.1 μm and below.

Focused ion beams: A small intense spot with a current density of 6.1–80 A/cm^2 is deflected over the surface [13].This is very similar to EBL as discussed earlier.

Ion beam proximity printing: It is also called masked ion beam lithography. Here, a stencil mask with open is held in close proximity to a resist covered sample and exposed to light ions such as H^+, H_2, or He. The ions can come from a standard ion implanter, and this technique of shadow printing can be very fast. The difficulties of making a mask at the final dimension that is either on transparent or open have limited the applicability of this technique. Lines and spaces down to 80 nm have been demonstrated as well as special applications which the limitations are less important.

Ion projection lithography: In this technique, a stencil mask is uniformly illuminated by a beam of light ions, and then by means of electrostatic optical elements, the image of the stencil mask is projected onto a sample surface. Recent progress in controlling the field distortion, in the image stability, and in achieving fine lines makes ion projection a promising lithography technique.

Laser lithography, multiphoton lithography, or direct laser writing (DLW) has been known for years by the photonic crystal community.

9.9.4 Nanoimprint lithography (NIL)/Hot Embossing

NIL/hot embossing is a method of fabricating nanometer-scale patterns. It is a simple nanolithography process with low cost, high throughput, and high resolution. It creates patterns by mechanical deformation of imprint resist, typically a polymer that is cured by heat or UV light. During the imprint step, resist is heated to a temperature above its glass transition temperature. Resist becomes viscous liquid and flows readily into the shape of the mold. Unlike conventional lithography methods, imprint lithography does not use energetic beams. Therefore, the resolution is not limited by effects of wave diffraction, scattering, and interference in a resist. Imprint lithography is more of a physical process than a chemical one. Figures 9.19 and 9.20 show essentials and classification of NIL.

9.9.5 Thermal Nanoimprint Lithography

Thermal imprint lithography, also referred to as NIL, was first introduced by Stephen Chou at the University of Minnesota in 1996. Figure 9.21 shows steps in thermal imprint lithography (Step 1: Imprinting using a mold to create a thickness contrast in a resist, Step 2: Mold removal, and Step 3: Pattern transfer using anisotropic etching to remove residue resist in compressed areas).

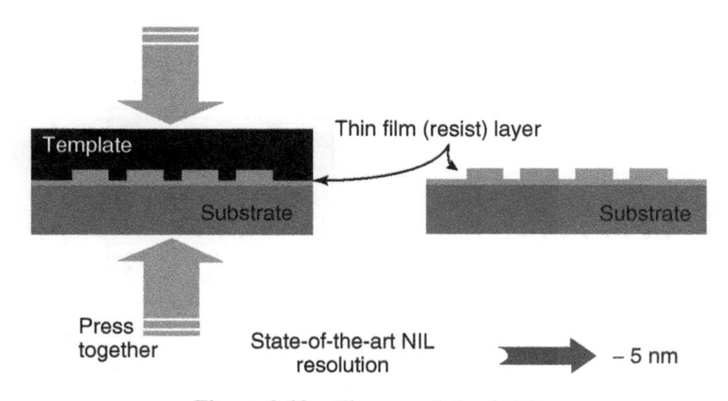

Figure 9.19 The essentials of NIL

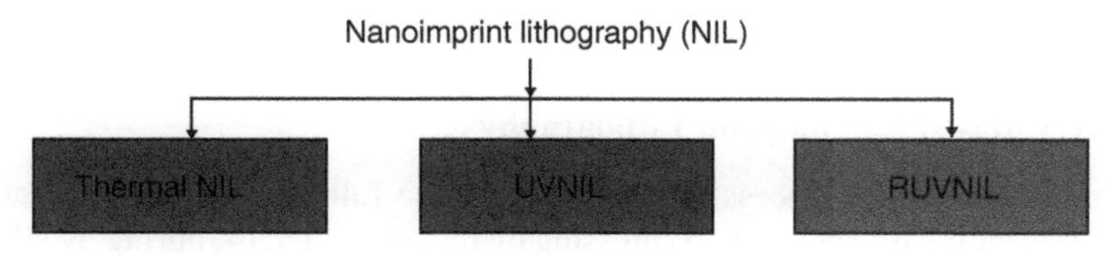

Figure 9.20 Classification of nanoimprint lithography (NIL)

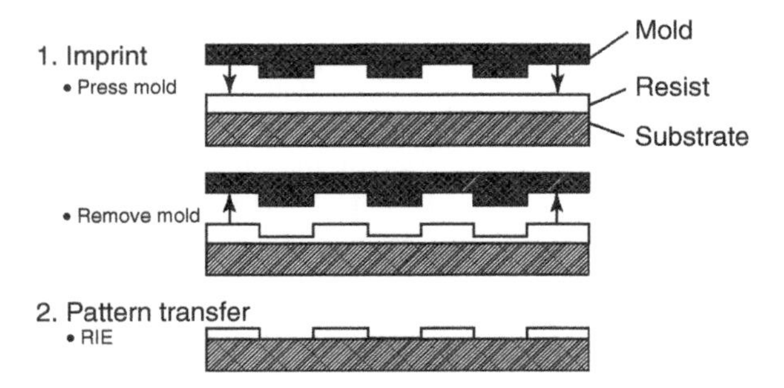

Figure 9.21 Steps in thermal imprint lithography (*Source:* Reproduced with permission from Ref. [18])

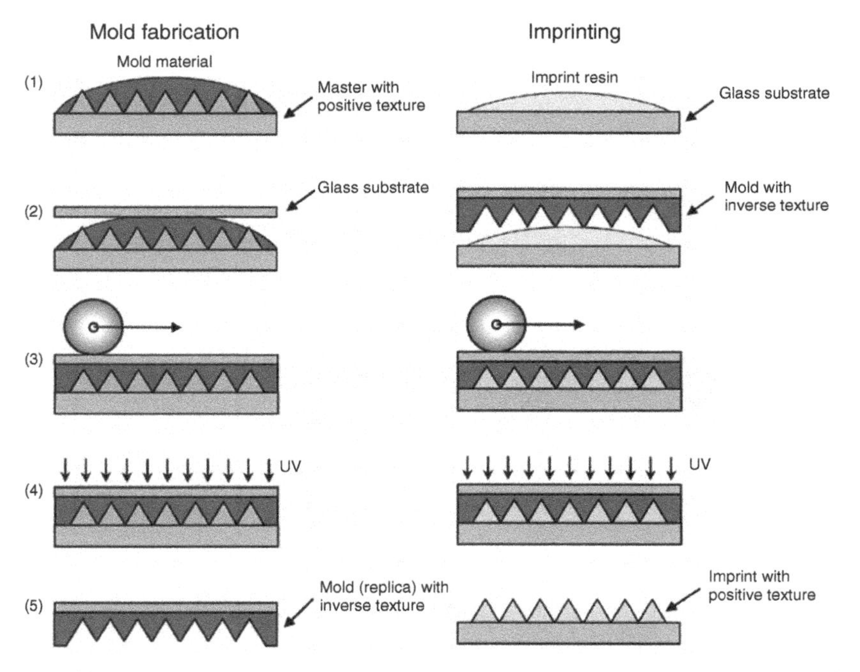

Figure 9.22 A basic scheme of UVNIL process (*Source:* Reproduced with permission from Ref. [19])

9.9.6 UV-Based Nanoimprint Lithography

Conventional UV NIL process scheme involves the following steps: (1) coating of resist, (2) bringing into contact, (3) pressing by moving roller, (4) curing by UV light, and (5) demolding, which is shown in Figures 9.22 and 9.23.

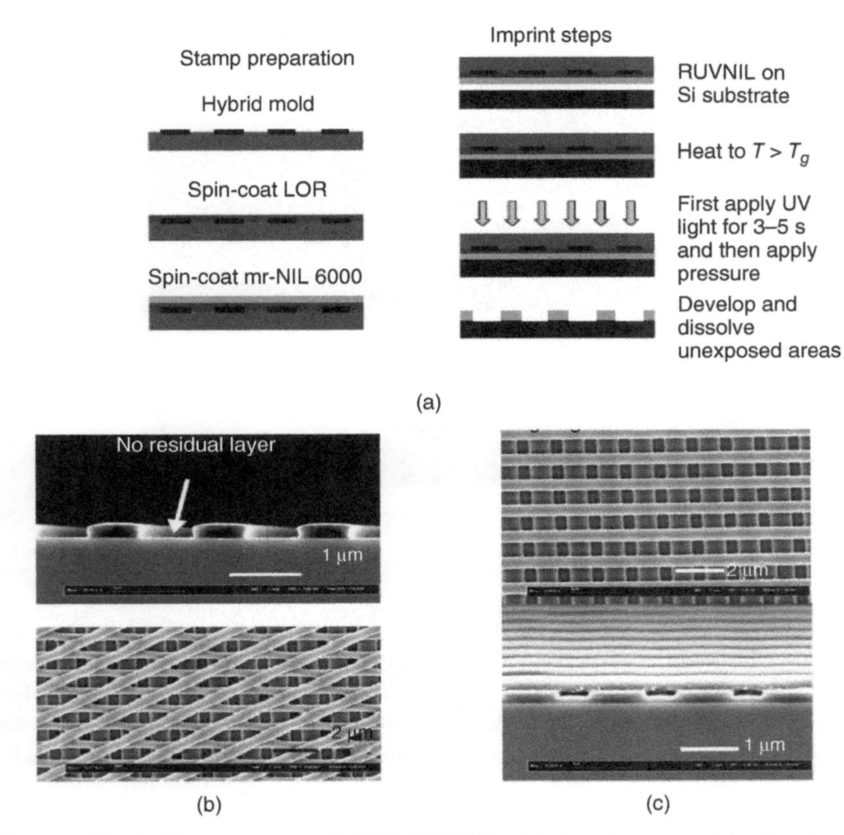

Figure 9.23 (a) Reverse contact UVNIL–RUVNIL, (b) three-layer woodpile-like structure by RUVNIL, (c) cross-section of a two-layer woodpile-like structure—no evidence of polymer flow in grooves. (*Source:* Reproduced with permission from Ref. [20])

9.9.7 Reverse Contact UVNIL–RUVNIL

Figures 9.24 and 9.25 show scanning electron microscope (SEM) images of the fabricated template and the silicon nanowire sensor for gas detection fabricated by nanoimprint on SU8/SiO_2/PMMA trilayer, respectively.

9.10 SURFACE OR BULK MICROMACHINING

The surface or bulk micromachining such as MEMS/NEMS that also can be used to develop the chipless RFID sensor. MEMS and nano-electro-mechanical systems (NEMS) rely on technologies of miniaturization. Watch makers have practiced the art of miniaturization since the thirteenth century. With the invention of the compound microscope in the 1600s and its subsequent improvement and later use to observe

Figure 9.24 SEM image of the fabricated template. The inset image shows the detail of the template

microbes, plant and animal cells and modern day, atomic force and electron microscopes that allow for observation at the molecular and atomic scale, there has been an interest to manipulate matter at a smaller and smaller scale. One success story has been the miniaturization of the modern era's transistor, which has allowed for the development of even smaller and more powerful gadgets and machines. The transistor in today's integrated circuits has a size of 0.18 μm in production and approaches 10 nm in research laboratories.

Surface micromachining provides a complementary metal-oxide semiconductor (CMOS) compatible fabrication technique for freestanding structures and allows a higher structure density than bulk micromachining. The basic idea of surface micromachining is the sacrificial layer concept as shown in Figure 9.26. The functional layer that will define the movable part of the device is deposited on a prestructured sacrificial layer. After appropriate structuring of the functional layer, the sacrificial layer is removed in an isotropic etch process. Figure 9.26 shows a novel capacitive-type humidity sensor using CMOS fabrication technology.

9.11 PRINTING TECHNIQUES

Printing is a fully additive process enabling direct deposition of ink on the substrate without the need to go through steps such as etching, stripping, and cleaning, which are most common in photolithography process. The ability to print a complete chipless RFID sensor on flexible substrate opens the door for its use in many low-cost

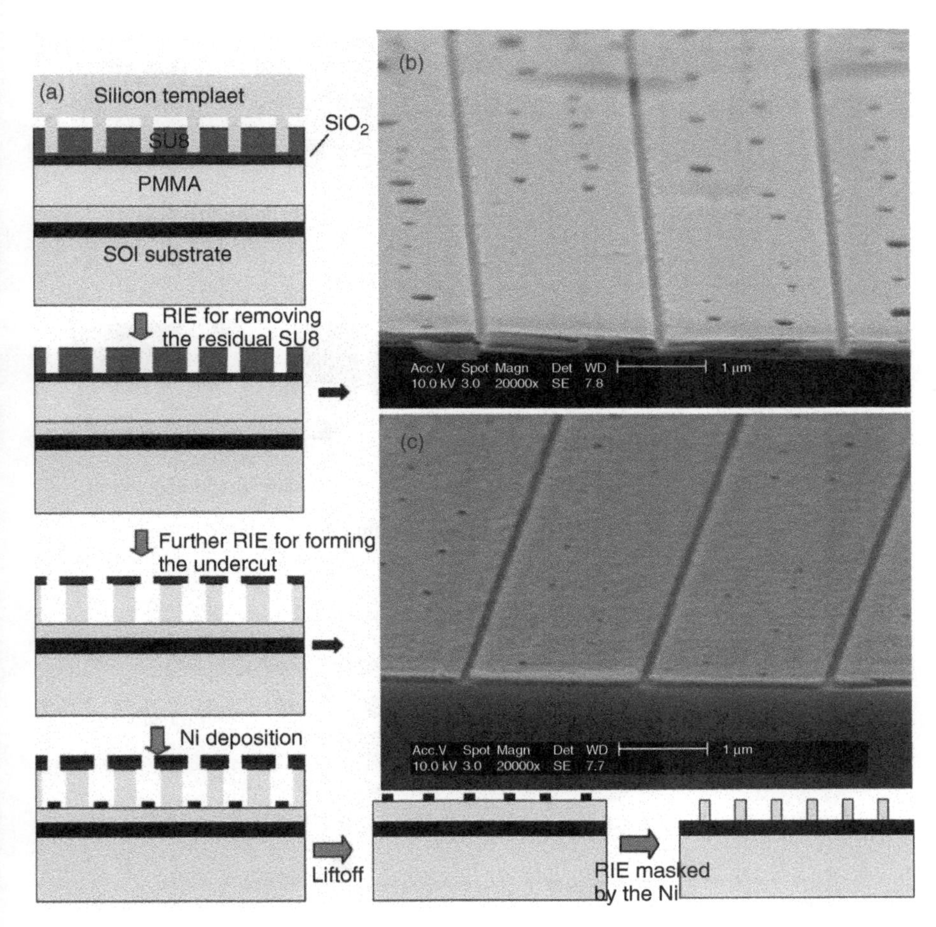

Figure 9.25 (a) Schematic view of the NW transfer steps by trilayer NIL on SU8/SiO_2/PMMA structure; (b) SEM image of the sample after imprint; and (c) SEM image of the sample after removing PMMA and formation of undercut

applications. The printing techniques that can be used to develop the chipless RFID sensor are screen, inkjet, thermal, flexographic, and gravure printing [14].

9.11.1 Screen Printing

In the rich palette of technologies nowadays available to print films for microelectronics and electronic-related systems, screen printing is the most simple, flexible, and cheap. Screen printing is a process used in the electronic industry, where films of functional materials are deposited on to a target (substrate) surface. as shown in Figure 9.27 [15].

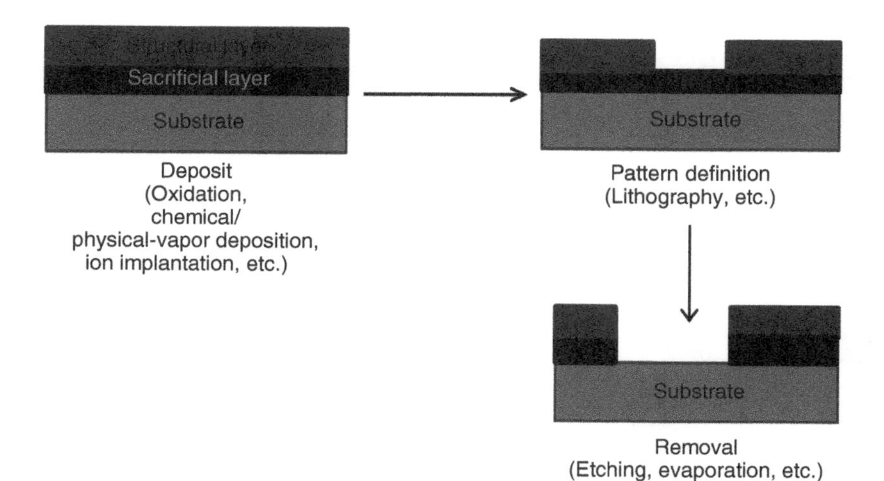

Figure 9.26 Surface micromachining techniques

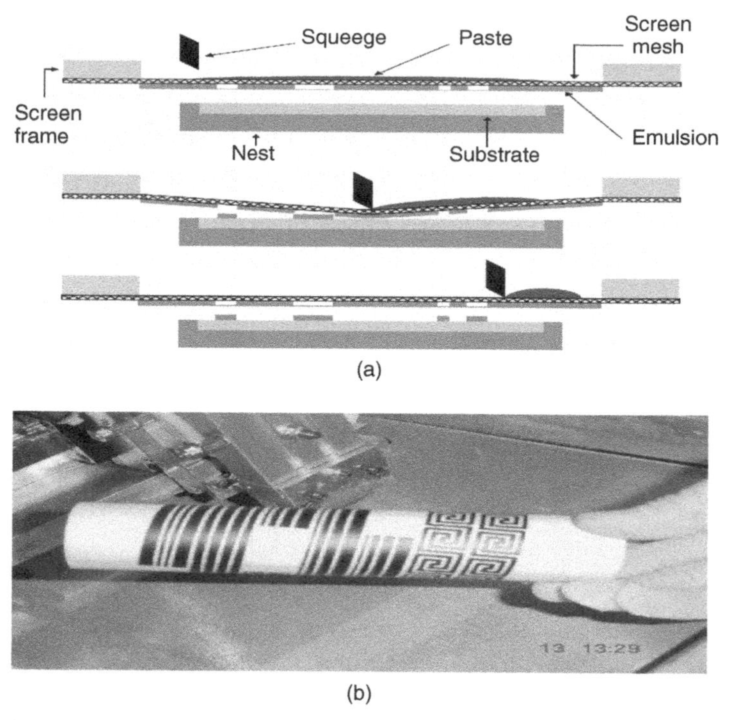

Figure 9.27 (a) The screen-printing process; (b) lines on an alumina tube screen-printed with IKO screen printer model T-620-80200 (*Source:* Courtesy of Institut für Sensor—und Aktuatorsysteme Technische Universität Wien; Reproduced with permission from Ref. [15])

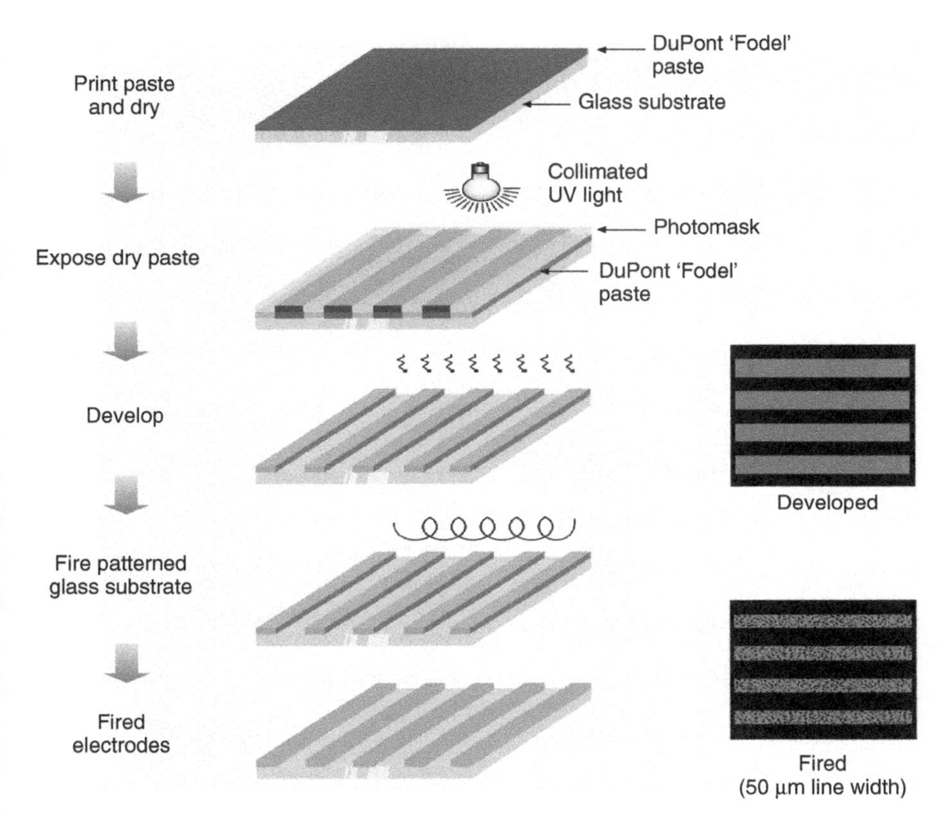

Figure 9.28 The process illustration for the screen-printing process (*Source:* Reproduced with permission from Ref. [15])

The process illustration for the screen-printing process is shown in Figure 9.28. The application of rotary screen printing will probably increase in future to get over the relatively modest throughput of conventional screen printing.

9.11.2 Inkjet Printing

Inkjet technology is a class of direct-write processes that share the appealing characteristic of unparalleled versatility: printing is possible onto virtually any substrate, of any size, with a resolution exceeding 200 lines/cm. As in home/small office desktop inkjet printers, tiny ink droplets of functional materials are reproducibly dispensed from inks of adequate viscosity, typically below 20 mPa s. Therefore, the inks may be liquid solutions, dispersion of small (preferably nano) particles, melts, or blends.

Droplets with diameters of 15–200 μm and volumes down to a few picoliters are deposited at a few kilohertz up to 1 MHz rates according to the equipment and inception of the deposition methods. Two different inkjet techniques have become mature

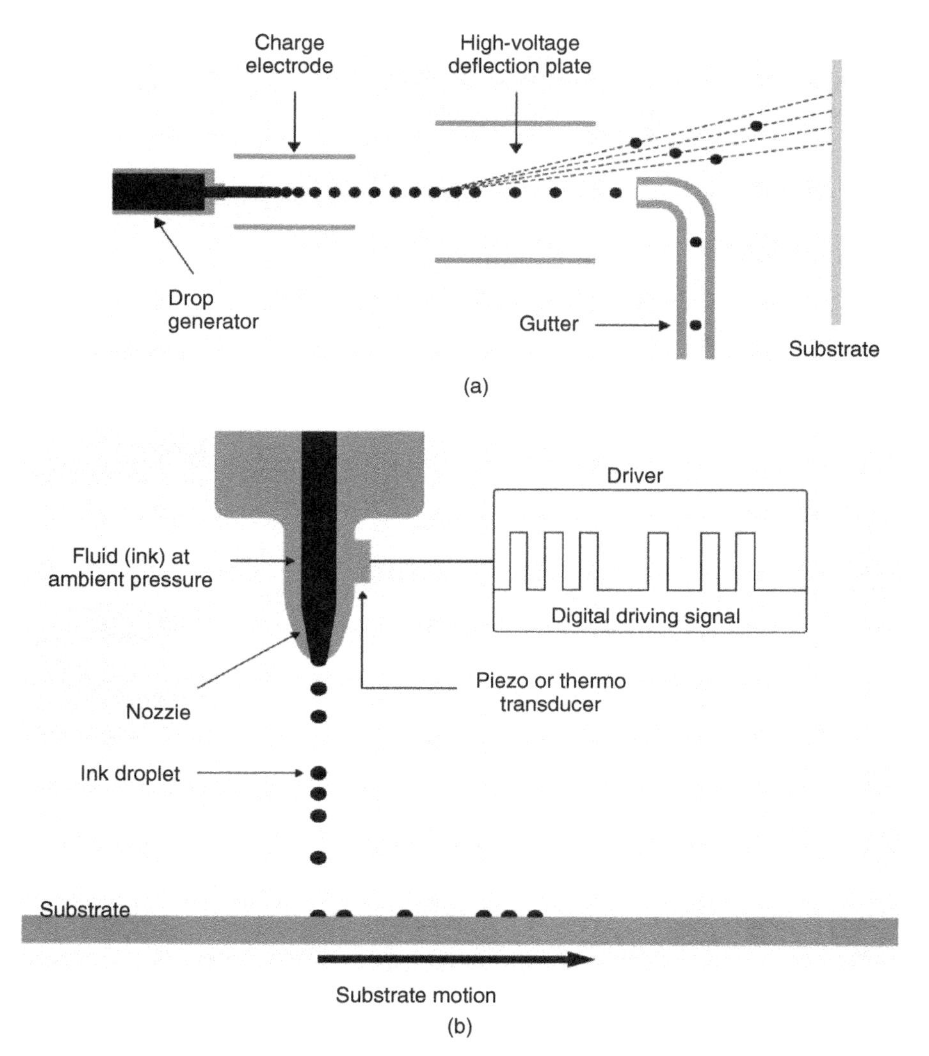

Figure 9.29 Inkjet printing: (a) continuous inkjet (CIJ) printing: multiple deflection system; (b) drop on-demand (DOD) system: single droplets ejected through an orifice at a specific point or time (*Source:* Reproduced with permission from Ref. [15])

and well understood in the printing industry for decades: drop-on-demand (DOD) and continuous inkjet (CIJ) as shown in Figure 9.29.

Figure 9.30 shows a single inkjet-printed humidity sensor tag from both sides. Printing was made using an iTi XY MDS 2.0 inkjet printer, equipped with Spectra S-class print heads. The printing resolution was set to 600 dpi. Using these parameters, the inkjet printing process should produce a conductor thickness ranging from 1.0 to 1.5 μm/printing layer. Samples were printed using two ink layers to reduce the

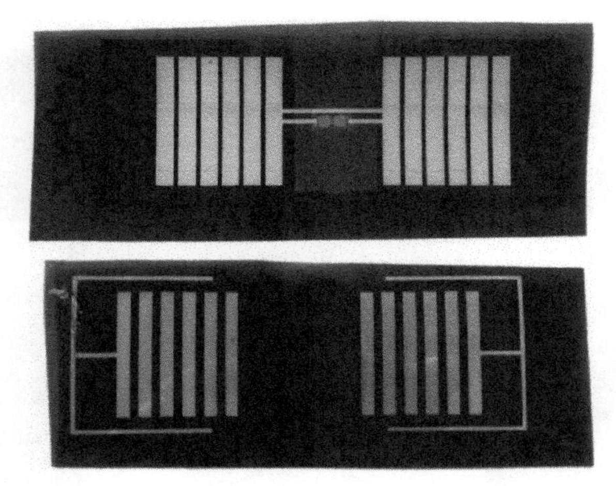

Figure 9.30 Inkjet-printed sensor tag (*Source:* Reproduced with permission from Ref. [16])

losses caused by the skin effect. The IC strap was attached by hands using conductive epoxy resin. Adhesive tape was laid on top of the IC to protect it from excessive humidity levels [16]. Figure 9.31 shows the fabrication of polyaniline-based gas sensors using piezoelectric inkjet and screen printing for the detection of hydrogen sulfide.

9.11.3 Laser Printing

The fabrication of chemical sensor arrays using laser-induced forward transfer (LIFT) as a means to deposit polymer layers to be used as chemically sensitive material is shown in Figure 9.32.

9.12 DISCUSSION ON NANOFABRICATION TECHNIQUES

Here, we have introduced the existing fabrication techniques and related techniques for sensing materials and chipless RFID sensors. Table 9.1 presents the advantages and disadvantages of these techniques for the synthesis of nanostructured sensing materials. A general survey and comparison of the fabrication techniques are given, including prototyping (hot embossing, surface/bulk micromachining, and printing) and direct fabrication (lithography and printing) techniques. Table 9.2 presents the advantages and disadvantages of these techniques for sensor development. Note that "photolithography" in this table refers to conventional photolithography that is mostly employed in IC industry for micrometer-scale patterning. In fact, in the past decade photolithography techniques have progressed to achieve smaller feature patterning ability and have been coupled to various plastic-/polymer-based techniques to better

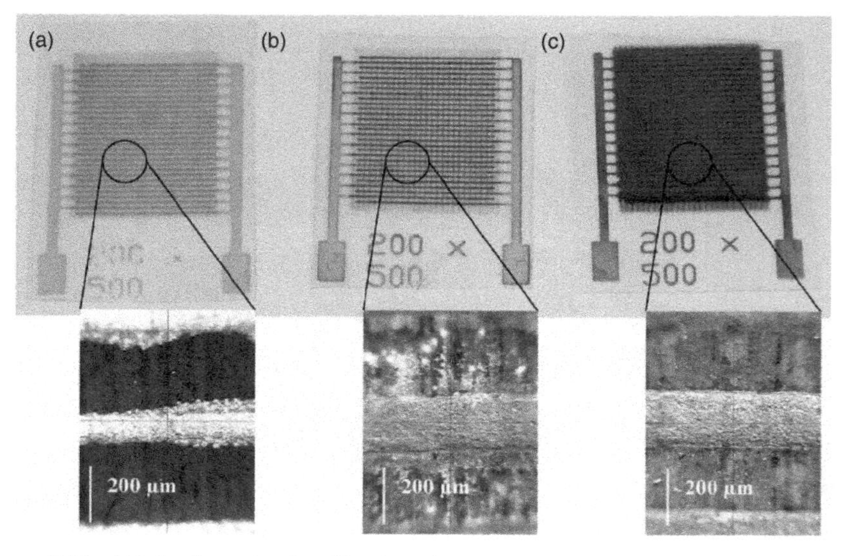

Figure 9.31 Fabrication of polyaniline-based gas sensors using piezoelectric inkjet and screen printing for the detection of hydrogen sulfide (interdigitated electrodes with inkjet-printed films of (a) PANI, (b) PANI-(preexposure), and (c) PANI-(postexposure). Insets compare optical microscopy of the electrode digits and sensor films (*Source:* Reproduced with permission from Ref. [21])

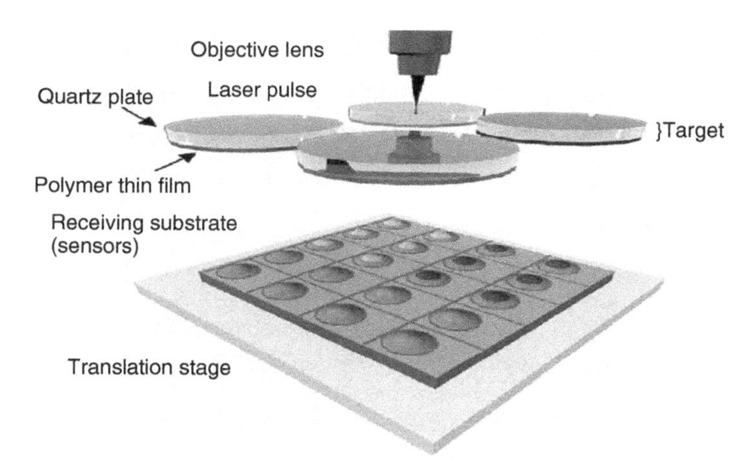

Figure 9.32 A sensor microarray realized by laser printing of polymers (*Source:* Reproduced with permission from Ref. [22])

TABLE 9.1 Different Fabrication Methods of Sensing Materials

Fabrication Methods	Advantages	Disadvantages
Random structures		
Deposition and annealing	Simple and capable of upscale for in-line mass production, inherent capability of producing hemispherical nanoparticles	Limited control of nanoparticle size, shape, and surface arrangement, incapable to fabricate complex nanostructures, vacuum is usually needed
Deposition through a patterned mask (e.g., anodic Al template)	Good control of nanostructures, no high temperature process involved	Poor reliability and low lifetime of the masks limit the mass production
Electrodeposition	High nanoparticle packing density on the surface, ignorable metal material wastes	Particle size relatively small, incapable of fabricating complex nanostructures, poor uniformity over a large surface
Chemical synthesis	Good and independent control of nanoparticle size, shape, and surface arrangement, capable of mass production	Limited capability of fabricating complex nanostructures
Laser ablation	Simple process	Incapable of fabricating complex nanostructures, low throughput
Ordered structures		
Direct pattern writing by electron beam lithography or ion beam lithography	Very fine control of nanostructures, 3-D nanostructures producible	Very low throughput, extremely expensive
Nanoimprint lithography	Fine control of nanostructures by replica, capable of roll-to-roll mass production	Patterns cannot be faithfully transferred for very rough device surfaces, medium cost level

suit lab-on-a-chip applications. To date, most of the current soft lithography processes still rely on modern photolithography techniques for master template/mask fabrication.

Consequently, the low-resolution ability of soft lithography can be gradually improved with high-quality masks by modern photolithography. Sub-100-nm fabrication resolution can also be achieved by composite layers of stamps. Other techniques were also used to obtain the soft lithography masters with nanometer-scale features below 5 nm such as to replicate these features from single-walled carbon nanotubes or from crystal fractures as soft lithography masters. For photolithography-made masters of the soft lithography process, the recently reported resolution limit has been pushed to around 20 nm. All these modern fabrication processes improve

TABLE 9.2 Different Fabrication Techniques of Sensors

Fabrication Methods	Advantages	Disadvantages
Conventional photolithography/ optical lithography	High-throughputs, ideal for microscale features	Usually requires a flat surface to start with, chemical posttreatment needed
Direct laser writing	Rapid, large format production	Multiple treatment sessions, limited materials
Hot embossing	Cost-effective, precise, and rapid replication of microstructures, mass production	Restricted to thermoplastics, difficult to fabricate complex 3-D structures
Electron beam lithography	Patterning is possible at a resolution of ~20 nm with e-beam lithography. This compares to a resolution of ~1 μm for conventional photolithography, no physical mask-plates are needed unlike photolithography, thus eliminating costs and time delays associated with mask production. Patterns can be optimized and changed very simply using flexible CAD software	The electron beam must be scanned across patterned areas pixel by pixel. Exposures can therefore take many hours to complete, conventional fabrication techniques such as metal lift-off and etching can become difficult at submicron length scales, EBL systems are generally expensive and highly complex machines requiring substantial maintenance
Ion beam lithography	High-exposure sensitivity, low backscattering from the substrate can be used as a physical sputtering etch or chemical-assisted etch	Lower throughput, extensive substrate damage
MEMS	The small size of MEMS is attractive for many applications because feature sizes are typically as small as 1 μm or less. MEMS offer opportunities to miniaturize devices, integrate them with electronics, and realize cost savings through batch fabrication. MEMS technology has enhanced many important applications and it holds great promise for continued contributions in the future	The performance of compliant mechanisms is highly dependent on the material properties, which are not always well known at this scale.

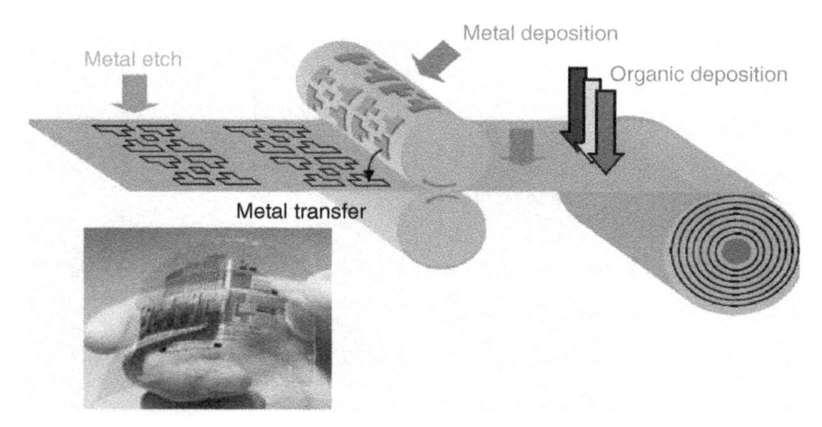

Figure 9.33 The process to print the conductive tracks and sensing materials simultaneously on a polymer substrate (*Source:* Reproduced with permission from Ref. [17])

the quality of the chipless RFID sensor and pave the way for their application in millimeter-wave frequency bands. These bands are less occupied and have promise to provide very compact sensors with high data capacity and functionalities.

9.13 CHIPLESS RFID SENSORS ON FLEXIBLE SUBSTRATES

There are also number of challenges in realizing a fully printable chipless sensor on flexible substrates such as plastic, polymer, or paper. Figure 9.33 shows the process to print the conductive tracks and sensing materials simultaneously on a polymer substrate [17].

Figure 9.34 shows a low-cost inkjet printing technique to print chipless RFID sensors directly on polymer and paper substrates. Here, the polymer substrate is patterned in an automated roller. In the first step, a dispenser delivers microliters of sensing material on the ELC resonator with the regular printing of the conducting tracks for ID generation. The polymer sheets are eventually coated to eliminate degradation from exposure to the environment and then packaged to suit the desired application.

9.14 CONCLUSION

This chapter has presented an overview of various fabrication techniques such as printing, micro-/nanofabrication and MEMS/NEMS that can be used for the development of various chipless RFID sensors. It includes a review on innovative micro- and nanofabrication technologies suitable for roll-to-roll chipless RFID sensor printing. A general survey and comparison of the different fabrication techniques were also given. The aim was to highlight the limitations of conventional fabrication process and

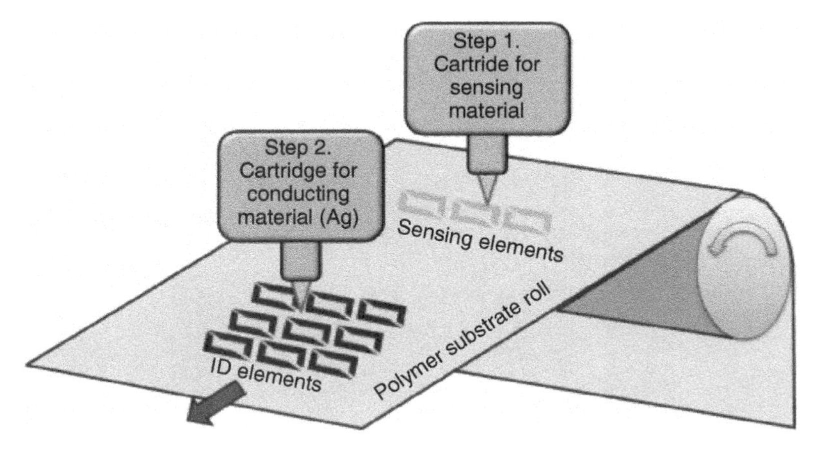

Figure 9.34 The roll-to-roll printable chipless RFID sensor

their solutions for on-demand, high-speed printing for flexible, robust, and mass production of chipless RFID sensor. Moreover, printing, imaging, and characterization procedures of micro- and nanostructures and their integration into RF sensing devices were presented. Printing of microwave passive design on polymers and organic materials has great research potential. From the above discussion, the following inferences are drawn.

(i) Micro-/nanofabrication techniques: The state-of-the-art micro-/nanofabrication processes such as electrodeposition, physical/chemical vapor deposition, laser ablation, direct pattern writing by photolithography/EBL/ion beam lithography, laser, NIL and etching, material modification by ion implantation, diffusion, doping, and thermal annealing can be used to develop the chipless RFID sensor. These nanofabrication processes are also categorized as additive, subtractive, patterning, and materials modification. The additive processes such as evaporation, CVD, oxidation, and plating are used for depositing materials on substrates. The subtractive processes such as plasma etching and reactive ion etching are used to subtract the materials or layers. The patterning processes such as deposition of photoresist, photolithography, and laser are used to develop the pattern according to the design. The materials modification processes such as ion implantation, diffusion, doping, and thermal annealing are used in order to achieve the certain or desired materials properties. Each process has its own distinctive attributes that can be used for the development of chipless RFID sensor.

(ii) MEMS/NEMS: The surface or bulk micromachining such as micro-/nanoelectromechanical systems (MEMS/NEMS) can be used to develop state-of-the-art chipless RFID sensor and is becoming emerging technologies for not only microwave/mm-wave but also sub-mm-wave and THz applications.

(iii) Printing techniques: The printing techniques such as screen, inkjet, thermal, flexographic, and gravure printing can be used to develop the chipless RFID sensor. Each printing technique has its own distinctive attributes that can be used for the development of chipless RFID sensors. Some of these techniques can be used for roll-to-roll printing of chipless RFID sensors.

(iv) RFID sensors fabrication techniques: The section discussed about the limitations of the conventional fabrication processes and the industrial solutions for on-demand and high-speed printing for flexible, robust, and mass productivity of chipless RFID sensors. It has been shown that roll-to-roll printing of the proposed chipless RFID sensor is possible with inkjet printing.

In the next chapter stand-alone reader architecture is discussed to read the data ID and sensing information from a chipless RFID sensor.

REFERENCES

1. J. C. A. Lwarence and J. Durney, *Science and Technology of Surface Coating*. London: Academic Press Inc., 1974.
2. M. M. J. Kimling, B. Okenve, V. Kotaidis, H. Ballot, and A. Plech, "Turkevich Method for Gold Nanoparticle Synthesis Revisited," *The Journal of Physical Chemistry B,* vol. 110, pp. 15700–15707, 2006.
3. A. Bogaerts, E. N. Neyts, R. Gijbels, and J. van der Mullen, "Gas discharge plasmas and their applications," Spectrochim. *Acta Part B*, vol. 57(4), pp. 609–658, pages 5–12, 2002.
4. N. Ohse, J. K. Saha, K. Hamada, K. Haruta, T. Kobayashi, T. Ishikawa, Y. Takemura, and H. Shirai, "Synthesis of Microcrystalline Silicon Films Using High-Density Microwave Plasma Source from Dichlorosilane," *Japanese Journal of Applied Physics*, vol. 46, pp. L696–L698, 2007.
5. D. W. Bäuerle, *Laser Processing and Chemistry*: Springer 2011.
6. D. W. Bäuerle, *Laser Processing and Chemistry*: Springer 1996.
7. K. B. Hermann Sachse, "Tandem-Photolithography Process for Thin-Film Sensors and Micromechanics," *Sensors and Actuators A*, vol. 42, pp. 569–572, 1994.
8. A. R. Mileham, L. B. Newnes, and A. Doniavia, "A Systems Approach to Photolithography Process Optimization in an Electronics Manufacturing Environment," *International Journal of Production Research*, vol. 38, pp. 2515–2528, 2000.
9. I. P. D. Gedamu, S. Kaps, O. Lupan,S. Wille, G. Haidarschin,Y. K. Mishra, and R. Adelung, "Rapid Fabrication Technique for Interpenetrated ZnO Nanotetrapod Networks for Fast UV Sensors," *Advanced Materials*, vol. 26, pp. 1541–1550, 2014.
10. M. Kawada, A. Tungkanawanich, Z. I. Kawasaki, and K. Matsu-Ura, "Detection of wide-band E-M signals emitted from partial discharge occurring in GIS using wavelet transform," *IEEE Transactions on Power Delivery*, vol. 15, pp. 467–471, 2000.
11. S. Qin and S. Birlasekaran, "The study of propagation characteristics of partial discharge in transformer," in *Electrical Insulation and Dielectric Phenomena, 2002 Annual Report Conference on*, pp. 446–449, 2002.
12. T. V. W.L. Brown, and A. Wagner, "Ion Beam Lithography," *Nuclear Instruments and Methods in Physics Research*, vol. 191, pp. 157–168, 1981.

13. J. Melngailis, "Focused Ion Beam Lithography," *Nuclear Instruments and Methods in Physics Research*, vol. 80/81, pp. 1271–1280, 1993.
14. S. M. Roy, R. E. A. Anee, N. C. Karmakar, R. Yerramilli, and G. F. Swiegers, "Printing techniques and performance of chipless tag design on flexible low-cost thin-film substrates," in *Chipless and Conventional Radio Frequency Identification: Systems for Ubiquitous Tagging*, Hoboken, USA: IGI Global, 2012.
15. J. H. M. Prudenziati, *Printed Films: Materials Science and Applications in Sensors, Electronics and Photonics (Woodhead Publishing Series in Electronic and Optical Materials)*: Woodhead 2012.
16. L. U. J. Virtanen, T. Björninen, L. Sydänheimo, and A. Z. Elsherbeni, "Inkjet-Printed Humidity Sensor for Passive UHF RFID Systems," *IEEE Transactions on Instrumentation and Measurement*, vol. 60, pp. 2768–2777, 2011.
17. S. R. Forrest, "The Path to Ubiquitous and Low-Cost Organic Electronic Appliances on Plastic," *Nature*, vol. 428, pp. 911–918, 2004.
18. L. J. Guo, "Nanoimprint Lithography: Methods and Material Requirements", *Advanced Materials*, vol. 19, pp. 495–513, 2007.
19. A. Bessonov, Y. Cho, S. J. Jung, E. A. Park, E. S. Wang, J. W. Lee, M. Shin, and S. Lee "Nanoimprint patterning for tunable light trapping in large-area silicon solar cells", *Solar Energy Materials & Solar Cells*, vol. 95, pp. 2886–2892, 2011.
20. N. Kehagias, V. Reboud, G. Chansin, M. Zelsmann, C. Jeppesen, C. Schuster, M. Kubenz, F. Reuther, G. Gruetzner, and C. M. Sotomayor Torres, "Reverse-Contact UV Nanoimprint Lithography for Multilayered Structure Fabrication", *Nanotechnology*, vol. 18(17), pp. 175303 (1–4), 2007.
21. A. M. Karl, H. Crowley, R. L. Shepherd, Marc in het Panhuis, G. G. Wallace, M. R. Smyth, and A. J. Killard, "Fabrication of polyaniline-based gas sensors using piezoelectric inkjet and screen printing for the detection of hydrogen sulfide," *IEEE Sensors Journal*, vol. 10, 2010.
22. C. Boutopoulos, V. Tsouti, D. Goustouridis, I. Zergioti, P. Normand, D. Tsoukalas, and S. Chatzandroulisa, "A Chemical Sensor Microarray Realized by Laser Printing of Polymers," *Sensors and Actuators B*, vol. 150, pp. 148–153, 2010.

10

CHIPLESS RFID READER ARCHITECTURE

10.1 INTRODUCTION

The final step of a complete chipless radio-frequency identification (RFID) sensor system is to include a stand-alone reader to read the data ID and sensing information from a chipless RFID sensor. An RFID reader interrogates a chipless RFID sensor using RF front-end circuitry. The backscattered or retransmitted signal from the sensor is captured and processed to extract data ID and sensing information. As discussed in the previous chapters, chipless sensors are passive circuitry and have no intelligence or wireless communication protocol. Therefore, in a chipless RFID system the reader performs all necessary data acquisition and signal processing tasks.

This chapter presents an overview of chipless RFID sensor reader architecture. Then it describes the operation and functionality of two primary sections of the reader: (i) RF section and (ii) digital control section. Finally, a detailed flowchart is presented to describe the step-by-step reading method.

10.2 READER ARCHITECTURE

An overall architecture of a chipless RFID reader is presented in Figure 10.1 [1]. The reader has three modules to carry out its operation. These are (i) RF module, (ii) digital module, and (iii) power module.

The *RF module* is designated for transmitting a UWB signal and it receives backscattered tag sensor response. Then it performs a down conversion of the signals and compares the transmitted and received signal using a gain phase detector (GPD). Finally, it sends the analog gain and phase data to the *Digital module*. The *Digital*

Chipless RFID Sensors, First Edition. Nemai Chandra Karmakar, Emran Md Amin and Jhantu Kumar Saha.

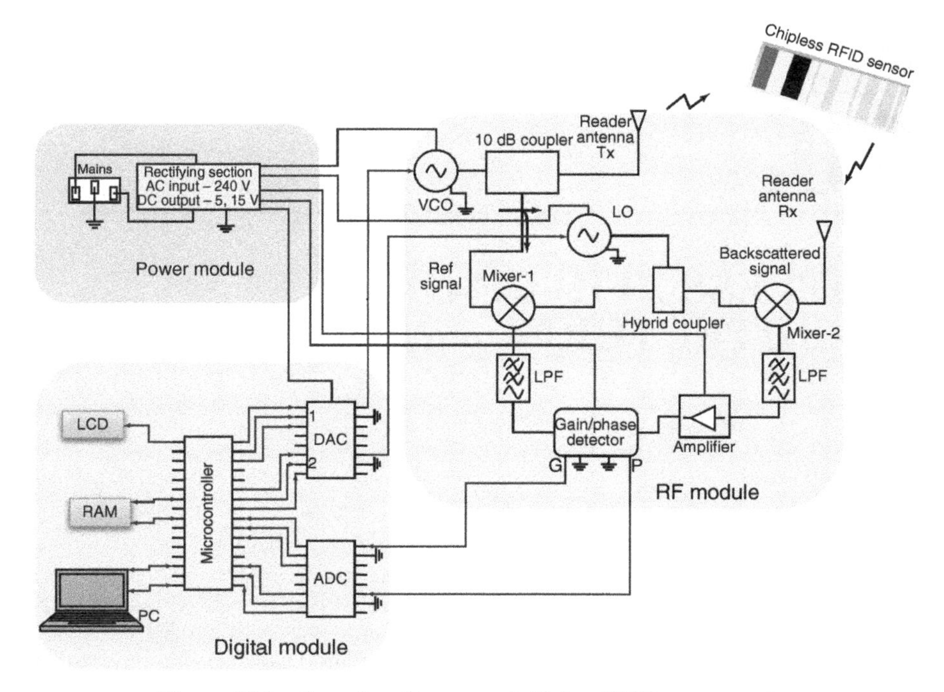

Figure 10.1 Overall architecture of chipless RFID sensor reader

module first converts the analog signal into low-frequency base band digital signal. Next, it performs necessary signal processing to decode the gain-phase information to user interface data. The *Power module* takes AC power from the line and converts into DC 5 and 15 V. It provides power to the active components in *RF module* and *Digital module*.

Detailed operation of the *RF module* and *Digital module* is discussed in the next section.

10.2.1 RF Module

RF module is the primary section of overall reader architecture. The reader architecture presented here uses continuous-wave sinusoidal signals for interrogation and is termed as frequency-modulated continuous-wave (FMCW) reader. Detailed architecture of RF module for an FMCW reader is shown in Figure 10.2. Also a photograph of reader RF module developed by Monash MMARS lab is shown in Figure 10.3.

The RF module generates a suitable interrogation signal to extract data from the tag sensor. The interrogation signal needs to be sufficiently wide band to cover all resonant frequencies. A voltage control oscillator (VCO) is used to generate such wideband interrogation signal. The VCO is the heart of RF module in this reader. Operation of a VCO is shown in Figure 10.4. Here, a voltage ramp is applied as the

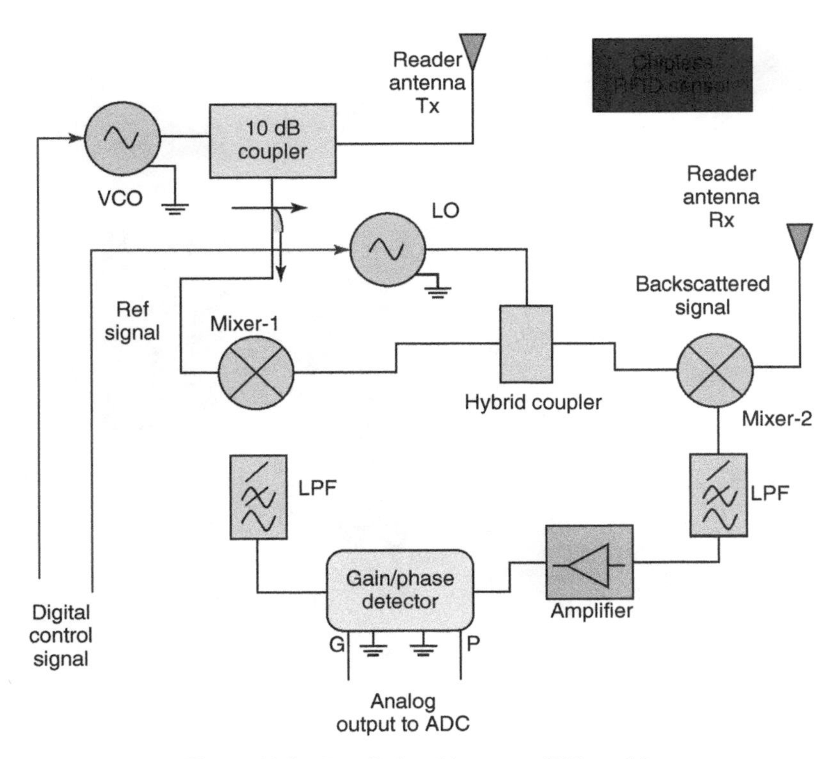

Figure 10.2 Detailed architecture of RF module

tuning voltage from V_0 to V_1 over time period T. This gives a controlled constant output spanning from frequency f_0 to f_1. In our reader, the voltage ramp signal is controlled from a digital to analog converter (DAC). Since the VCO input is discrete from the DAC, we get a stepped frequency sweep. However, by reducing the voltage step, we can approximate a linear voltage ramp resulting in a continuous frequency sweep.

At each voltage-level input for the VCO, its output is passed through a 10 dB coupler for reference signal (*Ref*). Concurrently, the VCO output signal is transmitted through antenna Tx and interrogates chipless sensor under test. The backscattered signal is captured by receiving antenna Rx. Both the *Ref* signal and backscattered signal are down converted to an intermediate frequency (IF) signal using two separate mixers. The IF signals are inserted into a GPD to determine the differences between the amplitude and phase of transmitted and backscattered signal. This is crucial information for reading the chipless sensor as position of resonance "dip" or "null" in the frequency spectra gives information about encoded data.

10.2.2 Digital Module

The digital module controls the overall operation of the reader and can be compared with human brain. Figure 10.5 illustrates a functional block diagram of digital module

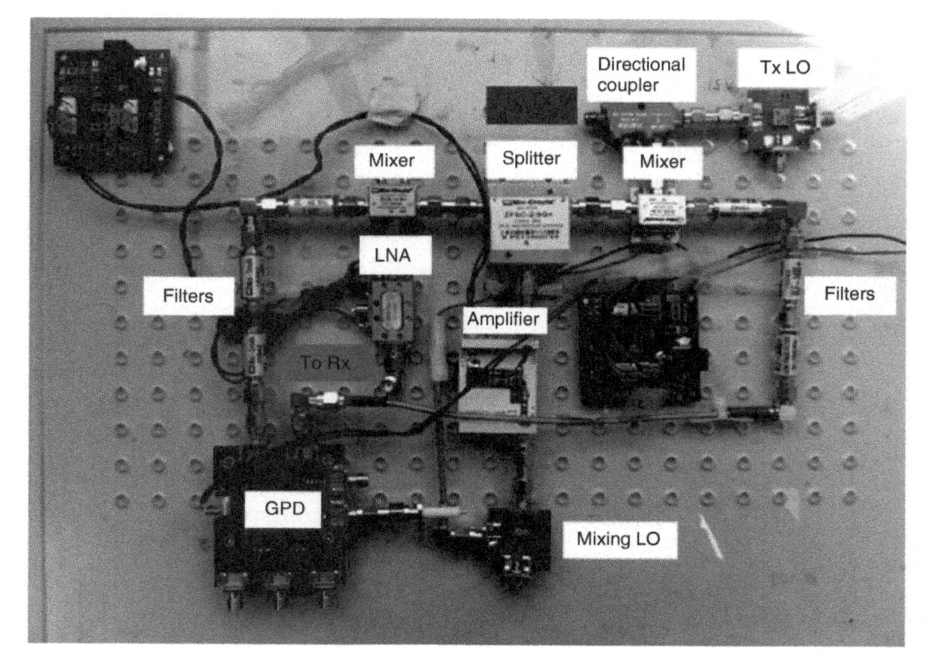

Figure 10.3 Photograph of RF module of chipless RFID reader developed by MMARS lab

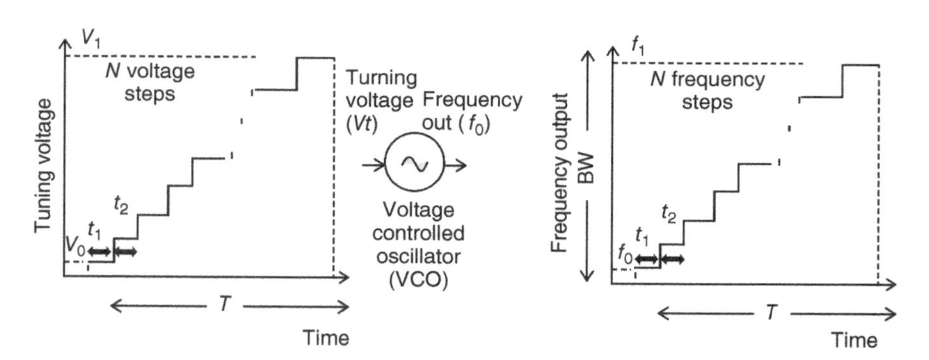

Figure 10.4 Operation of a VCO for discrete tuning voltage

in chipless RFID reader system [2, 3]. According to its functionality, digital module performs three primary tasks: (i) digital control, (ii) signal processing, and (iii) middleware implementation.

Digital control section sends control signals to the VCO and local oscillator (LO). In the previous section, we discussed about the tuning voltage of VCO for creating a voltage ramp. *Digital control section* sends the tuning voltage to the VCO for controlling the RF transmission from the reader. Also, the LO needs certain voltage

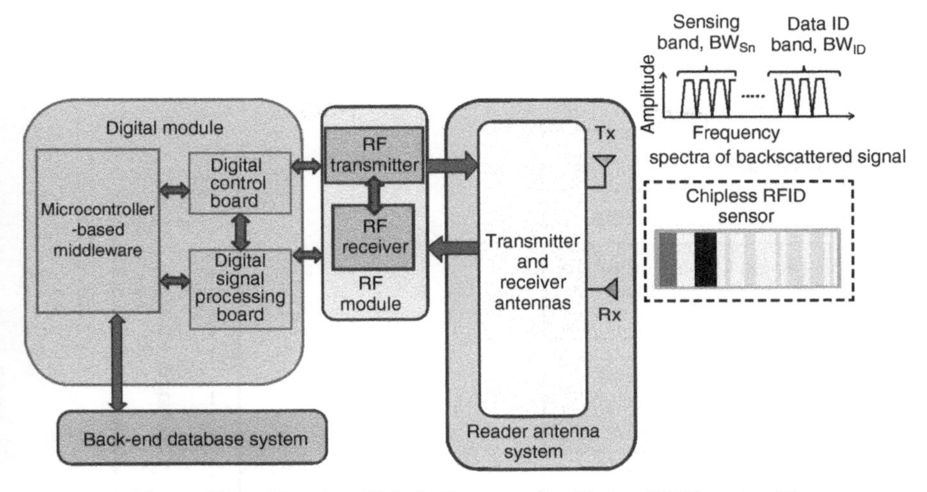

Figure 10.5 Functional block diagram of a chipless RFID reader [1]

input to create base band signal. This base band signal is used in mixer for down conversion.

Signal processing section is dedicated for digitizing the gain and phase signal sent from GPD. This section decodes the information embedded in backscattered signal. It implements necessary signal processing algorithm and denoising technique to identify encoded data bits and sensing parameter measure.

Microcontroller-based Middleware is dedicated for sending and receiving instructions from the user. The middleware receives instructions from the user or back-end database system and forward control and command signals to the digital control section or signal processing section. Similarly, the middleware sends the tag ID and sensing information to the database for further processing.

10.3 OPERATIONAL FLOWCHART OF A CHIPLESS RFID READER

This section presents the step-by-step operation of a chipless RFID reader for reading a tag sensor. Figure 10.6 shows a detailed flowchart of high-level functionality of the reader. This flowchart shows reading of the simplest tag sensor that uses binary encoding for ID generation (3-61) with single sensor. The reader operates in three main steps to decode a tag sensor. These are (i) reader calibration, (ii) real-time sensor data decoding, and (iii) tag ID decoding.

10.3.1 Reader Calibration

The first step of reading the sensor is calibrating the reader for accurate detection of spectral signature. To calibrate the reader across the bandwidth, two standard tags

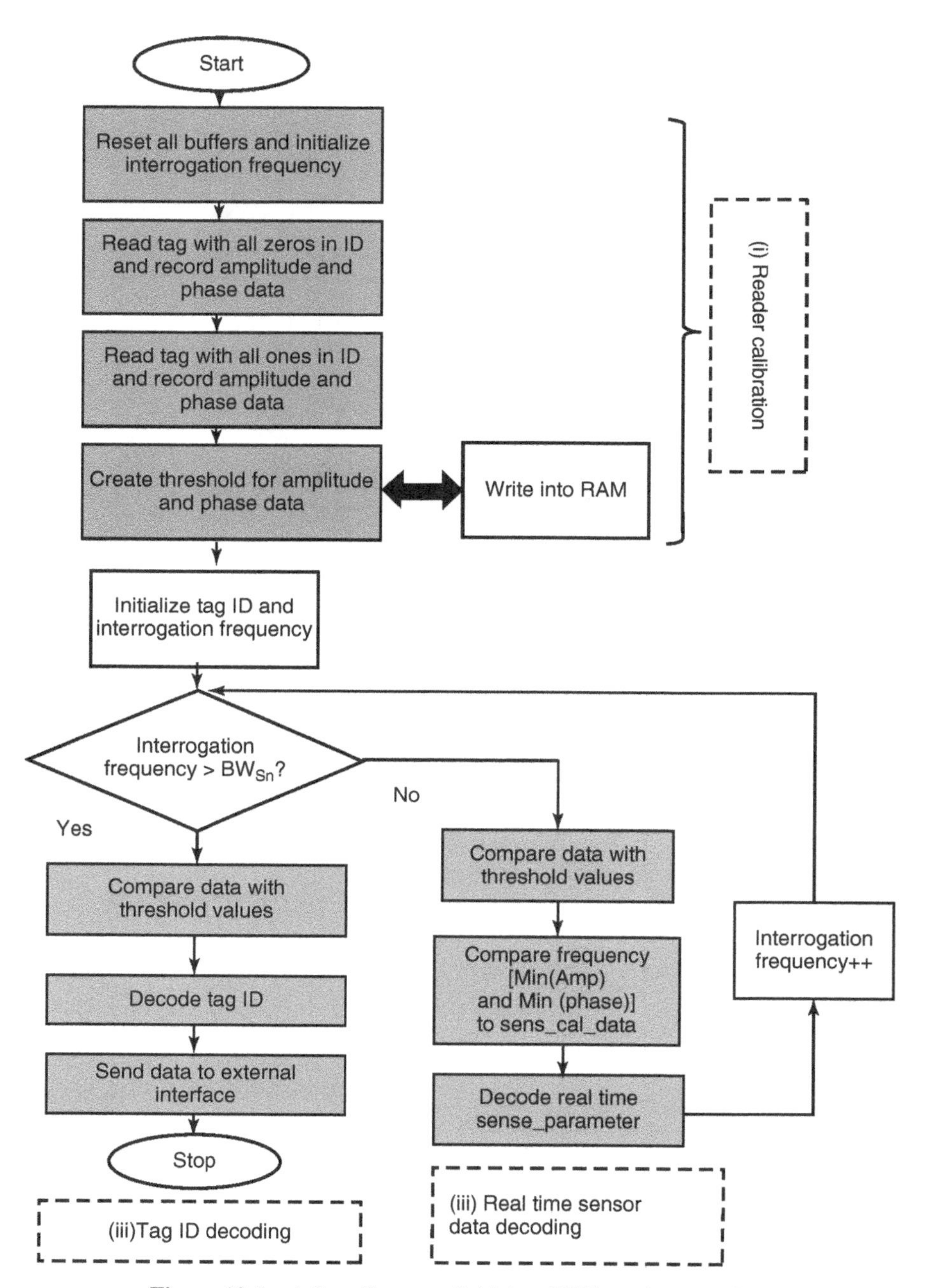

Figure 10.6 A flow diagram of chipless RFID reader operation

are used. One of the tags having all resonance dips (data ID "0") and another tag having no resonance dips (data ID "1"). The reader determines necessary amplitude and phase threshold for each data bit. Hence, the reader records in its memory the maximum difference of amplitude and phase for presence or absence of a resonance scatterer. This information is used in subsequent steps for data decoding.

10.3.2 Real-Time Sensor Data Decoding

In our chipless sensor, the overall bandwidth is divided into two subbands for sensing and data encoding. As shown in Figure 10.5, the lower bandwidth (BW_{Sn}) is dedicated for sensing information, whereas the high-frequency band (BW_{ID}) is dedicated for data ID. The signal processing section conducts decoding operation in the following order. Firstly, it separates the sensing information and data ID information from the backscattered signal by analyzing the frequency band. As shown in the flowchart (Figure 10.6), if the interrogation frequency is less than the sensing band, BW_{Sn}, the reader processes the information for sensing parameter calculation. Next, the signal processing unit correlates the frequency signature of sensing band, BW_{Sn} with previously calibrated data (Sens_cal_data) stored in memory. This gives the measured real-time sensing parameter.

10.3.3 Tag ID Decoding

The final step of chipless sensor reading is to decode tag ID. After recovering real-time sensing parameter, fixed threshold-based detection method [4] is carried out to decode ID information. In this method, the signal processing unit compares amplitude and phase of each scatterer response to the predetermined threshold value. Hence, the presence or absence of resonance dips can be located sequentially within the data ID band, BW_{ID}, which gives a "0"/ "1" pattern as tag ID. Finally, ID and sensing information is sent to the external user through middleware unit.

10.4 CONCLUSION

This chapter presents a guideline for chipless RFID sensor reader development. It gives an overview of reader architecture and operation. The reader architecture has two primary modules, namely, RF module and digital module. RF module is dedicated for RF signal transmission and reception at microwave frequency band. It also does signal down conversion and processing for detection and reading the chipless RFID sensors. On the other hand, digital module provides control instructions to the RF module and performs signal processing to interpret frequency signature. A detailed functional flow of chipless RFID sensor reader is presented to show its systemic approach in data decoding and implementation. There are two streams of data identification that provides the tracking and tracing information and sensing data that provide the information about the physical states of the tagged objects. This integrated sensing will revolutionize many fields of application in near future.

In the next chapter, various potential case studies of chipless RFID sensors are presented. These case studies highlight the potential application of food safety, pharmaceuticals, smart homes, agricultures, infrastructures condition monitoring, power industry, aged care, e-medicine/pills, emergency special services, luggage handling/logistic supports, and many more. The sensors have the ways for visions for Internet-of-things and smart cities.

REFERENCES

1. N. C. Karmakar, R. V. Koswatta, P. Kalansuriya, and R. E-Azim, *Chipless RFID Reader Architecture*. Boston: Artech House, 2013.
2. A. Lazaro, A. Ramos, D. Girbau, and R. Villarino, "Chipless UWB RFID Tag Detection Using Continuous Wavelet Transform," *IEEE Antennas and Wireless Propagation Letters*, vol. 10, pp. 520–523, 2011.
3. R. V. Koswatta and N. C. Karmakar, "Time domain response of a UWB dipole array for impulse based chipless RFID reader," in *Microwave Conference Proceedings (APMC), 2011 Asia-Pacific*, 2011, pp. 1858–1861.
4. S. Preradovic and N. C. Karmakar, "Design of short range chipless RFID reader prototype," in *Intelligent Sensors, Sensor Networks and Information Processing (ISSNIP), 2009 5th International Conference on*, 2009, pp. 307–312.

11

CASE STUDIES

11.1 INTRODUCTION

The chipless radio-frequency identification (RFID)-based wireless sensor network (WSN) is emerging as an inevitable technology in "our society to come" due to their reduced cost and simplicity of sensor nodes. The chipless RFID sensor has tremendous potential due to its robustness in operation and cost-effectiveness and special features that can withstand harsh environments such as extreme humid and dry conditions, temperature above 80 °C, and in cryogenic temperature close to absolute 0 K. It also has a number of innovative features as follows: it is fully printable, passive, disposable due to very low cost, and environmentally friendly. The potential advantages of these unique features permit chipless RFID sensor nodes in unique applications that could not be achieved previously with traditional RFID tags and sensors. The advancement of wireless communications, micro/nanofabrication, printing electronics, and integrated circuit technologies has enabled the development of low-cost chipless RFID sensor nodes. As these technologies are maturing, the potential of mass productions and developments of these sensor nodes become prevalent. New application areas are explored centering these innovation and ubiquitous sensor nodes. Figure 11.1 illustrates many visionary applications of wireless sensor nodes that can promulgate the vision for Internet of things (IoT) and smart cities.

The chipless RFID sensor can be used for various application areas such as food safety and retail, disaster management, health, military, homeland security, smart cities, agriculture, infrastructure, and environment monitoring. For each application area, there are different technical issues that researchers are currently resolving. However, many of them are trying to overcome the application-specific technological challenge. Specifically, RFID readers need to reconfigure to operate in that application

Chipless RFID Sensors, First Edition. Nemai Chandra Karmakar, Emran Md Amin and Jhantu Kumar Saha.

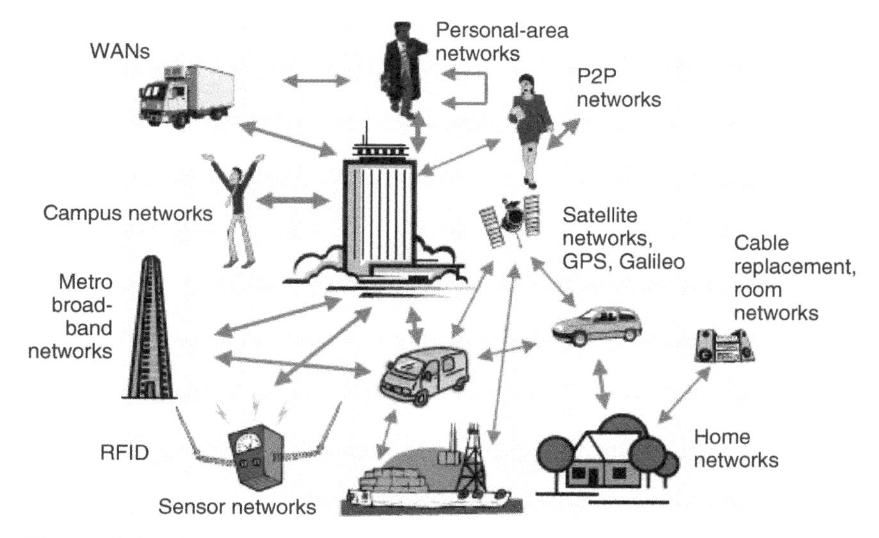

Figure 11.1 Illustration of wireless sensor network in multidimensional applications

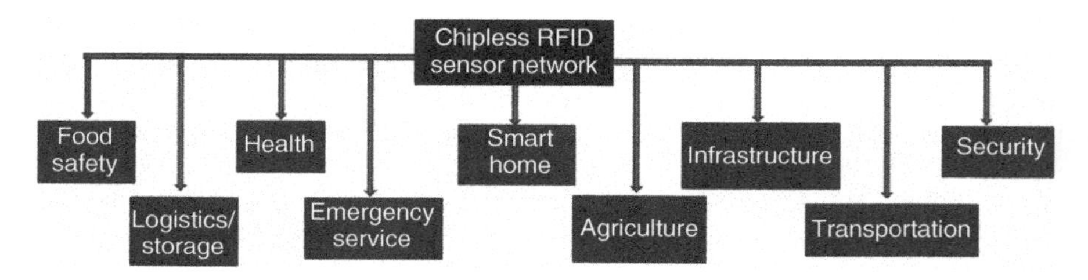

Figure 11.2 Application areas of chipless RFID sensor network

and RFID sensors position and orientation becomes vital for reading and interpolate results. In this regard, the effectiveness of some proposed approaches must be complemented by the supports of hardware design, implementation, and system integration. This chapter presents various application areas suitable for chipless RFID sensors that have been developed at the author's research laboratory at Monash University. Figure 11.2 shows a block diagram of application areas suitable for chipless RFID sensors.

11.2 FOOD SAFETY

Australia is a major producer of agricultural products. Food production and processing contributes 10.5% of Victoria's gross state product and the total sales in this sector were 23.7 billions in 2011–2012 (*Source:* Department of State Development Business and Innovation). Food manufacturing and process industry can be improved by

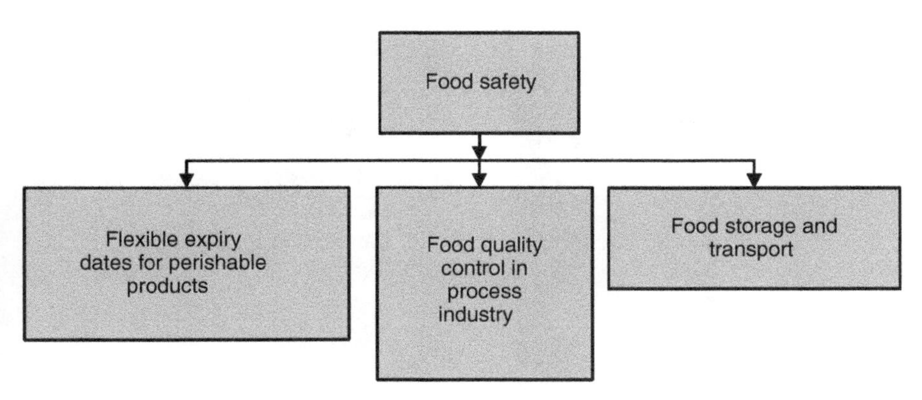

Figure 11.3 Chipless RFID sensor in food safety and quality control

utilizing advanced engineering principles and state-of-the-art technologies. In process industry, the quality of food needs to be monitored in item level at various stages. Moreover, transport and storage of food is an inevitable part of manufacturing industry (Figure 11.3). During transportation or storage, it is not cost-effective to monitor individual item for quality control. Therefore, industries use a central monitoring system for the whole storage room. However, chipless RFID sensors can provide critical physical parameter monitoring feature at subcent cost. This technology will transform the manufacturing and process industry through monitoring pH, temperature, or humidity for individual items.

Another vital area in food industry is to identify expired perishable products. Natural Resources Defense Council reports, "USA is losing up to 40% of its farm produced foods that is estimated to be about 165 billion dollars mostly because of sell-by date [1]." CBC report says-"360,000 tons of milk is wasted in the U.K. each year" [2]. The amount of wasted food and perishable products throughout the globe is astonishing, and, therefore, there is a huge economic aspect of perishable product condition monitoring and storage. Timely sensing and alerting product conditions not only reduces wastage but also improves their in-time intake by the consumer. This in turn immensely reduces health hazards caused by consuming expired unhealthy items such as food and medicine. RFID and sensor are two emerging technologies for tracking and condition monitoring of goods in retail chain [3]. But the cost of chipped RFID tags and sensors hinders their wide exposure to low-cost (one to few dollars) item tagging [3]. Perishable products such as milk, fruit juice, raw meat, and canned food in a supermarket can be tagged with our chipless RFID sensors as shown in Figure 11.4.

A recent study showed that a major retail chain could not maintain 4 °C in their frozen lines [4]. These short-term products exhibit certain chemical changes that can be sensed for status monitoring. The identification, collection, and dissipation of expired perishable products represent a huge international market, which will be immensely benefited by passive chipless sensors. Moreover, this would enhance product quality and customer satisfaction.

Figure 11.4 Coke can and milk carton with chipless RFID sensors

Figure 11.5 shows a conceptual design of chipless RFID pH sensor embedded inside typical milk carton. Refrigerated cartons have a number of polyethylene layers stuffed with paperboard. Our chipless sensor will be designed to utilize the paper board as substrate and polyethylene layer as protecting layer. Figure 11.5(b) shows a highly compact electric inductive–capacitive (ELC) resonator-based sensor design. In its cross-sectional view AA′ the multilayer materials are shown. Here, the pH sensing material will be exposed to the milk, whereas the conductors/resonators will be protected with the inner polyethylene. This will ensure the resonators' operation without being in contact with milk.

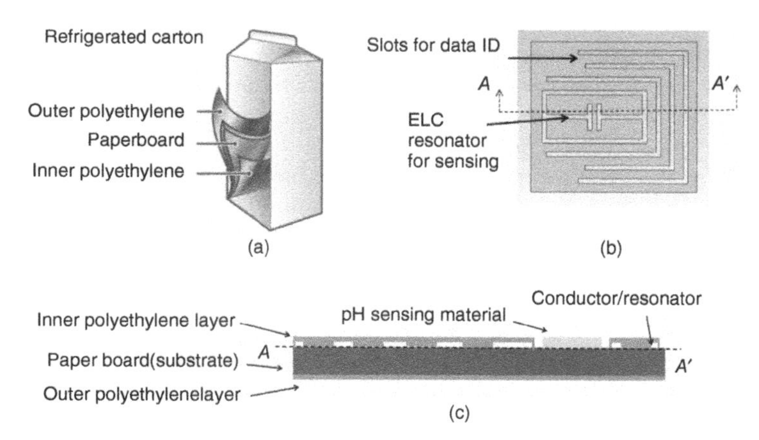

Figure 11.5 (a) A typical milk carton tagged with chipless RFID pH sensor, (b) highly compact chipless RFID sensor using ELC resonator, and (c) cross section of resonator structure showing different layer of materials

11.3 HEALTH

There is huge potential for chipless RFID sensors in shaping future healthcare system. As shown in Figure 11.6, there are a number of areas where chipless RFID sensors can be deployed.

Patient tracking in hospitals and clinics is an important part of hospital management. Patient's movement need to be monitored regularly within the hospital as well as authentication of a patient is required while surgery, prescribing medicine, or during pathological tests. Traditional wrist band type passive chipped RFID tags have limitations in accurate reading. To accurately read the tag, reader needs to wirelessly interrogate the tag IC without any blockage as shown in Figure 11.7.

In cases when a patient wearing a tag is sleeping or unconscious, RFID reader has less chance to read the tag IC as it is not in line of sight with the reader. The benefit

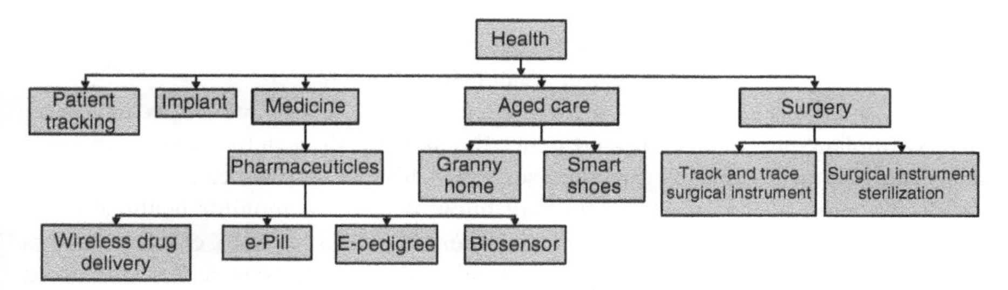

Figure 11.6 Application areas of chipless RFID sensors in healthcare

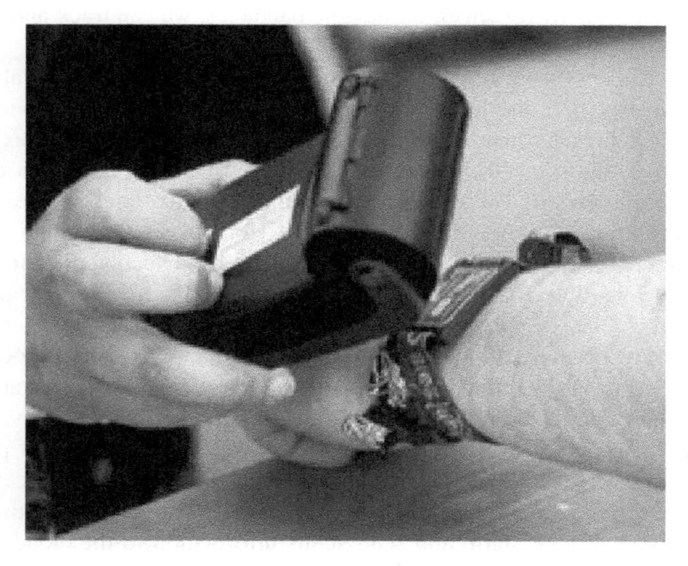

Figure 11.7 Traditional wristband type chipped RFID tag reading (*Source:* www.rfid-blog .com

Figure 11.8 Wristband type printed chipless RFID tag for patient tracking

of chipless RFID tag is it can be printed all-round the band as shown in Figure 11.8. Hence, for any orientation of the tag, it can be read by a reader.

Another important application for biodegradable chipless sensors is in bioimplants. Implants are devices placed under the skin for drug delivery or to monitor acute development of an organ after surgery. Chipless sensors can be embedded inside the skin during surgery. This will allow doctors to monitor healing of the affected organ wirelessly using a chipless reader (Figure 11.9). Also, chipless sensors can be embedded into medical instruments such as gauge bandages. Researchers at Seoul National University have developed printed bandages to monitor drug delivery (Figure 11.10). This shows the potential of printed circuits in medical diagnostics. By printing chipless RFID sensor on flexible bandages, we can track and monitor a number of biological parameters.

Chipless RFID sensors have tremendous potential in pharmaceutical industry. According to Ref. [3], the global market for RFID products and services in the pharmaceutical industry was valued at $112 million in 2008 and is expected to grow to $884 million in 2015. In the pharmaceutical industry, various chemicals and biomolecules require certain environmental conditions (temperature, humidity, pressure, pH level, etc.) for drug culture and storage [5]. A minute change in physical parameters can destroy the efficacy of a drug. In such applications, highly sensitive, flexible dielectric biosensors can improve production quality and reduce economic loss.

With the advancement of printing electronics and nanofabrication process, chipless RFID sensors can be printed on the surface of pills. These smart pills are referred as "e-pill." These pills can be used to monitor drug delivery wirelessly. For aged patients, it is necessary to routinely monitor whether they have taken prescribed medicine. An RFID reader can (Figure 11.11) detect e-pill presence at certain time frame. If the reader cannot detect the presence of the pill at prescribed time, it will automatically trigger an alarm and send status information to the cloud. This will reduce unwanted accidents for not taking regular medication.

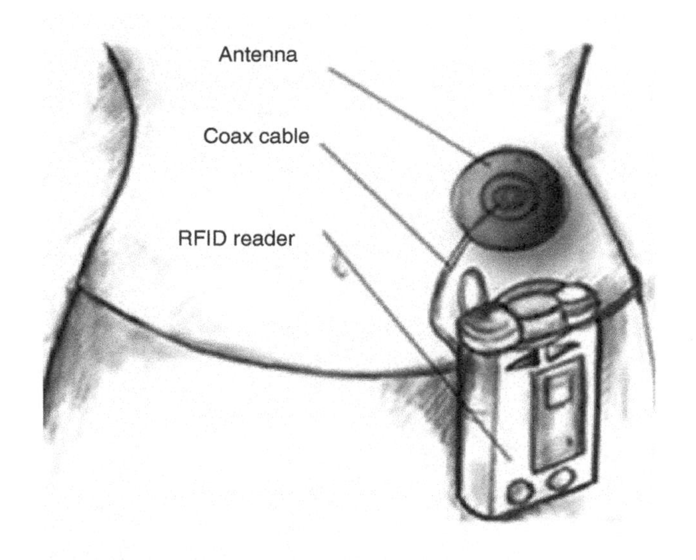

Figure 11.9 Monitoring implant healing after surgery

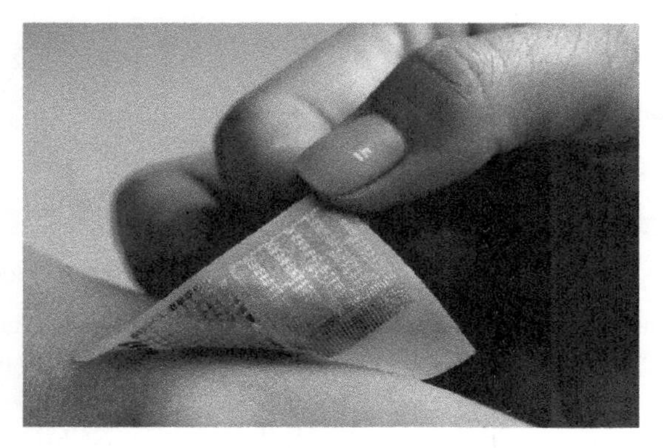

Figure 11.10 A smart bandage by researchers at Seoul National University can record muscle activity and trigger the release of a drug. (*Source:* Reprinted with permission from http://www.wsj.com/articles/SB10001424052702303825604579515704230508572)

Finally, chipless sensor-printed pills can revolutionize e-pedigree records and pharmaceutical supply chain management. e-pedigree is an electronic record for tracing the transport and storage of chemical drugs through the supply chain system. The global standard system GS1 has created a standard for supply chains to communicate product information through supply chain management data (SCMD) (Figure 11.12). In this, standard data from drug manufacturer is deposited through

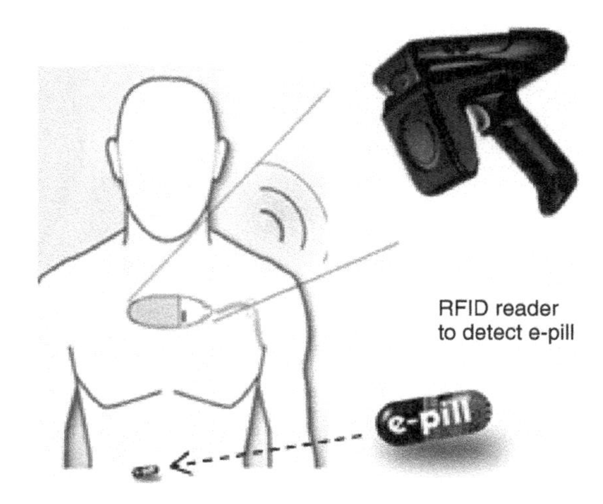

Figure 11.11 Wireless e-pill monitoring

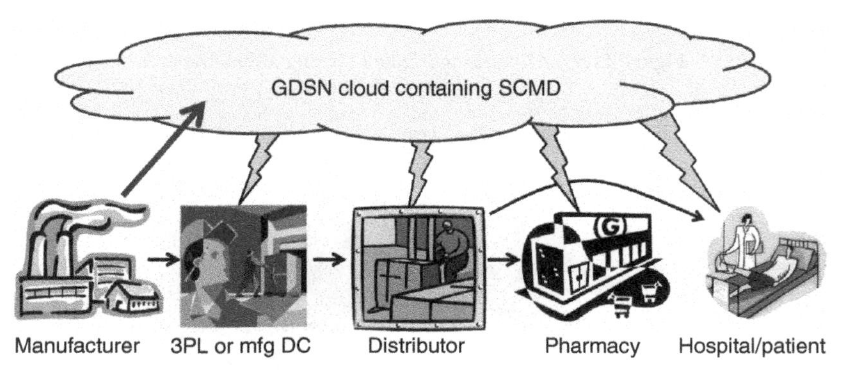

Figure 11.12 Illustration of e-pedigree proposed by GS1 (*Source:* Reprinted with permission from http://www.rxtrace.com/2012/04/gs1-standards-betcha-cant-use-just-one.html/)

global data synchronization network (GDSN) that contains SCMD. In this process, everyone in the supply chain until the end user can excess the SCMD whenever required. In this standard data management system, a low-cost chipless sensor can be added with the tagging feature to monitor the quality of drug.

11.4 EMERGENCY SERVICES

Chipless RFID sensors can be utilized in disaster management applications such as early warning of bushfires. In Australia, bushfires occur frequently due to its hot and dry weather conditions. Australia's Bureau of Meteorology (BOM) reported in 2002–2003 period about 21,241,000 hectares area (*Source:* Australia's Bureau

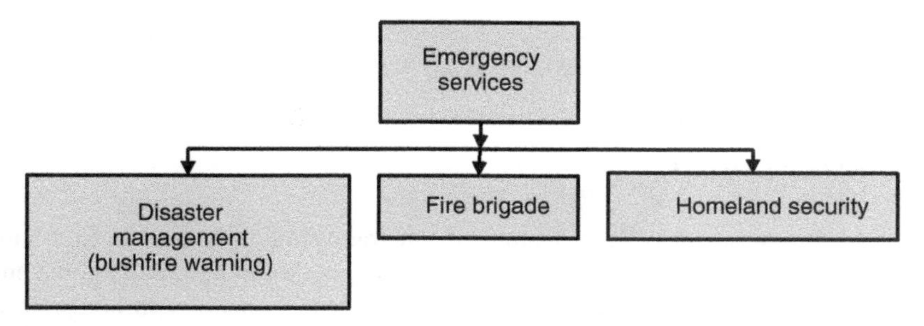

Figure 11.13 Various applications of chipless RFID sensors in emergency services

of Statistics, Year Book Australia, 2004) was burnt all over Australia causing catastrophic damage of property and infrastructure. Chipless RFID sensors are maintenance-free sensor nodes that can be deployed in remote bushlands for temperature and air dryness detection. The sensors can communicate through local hub to bushfire alarm stations for early warning. Similarly, chipless RFID sensors can be deployed in other emergency service management such as fire control and homeland security as shown in Figure 11.13.

1. *Bushfire warning via ecosystem monitoring*
   The developed chipless RFID sensors have the capability of multiparameter sensing. Deployment of such sensors in various points in the bushland and farming areas and monitoring continuous relative humidity, temperature, and stress of soil can provide ecological data for the land. The ranger and the farmer can collect ecological data and can take appropriate measures for the protection of fauna and flora.
2. *Fire brigade*
   Firefighters always work in emergency situations and need a large number of protective gears during their operations. The developed chipless RFID sensors can provide many useful pieces of information for firefighters. These gears are very expensive and need appropriate measures for their inventory control, condition monitoring, and disposal. For example, tagging of garments and protective gears help inventory checks during issuing new and used garments to individual firefighters, and real-time physical parameter monitoring helps detect stress level of the firefighters and warn them to take precautions.
3. *Police services*
   Police sergeants carry a large number of fire arms, protective gears, and uniforms. They need stringent checkout during the beginning of their operation. Counterfeiting and possession of these items can have significant consequences. Moreover, the law-enforcing personnel need continuous monitoring of their stress level during their operation. The developed chipless RFID sensors can provide physiological data of the personnel if smart uniforms engineered with the developed chipless RFID sensors are used. Therefore, the

developed chipless RFID sensors can ease the checkouts of these items and health monitoring of the law-enforcing personnel.

11.5 SMART HOME

Our low-cost sensor also has a feature of integrating multiple sensors in a single chipless RFID tag. This multiple parameter sensing node can be used in home environment monitoring. As for an example, CO sensor is mandatory in homes in the United Kingdom. Also, homes in arctic region need continuous temperature and humidity monitoring unit. In future smart homes [6], multiple sensor nodes will provide necessary information to enable pervasive condition monitoring of temperature, humidity, and presence of noxious gas. Using these sensors, various important parameters can be monitored in the home to give particular alarms or notifications. As shown in Figure 11.14, a smart home will have fire/gas alarm, temperature sensor, water-leak alarm, light sensor, and door security sensor. These sensors will bring ease of lifestyle as well as security and comfort.

11.6 AGRICULTURAL INDUSTRY

In modern agriculture, advanced technologies are being used to optimize the usage of water in farming. They are (i) monitoring local variation of soil moisture, (ii) drainage, and (iii) evaporation. Usually, agronomists visit the farming land and record

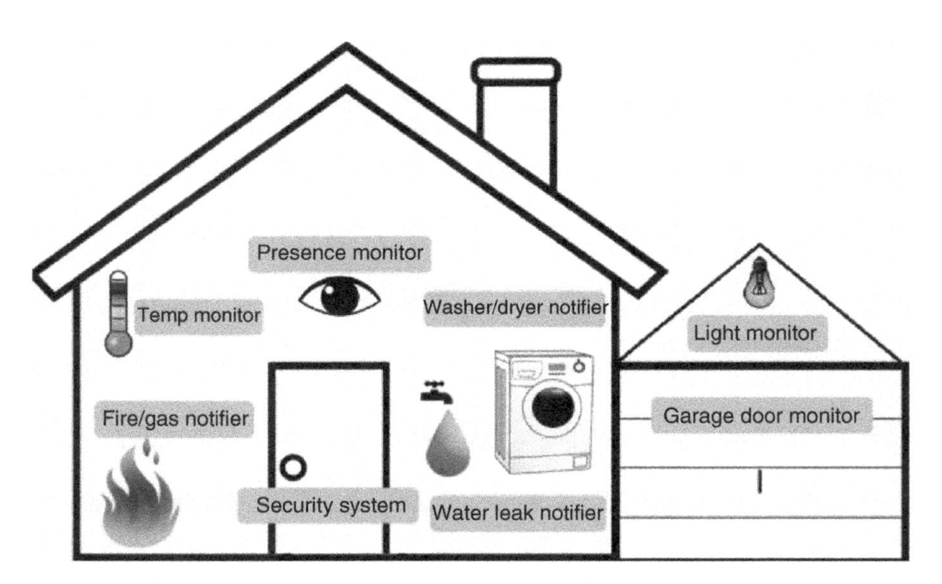

Figure 11.14 Future smart home embedded with wireless sensor

the data manually and later upload the information in a data management system. Based on these data, agronomists make management recommendations to the farmer. However, such manual collection of agronomic data is prone to large errors, data losses, and time delays. To alleviate the problem, RFID-based sensors are being introduced in farming for seamless information flow/management between various sectors of farming. However, these technologies are too expensive to be afforded by farmers in the developing and developed nations. Some of the activities such as irrigation and temperature control in greenhouse environment are done automatically with computerized remote terminal units. However, wireless sensor technology is still in early developmental stage and applications of wireless sensors in agriculture are still very rare.

Chipless RFID sensors can also wirelessly monitor soil moisture as well as the salinity of agricultural land. For plant growth, these two parameters are crucial, as farmers need to take action if the soil does not provide adequate nourishment. Moreover, passive, low-cost, chipless sensors can be deployed in agricultural land without maintenance. This will greatly contribute to agricultural production. Figure 11.15 shows a WSN comprising of N number of chipless sensors deployed in the field. Each sensor is capable of sensing soil moisture level and ionic concentration to predict the amount of fertilizer or irrigation required. Each sensor nodes directly communicate to a local reader or hub at certain interval. Also, the local hub transfers data to the farmhouse by using WLAN protocol.

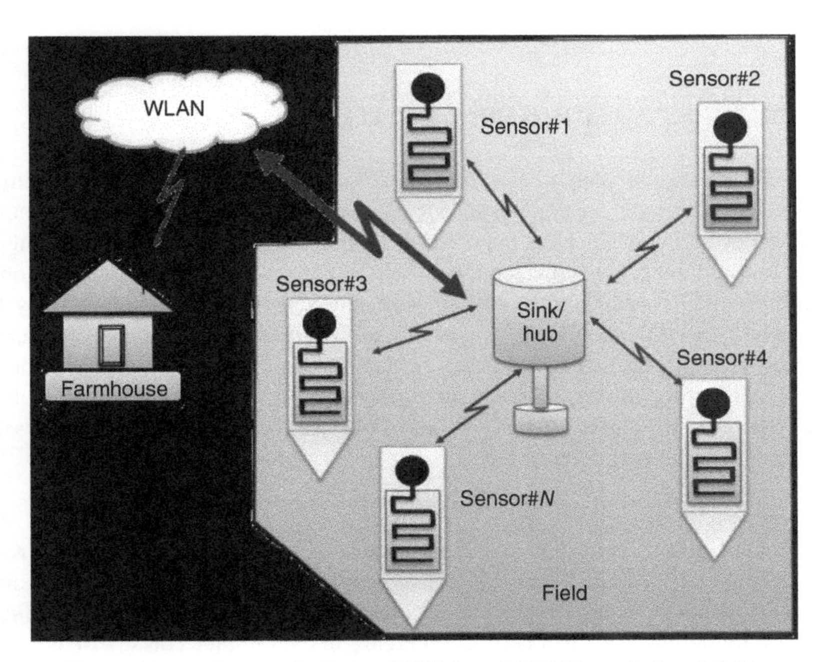

Figure 11.15 Proposed chipless RFID-based WSN in agricultural fields

11.7 INFRASTRUCTURE CONDITION MONITORING

Large and historic infrastructures such as bridges, towers, and buildings are frequently checked for structural health deterioration. To exploit detecting surface cracks in concrete and other construction materials, wireless sensors can be installed pervasively. Civil engineers have focused mainly on wired strain gauges and accelerometers to identify structural damages. The management of these sensors wiring can be exorbitant. Wireless complements are possible, but they are expensive hence do not give the feature of pervasive deployment. Moreover, chipless crack sensors have the potential to be installed inside walls during construction. This enables wireless monitoring of infrastructure for long periods of time without physical intrusion.

11.8 TRANSPORTATION AND LOGISTICS

In supply chain management and logistics, transportation is a crucial part, where items need to be monitored frequently. In Canada, there has been a major advancement in supply chain connectivity where 378 weight stations across 26 states are connected through WSN. Each weight station captures critical information of the cargo (i.e., temperature, humidity, gas, and pH) and sends it to the cloud. In such cases, information about the cargo environment is only available at the stations. Moreover, current technology cannot provide information about individual items. Chipless RFID sensor is low cost and passive. Hence, it can be tagged to individual items to track them and monitor physical parameters as needed.

11.9 AUTHENTICATION AND SECURITY

Australia has led the path to polymer banknotes since 1988 having the aim of enhanced security against counterfeiting, low-cost, durable, and recyclable notes. Today, about 23 countries have adopted this technology including Britain, Canada, Brazil, Romania and few other countries are in the stage of passing legislation to join the group. According to reports in Australia and Britain, this technology has drastically reduced the counterfeit instances. However, there are reports of practical problems of using polymer banknotes in daily transaction that raised skepticism in fully converting all paper notes into polymer. After reviewing articles from The Economist, BBC, This Day Live, and Polymer Notes Australia, two pressing problems are addressed here.

1. *Folding and tearing*

   One of the most prominent limitation of polymer notes are they resist attempts of folding. Therefore, consumers try to fold a note to put in pockets have difficulty compared to paper notes. Also, regular folding and unfolding of polymer note create a permanent crease in the middle. This crease becomes a weak point and has more chance to tear.

2. *Wet notes*

   As polymer absorbs more water than conventional paper, they get sticky in the presence of water. Hence, it becomes difficult to pull them apart. Also, counting wet notes becomes a problem.

3. *Storage of new notes*

   Newly printed polymer notes are to be kept in certain atmospheric conditions before releasing to the market. Therefore, continuous monitoring of relative humidity and temperature of the storage vaults is vital to keep the notes in appropriate shapes and conditions.

11.9.1 Solution

A distributive delay line-based RCS scatterer is designed to place in the critical region of banknote where folding may occur. In Figure 11.16(a), a 100 AUD with delay line and probable folding line is shown. Depending on the fold and unfold occurrence, the stress on the substrate polymer will change. This will also change the backscattered response of the RCS scatterer. This study first will be aimed to explore the

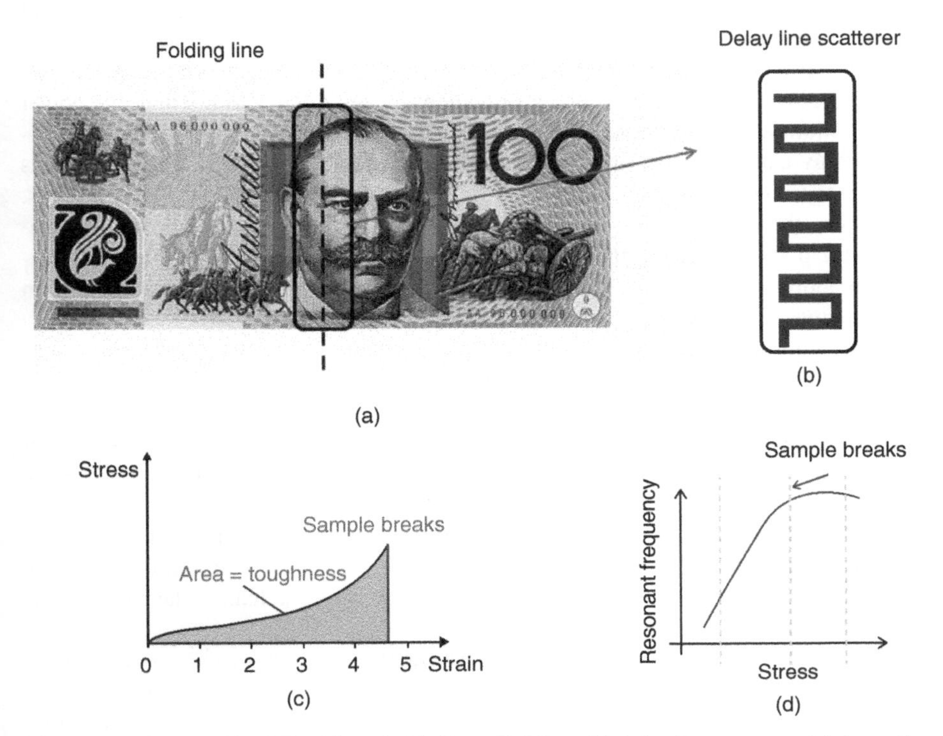

Figure 11.16 (a) 100 AUD with probable line of folding, (b) delay line scatterer, (c) theoretical stress–strain curve for polymer note, and (d) resonant frequency versus stress for delay-line scatterer fabricated on polymer notes

stress–strain relation of polymer banknotes (Figure 11.16(c)). From there, the toughness of polymer can be determined that indicates the breakdown stress for the note. Next, theoretical and experimental analyses of the resonant frequency of the delay line-based scatterer will be performed for various stress conditions. The qualitative relation between the resonant frequency versus stress/strain of the resonator indicates the break point of polymer substrate as shown in Figure 11.16(d). This relation can be a reference for banknote stress monitoring. By comparing the resonant frequency of an unknown note, indication of stress/strain level can be determined and possible breakage of the note can be predicted. This sensor can identify notes before they are torn apart for excessive usage. Thus, these stressed/stained notes can be culled from the market and new circulations can be made easier.

11.10 POWER INDUSTRY

The power industry has introduced RFID for asset management, inventory control, and equipment monitoring. In Ref. [7], five categories of problems in power system management that can be overcome through RFID and wireless technology are presented. These are (i) locating assets, (ii) identification and status of assets, (iii) tool tracking, (iv) fleet management, and (v) access management and infrastructure security. Power facility management using RFID integrated with WSN is proposed in Ref. [8]. Moreover, active RFID is a key ingredient for the Smart Grid, where the power distribution system automates the generation, delivery, and consumption of electrical energy using communication and information technologies [9]. Figure 11.17 is an illustration of a power distribution monitoring system using RFID tags and WSN. However, the cost is prohibitive and the industry looks for alternative cheap solutions for power facility management. Chipless RFID sensors can be highly potential to address these issues of low-cost faulty management and real-time condition monitoring.

The condition monitoring of high voltage (HV) equipment in a switchyard or substation involves routine checking throughout the whole lifespan of the equipment. The insulation in HV equipment may be damaged due to overvoltages, manufacturing defects, or aging. This can eventually result in electrical breakdown causing power failure, equipment failure, and loss of life and property. Partial discharge (PD) is the energy dissipation caused when the electric field across a dielectric exceeds a threshold breakdown value. PD detection is used in power systems to monitor the state of the dielectric insulation of HV equipment. As PD is an early warning of insulation deterioration, failure to detect PD efficiently and on time can lead to the catastrophic disruption of the equipment.

In this regard, chipless RFID sensors can be deployed in automated condition monitoring of HV equipment in a power substation by reducing costs and enhancing the dynamic range and longevity. Chipless RFID sensors can be utilized for online monitoring of high-voltage equipment in a switchyard without sacrificing reliability and capacity deployment.

Figure 11.17 Illustration of a power distribution monitoring system using RFID sensor network

11.11 CONCLUSION AND ORIGINAL CONTRIBUTIONS

This book has presented the novel frequency signature-based chipless RFID sensor system for low-cost item tagging and ubiquitous sensing. The motivation for this technology is explained as follows.

RFID technology is considered as the foundation for IoT and Smart Cities. It has got immense acceptability as a tracking/ID technology due to attractive features such as noninvasive operation, non-LOS reading, longer reading range, multiple tag detection, and environment sensing. However, RFID tags have not replaced the existing barcodes due to its comparatively high cost as well as electronic printing margins. The chipless RFID tag is a breakthrough to the limitations of conventional RFID technology as it cuts off the cost associated with the silicon IC chip in the tag circuit. Moreover, the tag is fully printable and passive and is thus resistant to harsh environments and weather conditions. The cost of a chipless RFID tag is reported to be lower than US $0.01. The potential advantages of this feature permit chipless RFID in applications such as low-cost item tagging such as for banknotes, ID cards, books, and consumer goods [10].

Apart from the tracking of an object, RFID tags can monitor the surrounding environmental conditions and act as a sensor. Thus, a sensor-enabled tag monitors physical parameters such as temperature, pressure, relative humidity, and presence of noxious gases in addition to its identification function. The upcoming research on RFID technology concentrates on integrating sensing mechanism to chipless RFID platform. Chipless RFID has already been regarded superior to conventional paper-based barcodes for its low cost and robustness. However, integration of sensing mechanism with the tag identification is analogous to adding "an extra flavor to the chipless RFID bar" as Mark Roberti describes in Ref. [11].

The novelty of a chipless RFID tag sensor has a remarkable impact in research publications, economic, and social aspects. To this date, a completely passive, printable, low-cost, mass deployable, environmentally friendly chipless RFID sensor node has not been reported for monitoring multiple sensing parameters. To address this challenge, we have developed a fully printable, high data density, compact, single-sided and high-sensitive chipless sensor for real-world applications. The original contributions portrayed in this book are as follows:

1. A comprehensive systematic review of RFID sensors has been performed. The review highlights the fundamental limitations of traditional RFID sensors and the potential of chipless RFID sensors as a solution to low-cost item tagging and condition monitoring
2. Design of a novel electromagnetic (EM) metamaterial structure for dielectric sensing
3. Design and optimization of compact, high Q stepped impedance resonators for chipless RFID PD detection
4. Design of a high data density, compact chipless RFID tag using multislot frequency selective surface (FSS) structure
5. Classification of smart sensing materials for sensing various physical parameters
6. Detailed review of microwave characterization techniques for smart material sensitivity analysis
7. A study of the effect of dielectric and conducting properties of superstrate material on RCS responses of EM metamaterial
8. Microwave characterization of humidity sensing materials polyvinyl alcohol (PVA) and Kapton for dielectric sensitivity
9. Comparative study of humidity sensing polymers for highly sensitive chipless RFID humidity sensors
10. Highly sensitive chipless RFID humidity sensor development for real-time environment monitoring
11. Dielectric study of temperature sensing material for irreversible temperature sensing properties
12. Novel nonvolatile memory sensor for event detection realized using a chipless RFID platform

13. Novel chipless RFID tag sensor development for both temperature and humidity sensing
14. Novel chipless RFID sensor system for partial discharge (PD) detection of high-voltage (HV) equipment
15. Time–frequency analysis and parameter optimization for simultaneous PD signal detection
16. Review of nanofabrication techniques for printable chipless RFID sensor realization on flexible substrates
17. Novel chipless RFID sensor reader architecture for decoding tag ID and sensing data
18. Investigate multidimensional real-world applications for the chipless RFID sensor

These original contributions are in line with the vision of MMARS chipless RFID sensor research. So far, the authors have demonstrated proof-of-concept prototypes of these low-cost sensors. There is ample scope of research and development in this area to deliver this technology to end users through commercialization. This book is a useful resource and gateway for the RFID sensors community to delve into the domain of low-cost tracking and sensing. Also, for retailers, government bodies, healthcare personnel, security force, and public management, it is an alternative technology to address various challenging real-world applications. The authors believe this book will contribute toward a new scientific breakthrough combing WSN, IoT, and green technology.

REFERENCES

1. R. Venkatesh and S. R. Kannan, "Detection of Faults and Ageing Phenomena in Transformer Bushings by Frequency Response Technique" Available: http://citeseerx.ist.psu.edu/viewdoc/summary?doi=10.1.1.549.7278 (accessed on 15 October 2015).
2. M. Kawada, A. Tungkanawanich, Z. I. Kawasaki, and K. Matsu-Ura, "Detection of Wide-Band E-M Signals Emitted from Partial Discharge Occurring in GIS Using Wavelet Transform," *IEEE Transactions on Power Delivery,* vol. 15, pp. 467–471, 2000.
3. P. Harrop and R. Das. *Printed and Chipless RFID Forecasts, Technologies & Players 2011-2021*. Available: http://www.idtechex.com/research/reports/printed-and-chipless-rfid-forecaststechnologies-and-players-2011-2021-000254.asp.
4. 2012. *Thin-Film Polymer Metamaterial Could be Used as Food Sensor.* Available: http://www.printedelectronicsworld.com/articles/thin-film-polymer-metamaterial-could-be-used-as-food-sensor-00004861.asp.
5. *RFID in Pharmaceuticals: Supply Chain Security Concerns Provide Impetus for RFID Adoption*. Available: http://www.rfidjournal.com/article/articleview/7708/1/1/
6. C. K. Abhaya. (2008). *Smart Homes, Intelligent Kitchens, Sensate Floors-Ambient Intelligence*. Available: http://www.eit.uni-kl.de/koenig/deutsch/TEKOGSYS_Abhaya_08.pdf.

7. D. Sen, P. Sen, and A. M. Das, *RFID for Energy and Utility Industries*: PennWell, 2009.
8. Y.-Il Kim, B.-J. Yi, J.-J. Song, J.-H. Shin, and J.-I. Lee, "Implementing a Prototype System for Power Facility Management using RFID/WSN," *International Journal of Applied Mathematics and Computer Sciences*, vol 16, pp. 781–786, 2006.
9. *RFID News Roundup* [Online]. Available: http://www.rfidjournal.com/article/view/8161.
10. S. Preradovic and N. C. Karmakar, "Chipless RFID: Bar Code of the Future," *IEEE Microwave Magazine*, vol. 11, pp. 87–97, 2010.
11. M. Roberti. *The Many Flavors of RFID* [Online]. Available: http://www.rfidjournal.com/article/purchase/7479.

INDEX

Chipless RFID Sensors, First Edition. Nemai Chandra Karmakar, Emran Md Amin and Jhantu Kumar Saha.